Lecture Notes in Mathematics

A collection of informal reports and seminars
Edited by A. Dold, Heidelberg and B. Eckmann, Zürich

243

Japan-United States Seminar on Ordinary Differential and Functional Equations

Held in Kyoto/Japan, September 6-11, 1971

Edited by Minoru Urabe, Kyoto University, Kyoto/Japan

Springer-Verlag
Berlin · Heidelberg · New York 1971

AMS Subject Classifications (1970): 28A65, 34Axx, 34Bxx, 34Cxx, 34Dxx, 34Exx, 34F05, 34Jxx, 34Kxx, 39Axx, 45J05, 45Mxx, 54H20, 90D25

ISBN 3-540-05708-0 Springer-Verlag Berlin · Heidelberg · New York
ISBN 0-387-05708-0 Springer-Verlag New York · Heidelberg · Berlin

Offsetdruck: Julius Beltz, Hemsbach

PREFACE

A Japan-United States Seminar on Ordinary Differential and Functional Equations was held on September 6 through 11, 1971 at Kyoto International Conference Hall, Kyoto, Japan. This SEMINAR as part of the Japan-United States Cooperative Science Program was held under the auspices of the Japan Society for the Promotion of Science and the United States National Science Foundation. The coordinators were John A. Nohel, University of Wisconsin, and Minoru Urabe, Research Institute for Mathematical Sciences, Kyoto University.

This SEMINAR succeeded the Seminar which has been held on June 26 through 30, 1967 at the University of Minnesota, entitled "United States-Japan Seminar on Differential and Functional Equations". The purpose of this SEMINAR was to stimulate further the cooperative efforts of research mathematicians in both countries who have contributed to the development of the theories of ordinary differential and functional equations.

These Proceedings are composed of two sections. Section I contains the texts of all major addresses and Section II contains the reports of short communications.

The coordinators of this SEMINAR owe a great debt of gratitude to the supporting organizations, the Japan Society for the Promotion of Science and the United States National Science Foundation.

Special thanks are due to the SEMINAR participants whose unhesitating and generous contribution of their time and work made possible the SEMINAR and these Proceedings.

CONTENTS

SECTION I

SECTION II

SECTION I

DIFFERENCE INEQUALITIES AND THEIR APPLICATIONS TO STABILITY PROBLEMS

Shohei Sugiyama

Introdcution

Difference equations are often used to analyze the so-called sampled-data systems, in which the stability problems are considered to be important (Cf. [8]). It seems, however, that, if we are concerned with the asymptotic behaviors of solutions of difference equations, not so many papers have been appeared so far. The purpose of this note is to show some results on certain types of the asymptotic behaviors of difference equations.

In order to discuss the asymptotic problems for difference equations, generally speaking, the analogous methods to those for differential and functional-differential equations are applicable, and the results obtained are also analogous to those for differential and functional-differential equations.

Furthermore, it is noted that some difference inequalities obtained for difference equations can be applied to the stability and boundedness problems for differential and functional-differential equations, and it is also expected that some results can be applied to the error estimations in the numerical analysis.

0. Preliminaries

We first summarize some lemmas which will often be used later. In the following, we denote by I_N a set of discrete points $t_0 + k$ $(k = 0, 1, \ldots, N \leq \infty)$, and the norms of matrices and vectors are suitably defined. The following result is easily proved by induction.

Lemma 1. Let $g(t, r)$ be defined for $t \in I_N$ and $0 \leq r < \infty$, and nondecreasing in r for any fixed t. Then, if u(t) and r(t) satisfy the relations

$$u(t+1) \leq g(t, u(t)), \quad r(t+1) = g(t, r(t)), \quad t \in I_N$$

respectively, there holds an inequality $u(t) \leq r(t)$, $t \in I_{N+1}$, provided $u(t_0) \leq r(t_0)$.

Corollary 1. Let $g(t, r)$ be defined for $t \in I_N$ and $0 \leq r < \infty$, and $r + g(t, r)$ be nondecreasing in r for any fixed t. Then, if $u(t)$ and $r(t)$ satisfy the relations

$$\Delta u(t) \leq g(t, u(t)), \quad \Delta r(t) = g(t, r(t)), \quad t \in I_N$$

respectively, where Δ is an operator such that $\Delta \varphi(t) = \varphi(t+1) - \varphi(t)$, there holds an inequality $u(t) \leq r(t)$, $t \in I_{N+1}$, provided $u(t_0) \leq r(t_0)$.

Corollary 2. If $k(t) > 0$ and $u(t)$ satisfies an inequality

$$u(t+1) \leq k(t)u(t) + p(t), \quad t \in I_N,$$

the following inequality is valid for $t \in I_{N+1}$:

$$u(t) \leq u(t_0) \prod_{s=t_0}^{t-1} k(s) + \sum_{s=t_0}^{t-1} p(s) \prod_{\tau=s+1}^{t-1} k(\tau).$$

Proof. It is sufficient to obtain a solution of the equation

$$r(t+1) = k(t)r(t) + p(t),$$

that is,

$$r(t) = r(t_0) \prod_{s=t_0}^{t-1} k(s) + \sum_{s=t_0}^{t-1} p(s) \prod_{\tau=s+1}^{t-1} k(\tau).$$

The direct proof is as follows. Multiplying $\prod_{s=t_0}^{t} k(s)^{-1}$, summing up from t_0 to $t-1$, and then multiplying $\prod_{s=t_0}^{t-1} k(s)$, we obtain the desired inequality.

Corollary 3. If $u_0 \geq 0$, $K(t) \geq 0$, $p(t) \geq 0$, and $u(t)$ satisfies an inequality

$$u(t) \leq u_0 + \sum_{s=t_0}^{t-1} (K(s)u(s) + p(s)), \quad t \in I_N,$$

the following inequalities are valid for $t \in I_N$:

$$u(t) \leq u_0 \prod_{s=t_0}^{t-1} (1 + K(s)) + \sum_{s=t_0}^{t-1} p(s) \prod_{\tau=s+1}^{t-1} (1 + K(\tau))$$

$$\leq u_0 \exp\left(\sum_{s=t_0}^{t-1} K(s)\right) + \sum_{s=t_0}^{t-1} p(s) \exp\left(\sum_{\tau=s+1}^{t-1} K(\tau)\right).$$

Proof. It is sufficient to obtain a solution of the equation

$$r(t) = u_0 + \sum_{s=t_0}^{t-1} (K(s)r(s) + p(s)),$$

that is,

$$r(t) = u_0 \prod_{s=t_0}^{t-1} (1 + K(s)) + \sum_{s=t_0}^{t-1} p(s) \prod_{\tau=s+1}^{t-1} (1 + K(\tau)).$$

The direct proof is similar to that of Corollary 2.

The following result is easily proved by induction.

Lemma 2. Let $f(t, x)$ and $g(t, |x|)$ be defined for $t \in I_N$ and $|x| < \infty$, $r + g(t, r)$ be nondecreasing in r for any fixed t. Then, if an inequality $|f(t, x)| \leq g(t, |x|)$ is satisfied, there holds an inequality $|x(t)| \leq r(t)$, $t \in I_{N+1}$, provided $|x_0| \leq r_0$, where $x(t)$ and $r(t)$ are the solutions of the equations respectively:

$$\begin{aligned} \Delta x(t) &= f(t, x(t)), \quad x(t_0) = x_0, \\ \Delta r(t) &= g(t, r(t)), \quad r(t_0) = r_o, \end{aligned} \qquad t \in I_N.$$

Lemma 3. Let $A(t)$ be an $n \times n$ matrix defined for $t \in I_\infty$ and $\det A(t) \neq 0$. If the trivial solution of the linear system $x(t+1) = A(t) x(t)$ is uniformly asymptotically stable, there exist two positive constants K and α such that $\|X(t, t_0)\| \leq K \exp(-\alpha(t-t_0))$, $t \geq t_o \geq 0$, where $X(t, t_0)$ is a fundamental matrix of the linear system.

1. Asymptotic behaviors

In this section, we are concerned with the asymptotic behaviors for nonlinear difference equations. In the following, I_∞ represents a set of nonnegative integers.

Theorem 1.1. Let $f(t, x)$ and $g(t, |x|)$ be defined for $t \in I_\infty$ and $|x| < \infty$, $r + g(t, r)$ be nondecreasing in r for any fixed t. Suppose that an inequality $|f(t, x)| \leq g(t, |x|)$ is satisfied for $t \in I_\infty$ and $|x| < \infty$. Then, if for any $t_0 \in I_\infty$ every solution $r(t)$ of

(1.1) $$\Delta r(t) = g(t, r(t)), \quad r(t_0) = r_0, \quad t \geq t_0$$

is bounded, any solution $x(t)$ of

(1.2) $$\Delta x(t) = f(t, x(t)), \quad x(t_0) = x_0, \quad t \geq t_0$$

such that $|x_0| \leq r_0$ has a limit as $t \to \infty$.

Proof. It follows by Lemma 2 that $|x(t)| \leq r(t)$, $t \geq t_0$, provided $|x_0| \leq r_0$. Since, for any integers $t_1, t_2 > t_0 (t_1 \leq t_2)$, we have

$$\begin{aligned} |x(t_2) - x(t_1)| &\leq \sum_{s=t_1}^{t_2-1} |f(s, x(s))| \\ &\leq \sum_{s=t_1}^{t_2-1} g(s, r(s)) = r(t_2) - r(t_1), \end{aligned}$$

by which we can easily obtain our result.

Corollary. Let $g(t, r)$ be of the form $g(t, r) = \lambda(t)M(r)$, where $M(r)$ is defined for $0 \le r < \infty$, nondecreasing, $M(r) > 0$ $(0 < r_0 \le r < \infty)$, $\lambda(t) \ge 0$, and $\sum_{s=0}^{\infty} \lambda(s) < \infty$, $\lim_{r\to\infty} r/M(r) = \infty$. Then the conclusion of Theorem 1.1 holds.

If $\lim_{r\to\infty} r/M(r) < \infty$, there are two cases that are bounded and unbounded.

Theorem 1.2. Let $g(t, r)$ be defined for $t \in I_\infty$ and $0 \le r < \infty$, and $r + g(t, r)$ be nondecreasing in r for any fixed t. Let $f_i(t, x)$ $(i = 1, 2)$ be defined for $t \in I_\infty$ and $|x| < \infty$, and satisfy an inequality

$$|x - y + f_1(t, x) - f_2(t, y)| \le |x - y| + g(t, |x - y|).$$

Then, if we denote by $x_i(t)$ $(i = 1, 2)$ and $r(t)$ the solutions of

$$\Delta x(t) = f_i(t, x(t)), \quad x(t_0) = x_i \quad (i = 1, 2),$$

$$\Delta r(t) = g(t, r(t)), \quad r(t_0) = r_0, \qquad t \in I_\infty$$

respectively, we have $|x_1(t) - x_2(t)| \le r(t)$, $t \in I_\infty$, provided $|x_1 - x_2| \le r_0$.

Proof. If we consider a function $u(t) = |x_1(t) - x_2(t)|$, we have from the hypothesis that

$$\Delta u(t) = |x_1(t) - x_2(t) + f_1(t, x_1(t)) - f_2(t, x_2(t))| - |x_1(t) - x_2(t)|$$

$$\le g(t, |x_1(t) - x_2(t)|) = g(t, u(t)).$$

Hence, it follows from Corollary 1 that $u(t) \le r(t)$, $t \ge t_0$, provided $|x_1 - x_2| \le r_0$.

Theorem 1.3. Let $g(t, r)$ be defined for $t \in I_\infty$ and $0 \le r < \infty$, $g(t, 0) \equiv 0$, and $r + g(t, r)$ be nondecreasing in r for any fixed t. Suppose that $f(t, x)$ is defined for $t \in I_\infty$ and $|x| < \rho$, $f(t, 0) \equiv 0$, and satisfy an inequality

$$|x + f(t, x)| \le |x| + g(t, |x|), \quad t \in I_\infty, \qquad |x| < \rho.$$

Then the stability properties of the trivial solution of (1.1) implies the corresponding stability properties of the trivial solution of (1.2).

Proof. Let the solution $r(t) \equiv 0$ of (1.1) be stable. Then, for any given $\varepsilon \in (0, \rho)$ and $t_0 \in I_\infty$, there exists a $\delta(\varepsilon, t_0) > 0$ such that

$r_0 < \delta$ implies $r(t, t_0, r_0) < \varepsilon$, $t \geq t_0$. It is claimed that, with these ε and δ, the trivial solution $x(t) \equiv 0$ of (1.2) is stable.

If this were false, there would exist a solution $x(t)$ of (1.2) and an integer $t_1 > t_0$ such that

$$|x(t_1+1)| \geq \varepsilon, \quad |x(t)| < \varepsilon, \quad t \in [t_0, t_1] \cap I_\infty.$$

For the above t's, setting $u(t) = |x(t)|$ and using Theorem 1.2, it follows that $|x(t)| \leq r(t, t_0, r_0)$, $t \in [t_0, t_1+1] \cap I_\infty$. At $t = t_1+1$, we arrive at the following contradiction:

$$\varepsilon \leq |x(t_1 + 1)| \leq r(t_1 + 1, t_0, r_0) < \varepsilon \quad .$$

Suppose that $r(t) \equiv 0$ is asymptotically stable. By means of the same reason as before, we have $|x(t)| < \varepsilon$, $t \geq t_0$, provided $|x_0| < \delta$ and also $|x(t)| \leq r(t, t_0, r_0)$, $t \geq t_0$, which implies the asymptotic stability of $x(t) \equiv 0$.

If $f(t, 0) \not\equiv 0$, we obtain the following

Theorem 1.4. Let $g(t, r)$ be defined for $t \in I_\infty$ and $0 \leq r < \infty$, and $r + g(t, r)$ be nondecreasing in r for any fixed t. Suppose that $f(t, x)$ is defined for $t \in I_\infty$ and $|x| < \rho$, and satisfy an inequality

$$|x - y + f(t, x) - f(t, y)| \leq |x - y| + g(t, |x - y|).$$

Then, for a given $t_0 \in I_\infty$, if any solution of

(1.3) $$\Delta r(t) = g(t, r(t)) + |f(t, 0)| \;, \quad r(t_0) = r_0 \geq 0$$

tends to zero as $t \to \infty$, every solution of (1.2) also tends to zero as $t \to \infty$.

Proof. Let $r(t) \equiv r(t, t_0, r_0)$ and $x(t) \equiv x(t, t_0, x_0)$ be solutions of (1.3) and (1.2) respectively. Then, by means of the same reason as before, if we choose r_0 sufficiently small, the inequality $\Delta|x(t)| \leq g(t, |x(t)|) + |f(t, 0)|$ implies $|x(t)| \leq r(t)$, $t \geq t_0$, provided $|x_0| \leq r_0$. Hence, we have from the hypothesis that $\lim_{t\to\infty} x(t) = 0$.

2. Perturbed systems

In this section, we shall consider the asymptotic behaviors of solutions of the perturbed difference equations such that

(2.1) $$x(t+1) = A(t)x(t) + F(t, x(t)), \quad x(t_0) = x_0$$

or

(2.2) $\Delta x(t) = A(t)x(t) + F(t, x(t)), \quad x(t_0) = x_0$

and show some relations between two solutions of (2.1) (or (2.2)) and the corresponding linear equations

(2.3) $x(t + 1) = A(t)x(t)$

or

(2.4) $\Delta x(t) = A(t)x(t).$

Theorem 2.1. Let $A(t)$ be an $n\times n$ matrix defined for $t\in I_\infty$, $\det A(t) \neq 0$, and $X(t)$, $X(t_0) = E$, be a fundamental matrix of (2.3). Let $F(t, x)$ be defined for $t\in I_\infty$ and $|x| < \infty$, $F(t, 0)\equiv 0$, and satisfy an inequality $|X(t+1)^{-1}F(t, X(t)y)| \leq g(t, |y|)$, where $g(t, r)$ is defined for $t\in I_\infty$ and $0\leq r < \infty$, and nondecreasing in r for any fixed t. Suppose that any solution of (1.1) is bounded. Then, if the trivial solution of (2.3) is exponentially asymptotically stable, the trivial solution of (2.1) is also exponentially asymptotically stable.

Proof. Since the transformation $x(t) = X(t)y(t)$ reduces (2.1) to $\Delta y(t) = X(t+1)^{-1}F(t, X(t)y(t))$, we have from the hypothesis that

$$\Delta |y(t)| \leq |X(t+1)^{-1}F(t, X(t)y(t))| \leq g(t, |y(t)|).$$

Then, for a solution $r(t)$ of (1.1), we have an estimate $|y(t)|\leq r(t)$, $t\geq t_0$, provided $|y(t_0)|\leq r(t_0)$. Hence, it follows from Lemma 3 that $|x(t)| \leq \|X(t)\|\, |y(t)| \leq K \exp(-\alpha(t-t_0))r(t)$, $t\geq t_0$, which implies our result, since $r(t)$ is bounded.

Note that the exponentially asymptotic stability is equivalent to the uniformly asymptotic stability for linear systems.

The following result is a direct consequence of Theorem 1.1, if we apply the same transformation $x(t) = X(t)y(t)$ as before.

Theorem 2.2. Let $A(t)$ be defined for $t\in I_\infty$ and $\det A(t) \neq 0$. Let $F(t, x)$ be defined for $t\in I_\infty$ and $|x|<\infty$, and satisfy an inequality $|X(t+1)^{-1}F(t, X(t)y)| \leq \lambda(t)|y|$, where $\sum_{s=0}^{\infty}\lambda(s)<\infty$. Then for any solution $x(t)$ of (2.1), the function $X(t)^{-1}x(t)$ has a finite limit as $t\to\infty$.

Theorem 2.3. Suppose that

(i) $A(t)$ is defined for $t \in I_\infty$, $\det A(t) > 0$, $\prod_{s=0}^{\infty} \det A(s) > 0$, and all solutions of (2.3) are bounded as $t \to \infty$;

(ii) $F(t, x)$ is defined for $t \in I_\infty$ and $|x| < \infty$, and satisfies an inequality $|F(t, x)| \leq \lambda(t)|x|$, where $\sum_{s=0}^{\infty} \lambda(s) < \infty$.

Then, for any solution $x(t)$ of (2.1), there exists a solution $y(t)$ of (2.3) such that $\lim_{t\to\infty} (x(t) - y(t)) = 0$.

Proof. Let $X(t)$ be a fundamental matrix of (2.3). Setting $x(t) = X(t)v(t)$, the function $v(t)$ satisfies an equation $\Delta v(t) = X(t+1)^{-1} F(t, X(t)v(t))$. It follows from (i) and (ii) that

$$\begin{aligned} |X(t+1)^{-1}F(t, X(t)v(t))| &\leq \|X(t+1)^{-1}\| \, \|X(t)\| \; \lambda(t)|v(t)| \\ &\leq K\lambda(t)|v(t)|, \end{aligned}$$

where K is a constant. Then, applying the preceding theorem, there exists a limit ξ of $v(t)$ as $t \to \infty$. Since a function $y(t) = X(t)\xi$ is a solution of (2.3) with an initial condition $y(t_0) = \xi$, it is a desired solution, because of the fact that

$$\lim_{t\to\infty} (x(t) - y(t)) = \lim_{t\to\infty} X(t)(v(t) - \xi) = 0.$$

Theorem 2.4. Suppose that

(i) $A(t)$ is defined for $t \in I_\infty$;

(ii) for any given $\varepsilon > 0$ there exists a $\delta(\varepsilon)$ such that $|F(t, x)| \leq \varepsilon|x|$ uniformly in t, provided $|x| < \delta(\varepsilon)$;

(iii) for a $t_0 \in I_\infty$, $\overline{\lim_{t\to\infty}} \frac{1}{t-t_0} \sum_{s=t_0}^{t-1} (\|E + A(s)\| - 1) < 0$.

Then the trivial solution of (2.2) is asymptotically stable.

It is noted that the quantity $\|E + A\| - 1$ corresponds to the logarithmic norm $\lim_{h\to 0} (\|E + Ah\| - 1)/h$ in differential equations.

Proof. For a sufficiently small ε, it follows from (iii) that there exists a constant K such that

$$\max_{t \geq t_0} (\exp(\varepsilon(t-t_0) + \sum_{s=t_0}^{t-1} (\|E + A(s)\| - 1))) \leq K$$

and we can choose a δ_1 such that $K\delta_1 < \delta(\varepsilon)$. Then, if $|x_0| < \delta_1$, we claim $|x(t)| < \delta(\varepsilon)$, $t \geq t_0$, where $x(t) \equiv x(t, t_0, x_0)$ is a solution of (2.2). If this were false, there would exist an integer $t_1 > t_0$ and a solution $x(t)$ with the property that

$$|x(t_1+1)| \geq \delta, \quad |x(t)| < \delta, \quad t \in [t_o, t_1] \cap I_\infty .$$

Setting $u(t) = |x(t)|$, we observe that

$$\text{(2.5)} \quad \begin{aligned} \Delta u(t) &= |x(t) + A(t)x(t) + F(t, x(t))| - |x(t)| \\ &\leq (\|E + A(t)\| - 1 + \varepsilon)u(t), \end{aligned}$$

which yields $u(t+1) \leq (\|E + A(t)\| + \varepsilon)u(t)$, $t \in [t_o, t_1] \cap I_\infty$. Then, choosing $u(t_o) = u_o$, we obtain

$$\text{(2.6)} \quad \begin{aligned} u(t) &\leq u_o \prod_{s=t_o}^{t-1} (\|E + A(s)\| + \varepsilon) \\ &\leq |x_o| \exp\left(\varepsilon (t-t_o) + \sum_{s=t_o}^{t-1} (\|E + A(s)\| - 1)\right), \end{aligned}$$

which leads to an absurdity:

$$\delta \leq u(t_1+1) \leq |x_o| \exp(\varepsilon(t_1+1-t_o) + \sum_{s=t_o}^{t_1}(\|E+A(s)\|-1)) < K \delta_1 < \delta .$$

Consequently, the inequality (2.5) and hence (2.6) holds for every $t \geq t_o$. Then, it follows that $\lim_{t\to\infty} x(t) = 0$, if ε is small enough.

Theorem 2.5. Suppose that

(i) $A(t)$ is defined for $t \in I_\infty$, and $\|E + A(t)\| - 1 < -\sigma$, $t \in I_\infty$, where σ is a positive constant;

(ii) $F(t, x)$ is defined for $t \in I_\infty$ and $|x| < \rho$, and satisfies the condition (ii) in Theorem 2.5;

(iii) $G(t, x)$ is defined for $t \in I_\infty$ and $|x| < \rho$, $G(t, 0) \equiv 0$, and

$$|G(t, x)| \leq \gamma(t), \quad t \in I_\infty , \quad |x| < \infty ,$$

where $\gamma(t) \to 0$ $(t \to \infty)$.

Then the trivial solution of

$$\text{(2.7)} \quad \Delta x(t) = A(t)x(t) + F(t, x(t)) + G(t, x(t))$$

is asymptotically stable.

Proof. For any given $\varepsilon \in (0, \min(\sigma, \rho))$, there exist $\delta(\varepsilon)$ $(\leq \varepsilon)$ and $T(\varepsilon)$ such that, for any $t_o \geq T(\varepsilon)$ an inequality

$$\sum_{s=t_o}^{t-1} \gamma(s) \exp(-(\sigma - \varepsilon)(t - s)) < \frac{\delta(\varepsilon)}{2} = \delta_1, \quad t \geq t_o$$

is valid. We claim that for a solution $x(t) \equiv x(t, t_o, x_o)$ of (2.1), $|x(t)| < \delta(\varepsilon)$, $t \geq t_o$, whenever $|x_o| < \delta_1$. If this were false, there would exist an integer $t_1 > t_o$ such that

$$|x(t_1+1)| \geq \delta(\varepsilon), \quad |x(t)| < \delta(\varepsilon), \quad t \in [t_o, t_1] \cap I_\infty .$$

Setting $u(t) = |x(t)|$, we obtain

$$\Delta u(t) = |x(t) + A(t)x(t) + F(t, x(t)) + G(t, x(t))| - |x(t)|$$
$$\leq (\|E + A(t)\| - 1 + \varepsilon)u(t) + \gamma(t),$$

which implies from Corollary 2 that

$$(2.8) \quad |x(t)| \leq |x_0| \exp(-(\sigma - \varepsilon)(t - t_0)) + \sum_{s=t_0}^{t-1} \gamma(s) \exp(-(\sigma - \varepsilon)(t - s - 1)).$$

Then, at $t = t_1 - 1$, there gives rise to a contradiction: $\delta(\varepsilon) \leq |x(t_1 + 1)| < \delta_1 + \delta_1 = \delta(\varepsilon)$. This proves that $|x(t)| < \delta(\varepsilon)$, $t \geq t_0$, whenever $|x_0| < \delta_1$. Hence, the inequality (2.8) shows the asymptotic stability of the trivial solution of (2.7).

Theorem 2.6. Suppose that

(i) $A(t)$ is defined for $t \in I_\infty$, $\det A(t) \neq 0$, and the trivial solution of $x(t+1) = A(t)x(t)$ is exponentially asymptotically stable;

(ii) for any given $\varepsilon > 0$ there exist $\delta(\varepsilon)$ and $T(\varepsilon)$ such that $|F(t, x)| \leq \varepsilon |x|$, provided $|x| < \delta(\varepsilon)$ and $t \geq T(\varepsilon)$;

(iii) there exists a constant $\eta > 0$ such that, if $|x| < \eta$ and $t \in I_\infty$, $|G(t, x)| \leq \gamma(t)$, where $\gamma(t) \to 0$ $(t \to \infty)$.

Then there exist T_0 and $\delta_1 > 0$ such that, for every $t_0 \geq T_0$ and $|x_0| < \delta_1$, a solution $x(t) \equiv x(t, t_0, x_0)$ of

$$x(t+1) = A(t)x(t) + F(t, x(t)) + G(t, x(t)), \quad x(t_0) = x_0$$

satisfies $\lim_{t\to\infty} x(t) = 0$.

It is noted that the type of stability corresponds to the eventual stability in differential equations.

Proof. It follows from (i) that for a fundamental matrix $X(t, s)$ there exist two constants K and α such that $\|X(t, s)\| \leq K \exp(-\alpha(t - s))$, $t \geq s \geq 0$. For any given $\varepsilon \in \min(\eta, \alpha/Ke^\alpha)$, let $T(\varepsilon)$ be so large that $t \geq t_0 \geq T(\varepsilon)$ implies

$$\sum_{s=t_0}^{t-1} \gamma(s) \exp(-(\alpha - Ke^\alpha \varepsilon)(t - s - 1)) < \frac{\delta(\varepsilon)}{2} = \delta_1 .$$

For $t \geq t_0$ and $|x_0| < \delta_1$, as long as $|x(t)| < \delta(\varepsilon)$, we have

$$|x(t)| = \left|X(t,t_0)x_0 + \sum_{s=t_0}^{t-1} X(t, s+1)(F(s, x(s)) + G(s, x(s))\right|$$

$$\leq Ke^{-\alpha(t-t_o)}|x_o| + \sum_{s=t_o}^{t-1} Ke^{-\alpha(t-s-1)}(\varepsilon|x(s)| + \gamma(s)).$$

If we put $u(t) = |x(t)|e^{\alpha t}$, the above inequality is reduced to

$$u(t) \leq Ku(t_o) + \sum_{s=t_o}^{t-1}(Ke^{\alpha}\varepsilon u(s) + Ke^{\alpha(s+1)}\gamma(s)).$$

Then, from Corollary 3 we obtain

$$u(t) \leq Ku(t_o)\exp(Ke^{\alpha}\varepsilon(t-t_o)) + K\sum_{s=t_o}^{t-1}\gamma(s)\exp(-(\alpha - Ke^{\alpha}\varepsilon)(t-s-1)),$$

which yields an inequality

$$|x(t)| \leq K|x_o|\exp(-(\alpha - Ke^{\alpha}\varepsilon)(t-t_o)) + K\sum_{s=t_o}^{t-1}\gamma(s)\exp(-(\alpha - Ke^{\alpha}\varepsilon)(t-s-1)). \tag{2.9}$$

By means of the same reason as before, we can prove that $|x(t)| < \delta(\varepsilon)$, $t \geq t_o$, whenever $|x_o| < \delta_1$. Since we can choose $\delta(\varepsilon)$ not greater than ε, we have $|x(t)| < \varepsilon$, $t \geq t_o$, whenever $|x_o| < \delta_1$, and the inequality (2.9) implies the desired result.

3. Applications of Lyapunov functions

As for the applications of Lyapunov functions to the stability problems of difference equations, we can find some results in [3, 4, 5, 6, 8], and in [7] concerning the criteria of Popov type for the absolute stability. In this section, we shall show some other results including a construction of Lyapunov function and the applications to perturbed systems. It is observed that the analogous methods as in differential equations can be applied to obtain the related results. We first begin with the following definition which will be found in [2].

Definition. Let $x(t, t_o, x_o)$ be a solution of

$$x(t+1) = f(t, x(t)), \quad x(t_o) = x_o, \quad t \geq t_o, \tag{3.1}$$

where $f(t, x)$ is defined for $t \in I_\infty$ and $|x| < \rho$, and $f(t, 0) \equiv 0$. Suppose that there exists a function $p(t)$ which is increasing for $t \in I_\infty$ and $p(t) \to \infty$ $(t \to \infty)$ such that

$$|x(t, t_o, x_o)| \leq K(t_o)|x_o|\exp(p(t_o) - p(t)), \quad t \geq t_o,$$

where $K(t) > 0$ is a function defined for $t \in I_\infty$. Then the trivial solution of (3.1) is said to be generalized exponentially asymptotically stable.

If $K(t) \equiv K$ is constant and $p(t) \equiv \alpha t$ for a constant $\alpha > 0$, we have the exponentially asymptotical stability.

If we define a function

$$V(t, x) = \sup_{\sigma \in I_\infty} |x(t + \sigma, t, x)| \exp (p(t+\sigma) - p(t))$$

for $t \in I_\infty$ and $|x| < \infty$, by means of just the same reason as in differential equations, we can easily obtain the following

Theorem 3.1. Suppose that $f(t, x)$ is defined for $t \in I_\infty$ and $|x| < \infty$, linear in x, and the trivial solution of (3.1) is generalized exponentially asymptotically stable. Then there exists a function $V(t, x)$ satisfying the following conditions:

(a) $V(t, x)$ is defined for $t \in I_\infty$ and $|x| < \infty$, and Lipschitzian in x for the function $K(t)$;

(b) $|x| \leq V(t, x) \leq K(t)|x|, \quad t \in I_\infty, \quad |x| < \infty$;

(c) for any solution $x(t)$ of (3.1),

$$\Delta V(t, x(t)) \leq -(1 - \exp(-\Delta p(t)))V(t, x(t)), \quad t \geq t_0.$$

Theorem 3.2. Suppose that

(i) $f(t, x)$ is linear in x for $t \in I_\infty$, and the trivial solution of (3.1) is generalized exponentially asymptotically stable;

(ii) $F(t, x)$ is defined for $t \in I_\infty$ and $|x| < \rho$, and $|F(t, x)| \leq g(t, |x|)$, $t \in I_\infty$, $|x| < \rho$, where $g(t, r)$ is defined for $t \in I_\infty$ and $0 \leq r < \infty$, nondecreasing in r for any fixed t, and $g(t, 0) \equiv 0$.

Then the stability properties of the trivial solution of

(3.2) $\Delta r(t) = -(1-\exp(-\Delta p(t)))r(t)+K(t+1)g(t, r(t)), \quad r(t_0)=r_0 \geq 0$

implies the corresponding stability properties of the perturbed system

(3.3) $$x(t+1) = f(t, x(t)) + F(t, x(t)).$$

Proof. From (i) there exists a function $V(t, x)$ fulfilling the conditions (a), (b), and (c) in Theorem 3.1. Let $x(t) \equiv x(t, t_0, x_0)$ and $y(t) \equiv y(t, t_0, y_0)$ be solutions of (3.3) and (3.1) respectively. Then, for any given $\varepsilon \in (0, \rho)$ and $t_0 \in I_\infty$, there exists a $\delta(\varepsilon, t_0)$ such that $|x(t)| < \rho$, $t \geq t_0$, whenever $|x_0| < \delta$. If this were false, there would exist an integer $t_1 > t_0$ such that, for a solution $x(t) \equiv x(t, t_0, x_0)$ of (3.3),

$$|x(t_1 + 1)| \geq \rho, \quad |x(t)| < \rho, \quad t \in [t_0, t_1] \cap I_\infty .$$

Setting $u(t) = V(t, x(t))$, whenever $|x(t)| < \rho$, it follows from Theorem 3.1 and (ii) that

$$\begin{aligned}\Delta u(t) &= V(t+1, x(t+1, t_0, x_0)) - V(t+1, y(t+1, t, x(t, t_0, x_0))) \\ &\quad + V(t+1, y(t+1, t, x(t, t_0, x_0))) - V(t, x(t, t_0, x_0)) \\ &\leq K(t+1)|x(t+1, t_0, x_0) - y(t+1, t, x(t, t_0, x_0))| \\ &\qquad\qquad - (1 - \exp(-\Delta p(t)))u(t) \\ &\leq K(t+1)|F(t, x(t, t_0, x_0))| - (1 - \exp(-\Delta p(t)))u(t) \\ &\leq K(t+1)g(t, u(t)) - (1 - \exp(-\Delta p(t)))u(t).\end{aligned}$$

Then, for a solution $r(t, t_0, r_0)$ of (3.2), it follows from Corollary 1 that $V(t, x(t, t_0, x_0)) \leq r(t, t_0, r_0)$, provided $V(t_0, x_0) \leq r_0$. Hence we have from (b) in Theorem 3.1 that

$$(3.4) \qquad |x(t, t_0, x_0)| \leq r(t, t_0, r_0), \quad t \in [t_0, t_1 + 1] \cap I_\infty .$$

If the trivial solution of (3.2) is stable or asymptotically stable, for any given $\varepsilon \in (0, \rho)$ and $t_0 \in I_\infty$, there exists a $\delta(\varepsilon, t_0)$ such that $r(t, t_0, r_0) < \varepsilon$, $t \geq t_0$, provided $r_0 < \delta$. Then the inequality (3.4) yields a contradiction: $\rho \leq |x(t_1+1)| \leq r(t_1+1, t_0, r_0) < \rho$, since $|x_0| \leq V(t_0, x_0) \leq r_0 < \delta$. Therefore, we have the validity of $|x(t, t_0, x_0)| \leq r(t, t_0, r_0)$ for any $t \geq t_0$, which implies our result.

Theorem 3.3. Suppose that

(i) $f(t, x)$ is linear in x for any $t \in I_\infty$, and the trivial solution of (3.1) is exponentially asymptotically stable;

(ii) $F(t, x)$ is defined for $t \in I_\infty$ and $|x| < \rho$, and satisfies an inequality $|F(t, x)| \leq c|x|$, $t \in I_\infty$, $|x| < \rho$ for sufficiently small c.

Then the trivial solution of (3.3) is also exponentially asymptotically stable.

Proof. A proof using not Lyapunov function, but Lemma 3 has been shown in [3]. The following is a direct application of Lyapunov function obtained above. Now, it follows from (i) that there exists a function $V(t, x)$ satisfying three conditions in Theorem 3.1. Let $x(t) \equiv x(t, t_0, x_0)$ be a solution of (3.3) such that $|x_0| < \rho/K$. Setting $u(t) = V(t, x(t, t_0, x_0))$, because of (b), we have $u(t_0) = V(t_0, x_0) \le K|x_0| < \rho$, if $|x_0| < \rho/K$. We claim that $u(t) < \rho$, $t \ge t_0$. If this were false, there would exist a $t_1 > t_0$ such that $u(t_1 + 1) \ge \rho$, $u(t) < \rho$, $t \in [t_0, t_1] \wedge I_\infty$. For these t's, by means of the same method as in the proof of the preceding theorem,

$$\Delta u(t) \le K|F(t, x(t,t_0,x_0))| - (1-e^{-\alpha})u(t) \le (Kc - (1-e^{-\alpha}))u(t).$$

For a constant $\sigma \in (0, 1)$, we choose c so small that $Kc - (1 - e^{-\alpha}) < -\sigma$. Then, from the above inequality, we obtain $\Delta u(t) \le -\alpha u(t)$, which yields $u(t) \le u(t_0) \exp(-\alpha(t-t_0))$. This inequality leads to a contradiction: $\rho \le u(t_1 + 1) \le K|x_0| < \rho$. Hence, for sufficiently small c, we arrive at the inequality

$$\Delta V(t, x(t, t_0, x_0)) \le -\sigma V(t, x(t, t_0, x_0)), \quad t \ge t_0.$$

Hence, we obtain

$$|x(t, t_0, x_0)| \le V(t, x(t, t_0, x_0)) \le V(t_0, x_0)e^{-\sigma(t-t_0)}$$

$$\le \rho e^{-\sigma(t-t_0)}, \quad t \ge t_0.$$

Theorem 3.4. Suppose that

(i) $f(t, x)$ and $F(t, x)$ is defined for $t \in I_\infty$ and $|x| < \infty$;

(ii) $V(t, x)$ is defined for $t \in I_\infty$ and $|x| < \infty$, Lipschitzian in x for a function $K(t)$, and

$$a(|x|) \le V(t, x) \; (\le b(|x|)), \quad t \in I_\infty, \quad |x| < \infty,$$

where $a(r)$ and $b(r)$ are continuous for $0 \le r < \infty$, strictly monotone increasing, and $a(0) = b(0) = 0$;

(iii) $g(t, r)$ is defined for $t \in I_\infty$ and $0 \le r < \infty$, $g(t, 0) \equiv 0$, and $r + g(t, r)$ is nondecreasing in r, and

$$\Delta V(t, x(t)) \le g(t, V(t, x(t))), \quad t \ge t_0,$$

where $x(t)$ is an arbitrary solution of

(3.5) $\qquad x(t+1) = f(t, x(t)), \quad x(t_0) = x_0, \quad t \ge t_0;$

(iv) $w(t, r)$ is defined for $t \in I_\infty$ and $0 \le r < \infty$, nondecreasing in r, $w(t, 0) \equiv 0$, and

$$|F(t, x)| \leq w(t, |x|), \quad t \in I \quad , |x| < \infty .$$

Then the stability properties of the trivial solution of

(3.6) $\Delta r(t) = g(t, r(t)) + K(t+1)w(t, a^{-1}(r(t)))$

implies the corresponding stability properties of (3.3).

Proof. Let $x(t) \equiv x(t, t_0, x_0)$ be a solution of (3.3). Then, setting $u(t) = V(t, x(t))$, by the same reason as before, it follows that

$$\Delta u(t) \leq g(t, u(t)) + K(t+1)w(t, a^{-1}(u(t))),$$

which implies $u(t) \leq r(t)$, $t \geq t_0$, where $r(t) \equiv r(t, t_0, r_0)$ is a solution of (3.6) such that $r_0 = V(t_0, x_0)$. Hence from (ii) we have $|x(t, t_0, x_0)| \leq a^{-1}(r(t, t_0, r_0))$, $t \geq t_0$. The conclusion can easily be obtained from the inequality. The function $b(r)$ is used for the uniformity.

Corollary. The functions $a(r) \equiv r$, $g(t, r) \equiv -\alpha r$, $0 < \alpha < 1$, $K(t) \equiv K$, $w(t, r) \equiv \lambda(t)r$, $\lambda(t) \geq 0$, and

$$\overline{\lim_{t \to \infty}} \frac{1}{t-t_0} \sum_{s=t_0}^{t-1} \lambda(s) < \frac{\alpha}{K}$$

are admissible in the preceding theorem to guarantee the uniformly asymptotic stability of the trivial solution of (3.3).

Theorem 3.5. Suppose that

(i) $A(t)$ is defined for $t \in I_\infty$;

(ii) $F(t, x)$ is defined for $t \in I_\infty$ and $|x| < \rho$, $F(t, 0) \equiv 0$, and for any given $\varepsilon > 0$ there exists a $\delta(\varepsilon)$ such that $|F(t, x)| \leq \varepsilon |x|$ uniformly in $t \in I_\infty$, provided $|x| < \delta(\varepsilon)$;

(iii) $V(t, x)$ is defined for $t \in I_\infty$ and $|x| < \infty$, Lipschitzian in x for a constant $K > 0$ and

$$|x| \leq V(t, x) \leq K|x|, \quad t \in I_\infty, \quad |x| < \infty \quad ;$$

(iv) for a solution $y(t)$ of $y(t+1) = A(t)y(t)$,

$$\Delta V(t, y(t)) \leq \alpha(t)V(t, y(t)), \quad t \in I_\infty ,$$

where

$$\overline{\lim_{t \to \infty}} \frac{1}{t-t_0} \sum_{s=t_0}^{t-1} \alpha(s) < 0.$$

Then the trivial solution of

(3.7) $\Delta x(t) = A(t)x(t) + F(t, x(t)), \quad x(t_0) = x_0, \quad t \geq t_0$

is asymptotically stable.

Proof. From (iv), for a sufficiently small $\varepsilon > 0$ we have

$$\lim_{t\to\infty} \exp\,(K\,\varepsilon(t-t_0) + \sum_{s=t_0}^{t-1} \alpha(s)) = 0.$$

Hence there exists a constant B such that

$$\max_{t \geq t_0} \exp\,(K\,\varepsilon(t-t_0) + \sum_{s=t_0}^{t-1} \alpha(s)) \leq B.$$

Then, if we choose $|x_0| < \delta_1$ such that $KB\,\delta_1 < \delta(\varepsilon)$, by means of the same reason as before, we have $|x(t)| < \delta(\varepsilon)$, $t \geq t_0$, where $x(t) \equiv x(t, t_0, x_0)$ is a solution of (3.7). Furthermore, we have

$$|x(t)| \leq V(t, x(t)) \leq V(t_0, x_0)\exp(K\,\varepsilon(t-t_0) + \sum_{s=t_0}^{t-1} \alpha(s)).$$

Thus, if ε is small enough, it is easily obtained that $\lim_{t\to\infty} x(t) = 0$.

References

[1] Halanay, A. Differential equations. Stability, Oscillations, Time-lags. Acad. Press, 1966.

[2] Lakshmikantham, V. and Leela, S. Differential and integral inequalities, I. Acad. Press, 1969.

[3] Sugiyama, S. Bull.Sci.Engr.Research Lab.Waseda Univ. 45, 140-144 (1969).

[4] Sugiyama, S. Proc. Japan Acad. 45, 526-529(1969).

[5] Sugiyama, S. Mem.School Sci.Engr.Waseda Univ. 32, 79-88(1969).

[6] Sugiyama, S. Bull.Sci.Engr.Research Lab.Waseda Univ. 47, 77-82 (1970).

[7] Szegö, G. and Kalman, R. C.R.Acad.Sci.Paris,257, 388-390(1963).

[8] Vidal, P. Non-linear sampled-data systems. Gordon and Breach, 1969.

SELECTED TOPICS IN DIFFERENTIAL DELAY EQUATIONS

James A. Yorke

On the occasion of this Japanese-American Conference, it seems perhaps appropriate that my paper on differential delay equations should not just describe my own work, but also should attempt to discuss some of the most recent American papers with which I am familiar. The choice of topics is determined largely from my own interests but is limited by papers of which I have preprints, (or other knowledge). Almost all of the papers are presently unpublished. In order to keep this report short and as understandable as possible, my emphasis is on new ideas which are easy to describe and so the reader should assume that only special cases of the several authors' results are stated here,--that complicated hypotheses have been replaced by simpler special cases of these hypotheses--in order to clarify the most important idea. I am indebted to the authors for discussion and hope I have stated their results accurately. In addition I am indebted to A. Halanay and G. Dunkel for discussion concerning this "selected survey". For another recent selected survey, see [HY] or parts of [Y]. For an elementary book, see [E]. For a more specialized book discussing several special topics see [H].

1. Notation

For $q > 0$, let C_q be the set of continuous functions from $[-q,0]$ to R^n (with the supremum norm). If x is a function defined on at least $[t-q,t]$, with values in R^n, write x_t for the function defined on $[-q,0]$ by

$$x_t(s) = x(t+s) \quad \text{for} \quad s \in [-q,0] . \tag{1.1}$$

Hence if x is continuous, then $x_t \in C_q$. The equation (where $g : R^n \to R^n$ is continuous)

$$x'(t) = g(x(t),x(t-1))$$

can then be written in the usual notation

$$x'(t) = g(x_t(0),x_t(-1)) .$$

For $\phi \in C_q$ (for $q \geq 1$) we have the continuous map F given by

---

This research was partially supported by the National Science Foundation under Grant NSF GP 27284.

$$\phi \overset{F}{\to} g(\phi(0),\phi(-1)) : C_q \overset{F}{\to} R^n .$$

In this notation Eq. (1.1) can be written

$$x'(t) = F(x_t) . \tag{1.2}$$

For non-autonomous equations like $x'(t) = g(t,x(t),x(t-1))$, we allow $F : R \times C_q \to R^n$ and

$$x'(t) = F(t,x_t) \tag{1.3}$$

For a solution of (1.1), (1.2), or (1.3) write $x(t) = x(t,t_o,\phi)$ if $x_{t_o} = \phi \in C_q$. The function ϕ is the initial function (at time t_o).

2. Results on Existence and Uniqueness

Delfour and Mitter [DM1, DM2, DM3, DM4] study equations which include the case

$$x'(t) = g(t,x(t-\theta_1),\ldots, x(t-\theta_N)) \tag{2.1}$$

where $0 \leq \theta_1 < \ldots < \theta_N \leq b$ and where $x(t) \in E$, a Banach space and $g : R \times E^N \to E$ (letting E^N denote $E \times \ldots \times E$, N times). The main novelty of the approach is to allow the initial condition $x_{t_o} = \phi$ where ϕ is an element of $L^p([-b,0];E)$ with the extra condition that $\phi(0)$ is specified.

Let $p \in [1,\infty)$ and let E be any Banach space. Define $M^p([-b,0];E)$ to be the space of measurable function ϕ from $[-b,0]$ to E, identifying functions ϕ_1 and ϕ_2 which are equal at 0 and are equal almost everywhere on $[-b,0]$. The norm for M^p is

$$||\phi||_{M^p} = (|\phi(0)|^p + ||\phi||_p^p)^{1/p}$$

where we let $|\cdot|$ denote the norm on E and let $||\phi||_p$ be the norm on $L^p([-b,0];E)$. Their initial condition ϕ is any element of M^p.

Consider the equation

$$x'(t) = x(t-1) .$$

This may be written

$$x'(t) = F(x_t)$$

where $F(\phi) = \phi(-1)$ if ϕ is assumed to be defined at -1. But the mapping $\phi \to \phi(-1)$ is not well defined for elements of M^p which are only defined almost everywhere. Yet they are able to conclude a solution $x(t,t_o,\phi)$ exists for each initial $\phi \in M^p$. Delfour and Mitter actually allow g in Eq. (2.1) to depend on an additional term x_t. Since x_t is measurable, it does not include the other terms of the form $x(t-\theta_N)$. We omit the x_t term to simplify the notation.

Their main results are Carathéodory-type existence theorems. The methods of proof are attributed to A. Bielecki and C. Corduneanu.

Theorem 2.1. Assume g has the following properties:

(CAR-1) The map $t \to g(t,z)$ is measurable for each $z \in E^N$;

(LIP) there exists a non-negative measurable function $m : R \to R$ which is L^q integrable on compact intervals, where $p^{-1} + q^{-1} = 1$, such that for z_1 and $z_2 \in E^N$

$$|g(t,z_1) - g(t,z_2)| \leq m(t)\ |z_1 - z_2|_{E^N} \quad \text{(for almost all } t)$$

where $|\cdot|_{E^N}$ is, say, the sum of norms of the components.

(BC) The map $t \to g(t,0)$ is L^1-integrable on compact intervals.

Then for each $\phi \in M^p$ there exists a unique global solution of (2.1) $x : [t_o-b,\infty) \to E$ such that $x_t \in M^p$ and x is absolutely continuous on each $[t_o,t]$ where $t_o < t$. Furthermore, the solution depends continuously on the initial function $\phi \in M_p$.

The above type of result would seem advantageous for a variety of problems, particularly those dealing with linear equations, to allow initial data to be chosen from a complete inner product space (as the above theorem allows for the case $p = 2$).

Let X be the space of all continuous $F : R \times C_q \to R^n$, and let X be given the topology of uniform convergence (or alternatively, the topology of uniform convergence on bounded subsets of $R \times C_q$). Then a metric can be given to X-consistent with the topology-so that it is a complete metric space. Let $\phi_o \in C_q$ and $t_o \in R$ and consider all of the equations

$$x'(t) = F(t,x_t)\ ,\ x_{t_o} = \phi_o\ . \tag{2.1}$$

of course examples can be given showing that F does not always uniquely determine the solution (for fixed ϕ_o and t_o). If $F(t,\phi)$ satisfies a Lipschitz condition in ϕ as in the above theorem (or if $F(t,\phi)$ is locally Lipschitzean in ϕ) then the solution of (2.1) is unique. But the set $L = L_{t_o,\phi_o}$ of F which are Lipschitzean in ϕ is some neighborhood N (depending on F) of $(t_o,\phi_o) \in R \times C_q$ is a small set in the sense of Baire category; that is, L is meager set (a set of the first category), the countable union of nowhere dense sets. Let $U = U_{t_o,\phi_o}$ be the set of $F \in X$ such that (2.1) has a unique solution. Since the set L_{t_o,ϕ_o} is "small", it might be expected that U_{t_o,ϕ_o} would be small. Costello [C] showed just the opposite.

Theorem. The set $X-U_{t_o,\phi_o}$ which is the set of continuous F which fail to have a unique solution of (2.1) is a set of the first category.

This result say that uniqueness of solutions is a generic property. This result generalizes a similar result of Orlicz [Or] using techniques of Lasota and Yorke [LY1] for ordinary differential equations in Banach spaces. Costello also proved similar results for uniqueness and existence for hyperbolic partial differential equations.

3. Linear Theory

Weiss [W] asked a question concerning the linear equation

$$x'(t) = Ax(t) + Bx(t-1) \tag{3.1}$$

where A and B are (constant) $n\times n$ matrices. Consider all initial conditions $\phi \in C_q$ at time 0 and define the reachable set (in R^n) at time t by

$$R(t) = \{x(t,0,\phi) : \phi \in C_q\} .$$

He asked if $R(t)$ is always equal to R^n (for $t > 0$) or if it can be a subspace of lower dimension. We say (3.1) is pointwise complete if $R(t) = R^n$ for all $t > 0$. V.M. Popov [P1] and A.M. Zverkin [Z] have both independently found an example for which $R^n = R^3$ and $R(t)$ is two-dimensional (for sufficiently large t). Both have discovered several theorems showing when $R(t)$ can and cannot be lower dimensional. Popov's example follows.

Consider $x = (\xi,\eta,\psi)$ and

$$\begin{aligned} \xi'(t) &= 2\eta(t) \\ \eta'(t) &= -\psi(t) + \xi(t-1) \\ \psi'(t) &= 2\eta(t-1) . \end{aligned} \tag{3.2}$$

Consider $x(t) = (\zeta(t), \eta(t), \psi(t))$ defined on $[-1,\infty)$, satisfying (3.2) for $t \geq 0$, where x on $[-1,0]$ is an arbitrary continuous initial function. It follows immediately that

$$\eta''(t) = 0 \quad \text{for} \quad t > 1 \tag{3.3}$$

and $\xi''' = 0$, so ξ is quadratic or linear for $t > 1$ and so

$$\xi(t) - \xi(t-1) = [\xi'(t) + \xi'(t-1)]/2 \quad \text{for} \quad t > 2 \tag{3.4}$$

so ξ is quadratic (or linear). Consider the inner product of x with $(1,-2,-1)$ in R^3. Write ζ, η and ψ for $\zeta(t)$, $\eta(t)$ and $\psi(t)$ and write $\overline{\zeta}$, $\overline{\eta}$ and $\overline{\psi}$ for $\zeta(t-1)$, $\eta(t-1)$ and $\psi(t-1)$. Then for $t > 2$ $\langle x(t), (1,-2,-1)\rangle$ is

$$\begin{aligned}\xi - 2\eta - \psi &= (\overline{\xi} + \frac{\overline{\xi'+\xi'}}{2}) - 2\eta - \psi \quad \text{(from Eq. (3.4))}\\ &= (\overline{\zeta} + \overline{\eta} + \eta) - 2\eta - \psi \\ &= \overline{\xi} - \eta' - \psi \quad \text{(from Eq. (3.3))}\end{aligned}$$

which is identically 0 from the center line of the system (3.2). Hence for $t \geq 2$, each point of $R(t)$ is perpendicular to $(1,-2,-1)$ so $R(t)$ has dimension less than 3. This derivation showing that (3.2) is not pointwise complete is based on a suggestion of R. Driver and is different from the derivation in [Pl].

This startling example was the start of a complete theory of such behavior. Rather than discuss the detailed necessary and sufficient conditions for the autonomous case, a simpler sufficient condition for pointwise completeness is now stated which is valid for the non-autonomous system

$$x'(t) = A(t)x(t) + B(t)x(t-1) \tag{3.5}$$

(where $A(t)$ and $B(t)$ are continuous matrix functions of t).

<u>Theorem</u> (Zverkin). For $t_1 > t_0$ the reachable set for (3.5)

$$R(t_0,t_1) = \{x(t_1,t_0,\phi) : \phi \in C_q\}$$

equals R^n if $\{t \in [t_0,t_1] : \det B(t) = 0\}$ contains no intervals.

Brooks and Schmitt [BS] proved that the equation (where A_i are $n\times n$ matrices and $\tau_i > 0$)

$$x' = \Sigma A_i x(t-\tau_i)$$

is pointwise complete if $A_iA_j = A_jA_i$ for all i and j. Their methods extend to certain neutral equations. This result is proved by using the Banach algebra of all polynomials generated by the matrices A_i and studying the multiplicative (algebra) homomorphisms on this space. These functionals h are complex-valued maps which evaluate the eigenvalue of A on some subspace $L \subset R^n$ which is an eigenspace for all A_i. The Gelfand transform is used to give equations which they know can be solved.

Grainger Morris, Ernest W. Bowen and Alan Feldstein have studied the equation

$$x'(t) = -x(t/k) \qquad x(0) = 1$$

for $k > 1$, which is a differential delay equation on $[0,\infty)$. They show that the unique solution of this equation is unbounded (and it follows that x has infinitely many zeros). The study of the solution in the complex plane,

$$x(t) = \sum_{n=0}^{\infty} \frac{(-1)^n t^n}{n!k^{n(n-1)/2}}$$

which converges for all t and is entire. Their main tool is the Phragmen-Lindeloff Theorem, which enables them to prove that if x is bounded on $[0,\infty)$ then it is bounded in the complex plane and so x would have to be constant, (which it clearly cannot be). Their methods extend to certain scalar equations of the form

$$x'(t) = \sum_{i=1}^{N} a_i x\left(\frac{t}{k_i} - b_i\right), \quad x(0) = 1 .$$

4. The Theory of Disease

These are several approaches in attempting to find an equation to describe how a disease spreads through a population as a function of time. For example, we must choose between deterministic or probabilistic models. Probabilistic or stochastic models are necessary when dealing with diseased populations which have very few individuals sick at various times, but such models have the drawback that they are very difficult to analyze since there is no single solution $x : [t_o, t_1] \to R$ to analyze. It is also necessary to decide whether to use an equation with spacial dependence. Allowing spacial dependence can lead to partial differential equations of diffusion type (and these models are not very believable) or if a finite number of point locations are allowed, a system of ordinary differential equations for which there are many "free" parameters which must be chosen without much justification. This section discusses only deterministic spacially homogeneous (allowing for no spacial variation in the numbers of sick individuals).

For simplicity, at any time t the population is divided into four disjoint subpopulations:

$S(t)$ = the number of susceptible individuals

$E(t)$ = the number of individuals exposed to the infection, who will as a result become infections (but are not yet infectious)

$I(t)$ = the number of infectious individuals

$R(t)$ = the number of individuals who are immune to the infection .

Deterministic models treat these variables as being continuously varying, though the possible values actually are discrete.

Certain disease changes occur after fairly specific intervals and this effect causes time lags to enter into the equation. For example, an individual who is exposed to measles at time t (measured in days) can be assumed to be infectious for the period $[t + 12, t + 14]$, that is for the period from 12 to 14 days following exposure. We now describe a measles model. Measles is a very infectious disease and in large population centers almost all people are infected by age 15. It is also extremely rare (if it ever occurs) for individuals to contract the disease twice. Hence descriptively we have

births → susceptibles → exposed → infectious → immune .

The disease is spread from an infectious individual to a susceptible individual and the number of contacts capable of spreading the disease therefore depends linearly on $S(t)$ and $I(t)$, that is,

$$\text{rate of new infection} = \beta(t)S(t)I(t)$$

where $\rho(t)$ is a proportionality constant that is seasonally dependent, having a one-year period. The susceptibles $S(t)$ are increased by the rate γ at which individuals enter the population (through net immigration or birth), so

$$S'(t) = -\beta(t)S(t)I(t) + \gamma \quad . \tag{4.1}$$

The number of infectious individuals is precisely those individuals who were infected βSI between 12 and 14 days earlier,

$$I(t) = \int_{t-14}^{t-12} \beta(x)S(x)I(x)dx$$

$$= \int_{t-14}^{t-12} \gamma - S' = 2\gamma + S(t-14) - S(t-12) \quad .$$

Substituting this equation into (4.1) yields the differential equation

$$S'(t) = -\beta(t)S(t)[2\gamma + S(t-14) - S(t-12)] + \gamma .$$

Equation (4.2) has been investigated by W. London, M.D. and J.A. Yorke using actual monthly data of the past 30 years from New York City and Baltimore, to determine how closely the data satisfies the model. Using the function $\beta(t)$ calculated from the data, Eq. (4.2) has periodic solutions (according to computer simulations) with a two year period and another with a three year period. The two year period-a high incidence year followed by a low incidence year-has been observed in a number of large cities, and has been commented on in numerous papers. Two other contagious virus diseases are mumps and chickenpox, and these will be compared with measles since they do not exhibit two year cycles though they also should satisfy (4.2)-though the crucial function β should be different.

There is a tendency for people to use models without delays. Differentiating the first $I(t)$ equation yields

$$I'(t) = \beta(t-12)S(t-12)I(t-12)-\beta(t-14)S(t-14)I(t-14) \quad . \tag{4.3}$$

It is sometimes assumed that for diseases that (a) exposed individuals immediately become infectious (i.e. the incubation period of 12 days is changed to 0) and (b) that the recovery rate is purely proportional to $I(t)$ with some constant α (which must be guessed), so we would have

$$I'(t) = \beta(t)S(t)I(t) - \alpha I(t) .$$

For many diseases assumption (a) is justified since the incubation period may be short compared to the infectious period, but assumption (b) seems to be never justified. It does not correspond to plausible biological hypotheses.

Hoppensteadt and Waltman [HW] have investigated existence and uniqueness for disease models which exhibit time delays. The principle feature of their models is that they allow for certain threshold phenomena involved in the spread of infection. No diseases are mentioned which require their general threshold effect but cholera is a possibility. (Actually they need not have proved existence, since their equations are differential delay equations, even though the lags are complicated and are implicitly defined. The second of their papers allows for a period of immunity after recovery and then the individuals again become susceptible. They have included some numerical work.

Certain infectious diseases are more easily studied than others. Measles, Mumps and Chickenpox are relatively simple because a recovered individual can be ignored in the model since he is immune to further infection.

Another disease is relatively simple for a different reason. The disease gonorrhea has the property that people recover only after drug therapy and do not develop any observable resistence to the disease. After recovery, they immediately become susceptible. (Of course some individuals are more liable to get gonorrhea than others.) This disease has been studied by Cooke and Yorke [CY]. The gonorrhea model assumes the population is of constant size and that it is composed of two groups: susceptibles and infectious individuals. Also it is not a seasonal disease and the coefficients are time independent. The rate of new infection depends only on $S(t)$ and $I(t)$, but since $S(t)$ = total population - $I(t)$, it in effect depends only on $I(t)$ and can be written $g(I(t))$ for some continuous g. It may be assumed that either there is a single infectious period L (the time it takes the infectious individual to seek out and receive treatment) or a distributed lag. Considering here only the single lag we have

$$I'(t) = g(I(t)) - g(I(t-L)) \tag{4.5}$$

or considering an integrated form of this equation,

$$I(t) = \int_{t-L}^{t} g(I(x))dx \, . \tag{4.6}$$

Let the probability of still having the disease time s after initial infection is denoted $P(s)$, (so we assume for Theorems 4.1 and 4.2 that $P(0) = 1$ and $P(L) = 0$ and P is monotonically decreasing,) we have the more general equations

$$I'(t) = g(I(t)) + \int_{0}^{L} g(I(t-s))dP(s) \tag{4.7}$$

or the integrated form is

$$I(t) = \int_{t-L}^{t} g(I(s))P(t-s)ds + c \, . \tag{4.8}$$

This model is also useful for population growth models and economic growth models, so P and g are kept as general as possible. For gonorrhea, we may let $c = 0$. The following results are proved.

<u>Theorem</u> 4.1. Let $g : R \to R$ be locally Lipschitzean. Let I be a solution of (4.8) on $[0,\infty)$ for some c. Assume for some t_0 that

$$g(I(t_0)) \geq g(I(t)) \quad \text{for} \quad t \in [t_0-L, t_0] \, .$$

Then $I(t_0) \leq I(t)$ for all $t \geq t_0$.

This says that if the rate of new infection $g(I(t_o))$ at time t_o, is at least as high as during the previous period of length L, then in the future, the level of infection $I(t)$ will never be smaller than the present level $I(t_0)$.

Theorem 4.2. Let $g : R \to R$ be locally Lipschitzean and let I be a solution of (4.8) for all $t \geq 0$ for some c. If $|g|$ is bounded (as in the case with diseases) then

$$I(t) \to \text{constant} \qquad \text{as} \quad t \to \infty . \tag{4.9}$$

If g is unbounded either (4.9) occurs of

$$x(t) \to \infty \quad \text{or} \quad x(t) \to -\infty \quad \text{as} \quad t \to \infty .$$

Notice that even if g is unbounded all bounded solutions must go to a constant. For gonorrhea this says that the level of infection will tend to a constant and there can be no periodic oscillation (as occurs in epidemic diseases like measles). Of course the model ignores social changes in morals or changes in the ability of the disease to resist various treatments and ignores the possibility of new treatments. A more detailed discussion of the hypotheses appears in [CY].

Equations (4.7) and (4.8) can allow for immigration and emmigration. Suppose we wish to model gonorrhea in a city in which we assume that any individual in the city at time s will be in the city at $t > s$ with probability $e^{\alpha[s-t]}$ for some $\alpha \geq 0$. For simplicity, assume everyone who gets the disease has it for time precisely L (yielding Eq. (4.5) or (4.6) if $\alpha = 0$). Assume people entering the city do not have gonorrhea. Then the number of infections $I(t)$ in the city is the integrated sum of those individuals who contracted the disease, multiplied by the probability that they are still in the city, hence

$$I(t) = \int_{t-L}^{t} g(I(s))e^{\alpha[s-t]}ds ,$$

which is of the form (4.8), letting $P(t) = e^{-\alpha t}$ for $t \in [0,L)$ and $P(L) = 0$. Hence again (for g bounded), $I(t) \to$ constant as $t \to \infty$.

5. Additional Problems

Levin and Shea have studied the asymptotic behavior of the equations

$$x'(t) + \int_{-\infty}^{\infty} g(x(t-\zeta))dA(\zeta) = f(t) \tag{5.1}$$

$$x(t) + \int_{-\infty}^{\infty} g(x(t-\zeta))dA(\zeta) = f(t) \tag{5.2}$$

where A is of bounded variation, f is in L^∞ and $f(t) \to$ constant as $t \to \infty$. The following problem is studied in both (5.1) and (5.2). They study the asymptotic behavior of (locally absolutely continuous) bounded solutions $x(t)$ of these equations as $t \to \infty$ which can always be written in the form

$$x(t) = \sum_{m=1}^{\infty} \psi_m(t) y_m(t) + \eta(t) \quad t \geq 0$$

where $\eta(t) \to 0$, and each y_m is a solution of the same equation as x, except with f constant $(= \lim_{t\to\infty} f(t))$, and $\psi_m : [0,\infty) \to R$ is a set of weighting functions with $\Sigma\psi_m \equiv 1$. The objective of their study is to show when the functions ψ_m can be given special properties, each ψ_m being almost 1 over a long interval of time, an interval which increases in length as $m \to \infty$. Equation (5.1) includes certain ordinary differential equations and their results (and their approach) seem to be new even for that case. Their methods are based on fourier transforms (and they are forced to place hypotheses on $A(\zeta)$ in order to be able to apply transform theorems-including the Wiener-Pitt Theorem).

A related problem which they do not study would be to replace x and f by sequences of functions x_i and f_i for $i = 1,2,\ldots$ and assume that $\sup_{t\in R}|f_i(t)| \to c$ as $i \to \infty$. Then ask how x_i "approaches" solutions of (5.1) (with $f = c$) as $i \to \infty$.

BIBLIOGRAPHY

Many of the results quoted in this paper are based on papers which have not yet been published.

[HY] A. Halanay and J.A. Yorke, Some new results and problems in the theory of differential-delay equations, SIAM Rev., 13(1971) 55-80.

[Y] T. Yoshizawa, Stability by Lyapunov's second method, Math. Society of Japan, 1966.

[E] L.E. El'sgol'ts, Differential Equations with deviating argument, Holden Day, San Francisco, 1966. MR. 36#5464.

[H] J. Hale, Functional Differential Equations, Springer-Verlag, 1971.

[BC] R. Bellman and K. Cooke, Differential-Difference Equations, Academic Press, 1963.

[DM1] M. Delfour and S. Mitter, Systèmes d'équations différentielles héréditaires à retards fixes. Théorèmes d'existence et d'unicité. An announcement.

[DM2] M. Delfour and S. Mitter, Systèmes d'équations différentielles héréditaires à retards fixes. Une classe de systèmes affines et le problème du système adjoint. C.R. Acad. So. Paris, T. 272, 1971 (to appear). An announcement.

[DM3] M. Delfour and S. Mitter, Hereditary differential systems with constant delays. I-General Case, to appear.

[DM4] M. Delfour and S. Mitter, Hereditary differential systems with constant delay, II- A class of affine systems and the adjoint problem.

[P1] V.M. Popov, Pointwise degeneracy of linear time-invariant delay differential equations.

[P2] V.M. Popov, On the property of reachability for some delay-differential equations, E.E. Tech. Report R-70-08, U. of Maryland.

[Z] A.M. Zverkin, a communication at a meeting at the Univ. of Friendship of Peoples, May 1971.

[BS] R.M. Brooks and K. Schmitt, Pointwise completeness of differential-difference equations.

[C] T. Costello, On the fundamental theory of differential-delay equations.

[Or] W. Orlicz, Zur Theorie der Differentialgleichung $y' = f(x,y)$, Bull. Akad. Polon. Sci., Sér A(1932), 221-228.

[LY] A. Lasota and J.A. Yorke, The generic property of existence of solutions of differential equations in Banach space.

[HW1] F. Hoppensteadt and P. Waltman, A problem in the theory of epidemics, I and II.

[HW2] F. Hoppensteadt and P. Waltman, A system of integral equations describing a deterministic epidemic model, in Seminar on Differential Equations and Dynamical Systems III, (ed. by D. Sweet and J.A. Yorke), to appear.

[CY] K.L. Cooke and J.A. Yorke, Equations modelling population growth, economic growth, and gonorrhea epidemiology.

New Papers Not Referred To In The Text

R. Datko, Criteria for exponential stability of linear semi-groups and difference schemes,

R.E. Fennell, Periodic solutions of functional differential equations, submitted to Proc. AMS.

R.B. Grafton, Periodic solutions of certain ienard equations with delay, March 1971.

S.E. Grossman, Stability and asymptotic behavior of differential-delay equations, J. Math. Anal. Appl., to appear.

S.E. Grossman and J.A. Yorke, Asymptotic behavior and stability criteria for delay differential equations, J. Differential Equations, to appear.

T.G. Hallam, On nonlinear functional perturbation problems for ordinary differential equations, Univ. of R.I. Technical Report, January 1971.

J. Kaplan and J.A. Yorke, On the stability of a periodic solution of a differential equation with a time lag.

J. Kurzweil, On solutions of nonautonomous linear delayed differential equations which are defined and bounded for $t \to -\infty$, Com. Math. Univ. Car., 12, 1 (1971) p.69-72.

G. Ladas, Oscillation and asymptotic behavior of solutions of differential equations with retarded argument, Univ. of R.I. Technical Report, July 1970.

G. Ladas and V. Lakshmikantham, Oscillations caused by retarded actions, Univ. of R.I. Technical Report, October 1970.

A. Lasota and J.A. Yorke, Bounds for periodic solutions of differential equations in Banach space, J. Differential Equations, to appear.

R.C. MacCamy and J.S.W. Wong, Stability theorems for some functional differential equations, Carnegie-Mellon Univ. Technical Report, March 1970.

R.K. Miller, Asymptotic stability properties of linear Volterra integrodifferential equations.

K. deNevers and K. Schmitt, An application of the shooting method to boundary value problems for second order delay equations, J. Math. Anal. Appl., to appear.

K. Schmitt, Differential inequalities and boundary value problems for differential equations with deviating argument, in Seminar on Differential Equations and Dynamical Systems III, ed. by D. Sweet and J.A. Yorke, to appear.

K. Schmitt, Comparison theorems for second order delay-differential equations, Rocky Mountain J., to appear.

P. Waltman and J.S.W. Wong, Two point boundary value problems for nonlinear functional differential equations, Trans. Amer. Math. Soc., to appear.

STABILITY FOR ALMOST PERIODIC SYSTEMS

Taro YOSHIZAWA

By assuming the boundedness of solutions, the existence of periodic solutions has been discussed by many authors. However, the boundedness of solutions does not necessarily imply the existence of almost periodic solutions even for scalar equations (cf.[4], [11]). In discussing the existence of an almost periodic solution, several authors have been assumed that the almost periodic system has a bounded solution which has some kind of stability properties, and many interesting results have been obtained. For the references, see [20] and [22]. Deysach and Sell [3] discussed the existence of an almost periodic solution under the assumption that the periodic system has a bounded solution which is uniformly stable. Miller [9] and Seifert [14] considered an almost periodic system. Miller assumed that the bounded solution is totally stable, and Seifert assumed the Σ-stability of the bounded solution which is equivalent to the stability under disturbances introduced by Sell [17]. All of them used the theory for dynamical systems, and hence the uniqueness of solutions is assumed. Coppel [2] and the author [22] have shown that those results can be obtained by using the property of asymptotically almost periodic functions without assuming the uniqueness of solutions. Stability properties which have been assumed are uniform stability, uniformly asymptotic stability, total stability and stability under disturbances from the hull. Relationships between these stability properties are simple for periodic systems, but not for almost periodic systems.

In this article, we shall show relationships between several kinds of stability properties of a bounded solution of an almost periodic system. Let $f(t,x)$ be a continuous function defined on $I\times S_{B*}$ with values in R^n, where $I=[0,\infty)$ and $S_{B*}=\{x;\ x<B^*\}$, and consider a system of differential equations

$$(1) \qquad x' = f(t,x) \qquad (' = \frac{d}{dt}).$$

In the case where $f(t,x)$ is defined on $R\times S_{B*}$, $R=(-\infty,\infty)$, and is periodic in t of period ω, $\omega>0$, that is, $f(t+\omega,x)=f(t,x)$ for all $t\in R$ and $x\in S_{B*}$, we shall denote by (P) the system (1), and in the case where $f(t,x)$ is defined on $R\times S_{B*}$ and is almost periodic in t uniformly for $x\in S_{B*}$, we shall denote by (AP) the system (1).

For the periodic system (P) defined on $R\times R^n$, the following existence theorem for a periodic solution of period ω is known [19]. For a more general dissipative system, see [5].

Proposition 1. Assume that the solution of (P) is unique for the initial value problem. If the solutions of (P) are equiultimately bounded for bound B (consequently uniformly ultimately bounded), then the system (P) has a periodic solution $p(t)$ of period ω such that $\|p(t)\| \leqq B$ for all $t \in R$.

Recently, Paval [12] has shown that if the solution of (P) is unique for the initial condition and the solutions of (P) are ultimately bounded, then the solutions of (P) are uniformly bounded and consequently the solutions are uniformly ultimately bounded. Without uniqueness we have the following result.

Proposition 2. For the periodic system (P) defined on $R \times R^n$, if there is a constant $B>0$ and for any $t_0 \varepsilon I$ and any $x_0 \varepsilon R^n$, there exists a $T(t_0,x_0) \geqq 0$ such that $\|x(t;x_0,t_0)\| < B$ for all $t \geqq T(t_0,x_0)$, where $x(t;x_0,t_0)$ is a solution of (P) through (t_0,x_0), then the solutions of (P) are uniformly bounded.

Proof. It is sufficient to show that the solutions of (P) are equi-bounded. Suppose that the solutions are not equi-bounded. Then there are an $\alpha>0$, $t_0 \geqq 0$, $\{x_k\}$ and $\{\tau_k\}$ such that $\|x_k\| \leqq \alpha$, $\tau_k \geqq t_0$ and $\|x(\tau_k;x_k,t_0)\| \geqq k$, where we can assume that $\alpha>B$ and $k>\alpha$. There exists a t_k such that

$$\|x(t_k;x_k,t_0)\| = \alpha, \quad \alpha < \|x(t;x_k,t_0)\| \quad \text{for } t_k < t \leqq \tau_k.$$

Let $m_k \geqq 0$ be an integer such that $t_k = m_k\omega + \sigma_k$, $0 \leqq \sigma_k < \omega$, and set $\tau_k = m_k\omega + \tau_k'$ and $\bar{x}_k = x(t_k;x_k,t_0)$. Then $\|\bar{x}_k\| = \alpha$ and $x(t;x_k,t_0) = x(t;\bar{x}_k,t_k)$ for $t \geqq t_k$. Since the system is periodic of period ω, there is a solution of (P) such that $x(t;\bar{x}_k,\sigma_k) = x(t+m_k\omega;\bar{x}_k,t_k)$ for $t \geqq \sigma_k$. Thus $\|x(\tau_k';\bar{x}_k,\sigma_k)\| \geqq k$, $\|x(\sigma_k;\bar{x}_k,\sigma_k)\| = \alpha$ and $\alpha < \|x(t;\bar{x}_k,\sigma_k)\|$ for $\sigma_k < t \leqq \tau_k'$. There are subsequences $\{\bar{x}_{k_j}\}$ and $\{\sigma_{k_j}\}$ such that $\bar{x}_{k_j} \to x_0$, $\sigma_{k_j} \to \sigma_0$ as $j \to \infty$, and we have $\|x_0\| = \alpha$, $0 \leqq \sigma_0 \leqq \omega$. Let $x(t;x_0,\sigma_0)$ be a solution of (P) through (σ_0,x_0). Then $\|x(t;x_0,\sigma_0)\| < B$ for all $t \geqq T(\sigma_0,x_0)$. Since every solution starting from $0 \leqq t \leqq T(\sigma_0,x_0)$, $\|x\| \leqq \alpha$ is continuable to $T(\sigma_0,x_0)$, there is a $K>0$ such that

$$\|x(t;\bar{x}_{k_j},\sigma_{k_j})\| < K \quad \text{for large } j \text{ and } \sigma_{k_j} \leqq t \leqq T(\sigma_0,x_0).$$

These solutions are uniformly bounded and equicontinuous, and hence there is a subsequence, which we shall denote by $x(t;\bar{x}_{k_j},\sigma_{k_j})$ again, such that $x(t;\bar{x}_{k_j},\sigma_{k_j})$ tends to a certain solution $x(t;x_0,\sigma_0)$ uniformly on $\sigma_0 \leqq t \leqq T(\sigma_0,x_0)$ as $j \to \infty$. For large j, τ_k' can not be less than $T(\sigma_0,x_0)$.

Since $\alpha<\|x(t;\bar{x}_{k_j},\sigma_{k_j})\|$ for $\sigma_{k_j}<t\leqq\tau'_{k_j}$ and $\|x(T;\bar{x}_{k_j},\sigma_{k_j})\|<B<\alpha$ for large j, there arise a contradiction if $\tau'_{k_j}\geqq T(\sigma_0,x_0)$. This proves the proposition.

Remark 1. Even for a linear scalar almost periodic system, ultimate boundedness does not necessarily imply uniform boundedness as Conley and Miller's example [1] shows. In their example, the zero solution is asymptotically stable, but not uniformly stable.

Remark 2. Fink and Frederickson [4] have constructed an almost periodic differential equation, for which the solutions are uniformly ultimately bounded, but none are almost periodic.

The following discussion is based on the papers by Kato and the author [6], [7], [23]. Let $\phi(t)$ be a solution of the system (1) defined on I and assume that $\|\phi(t)\|\leqq B$ for all $t\geqq 0$, where $B<B^*$. We denote by $x(t;x_0,t_0)$ a solution of (1) through (t_0,x_0). Sell [16] has introduced the following stability.

Definition 1. $\phi(t)$ is weakly uniformly asymptotically stable, if it is uniformly stable and, in addition, there is a $\delta_0>0$ such that if $t_0\varepsilon I$ and $\|\phi(t_0)-x_0\|<\delta_0$, then $\|\phi(t)-x(t;x_0,t_0)\|\to 0$ as $t\to\infty$.

The weakly uniformly asymptotic stability of $\phi(t)$ implies the equiasymptotic stability of $\phi(t)$, and for the periodic system (P), the weakly uniformly asymptotic stability of $\phi(t)$ is equivalent to the uniformly asymptotic stability of $\phi(t)$, see [21]. For an almost periodic system, Seifert [15] has shown that if the system has the zero solution which is weakly uniformly asymptotically stable, then the zero solution is uniformly asymptotically stable.

Remark 3. For the periodic system, even if $\phi(t)$ is uniformly asymptotically stable, there does not necessarily exist a periodic solution of period ω, but there exists a periodic solution of period $m\omega$ for some integer $m\geqq 1$ which is uniformly asymptotically stable.

Definition 2. $\phi(t)$ is totally stable, if for any $t_0\geqq 0$ and any $\varepsilon>0$ there exists a $\delta(\varepsilon)>0$ such that if $h(t)$ is continuous on $[t_0,\infty)$ and satisfies $\|h(t)\|<\delta(\varepsilon)$ for all $t\geqq t_0$ and if $y_0\varepsilon S_{B^*}$ satisfies $\|\phi(t_0)-y_0\|<\delta(\varepsilon)$, then any solution $y(t)$ through (t_0,y_0) of the system

$$(2)\qquad x' = f(t,x) + h(t)$$

satisfies $\|\phi(t)-y(t)\|<\varepsilon$ for all $t\geqq t_0$.

This definition is equivalent to the usual definition of total stability, see [7]. Clearly the total stability of $\phi(t)$ implies the uniform stability of $\phi(t)$, but the converse is not true as a simple example $x'=0$ shows. Moreover, total stability does not necessarily imply uniformly asymptotic stability even for a scalar autonomous equation, see [8]. However, for a linear system, the total stability of $\phi(t)$ is equivalent to the uniformly asymptotic stability of $\phi(t)$. If $f(t,x)$ in (1) satisfies

$$\|f(t,x) - f(t,y)\| \leq L\|x-y\|$$

for $t\varepsilon I$, $x\varepsilon S_{B*}$, $y\varepsilon S_{B*}$ and if the bounded solution $\phi(t)$ of (1) is uniformly asymptotically stable, then it is totally stable.

Now consider the almost periodic system (AP) and let S be a given compact set in S_{B*}. Let $\phi(t)$ be a solution of (AP) such that $\phi(t)\varepsilon S$ for all $t\geqq 0$. We shall denote by H(f) the hull of f(t,x). For $g\varepsilon H(f)$ and $p\varepsilon H(f)$, define $\rho(g,p;S)$ by

$$\rho(g,p;S) = \sup\{\|g(t,x)-p(t,x)\|;\ t\varepsilon R,\ x\varepsilon S\}.$$

Definition 3. $\phi(t)$ is stable under disturbances from H(f) with respect to S for $t\geqq 0$, if for any $\varepsilon>0$ there exists a $\delta(\varepsilon)>0$ such that $\|\phi(t+\tau)-x(t;x_0,0,g)\|\leqq\varepsilon$ for $t\geqq 0$, whenever $g\varepsilon H(f)$, $\|\phi(\tau)-x_0\|\leqq\delta(\varepsilon)$ and $\rho(f_\tau,g;S)\leqq\delta(\varepsilon)$ for some $\tau\geqq 0$, where $f_\tau=f(t+\tau,x)$ and $x(t;x_0,0,g)$ is a solution of

$$x' = g(t,x) \tag{3}$$

such that $x(0;x_0,0,g)=x_0$ and $x(t;x_0,0,g)\varepsilon S$ for all $t\geqq 0$.

For the solution $\phi(t)$ of (1) such that $\|\phi(t)\|\leqq B$ for $t\geqq 0$, let K be the set of x such that $\|x\|\leqq B_1$, where $B<B_1<B*$. Then K is a compact set in S_{B*} and $\phi(t)\varepsilon K$ for all $t\geqq 0$. Clearly, if the solution $\phi(t)$ of (AP) is stable under disturbances from H(f) with respect to K, then $\phi(t)$ is uniformly stable. For the periodic system (P), we have the following result.

<u>Proposition 3</u>. Let $\phi(t)$ be a solution of (P) such that $\|\phi(t)\|\leqq B$ for all $t\geqq 0$. If $\phi(t)$ is uniformly stable, then $\phi(t)$ is stable under disturbances from H(f) with respect to K.

Proof. In the case where f is not autonomous on R×K, there is a smallest positive period ω^* and we can see that for any $g\varepsilon H(f)$ and any $\tau\geqq 0$ there is a $\sigma(\tau,g)$ such that $\tau-\frac{\omega^*}{2}\leqq\sigma(\tau,g)\leqq\tau+\frac{\omega^*}{2}$ and $g(t,x)=f(t+\sigma,x)$ on R×K. For such a $\sigma(\tau,g)$, corresponding to any $\varepsilon>0$, there exists a $\gamma(\varepsilon)>0$ such that if $\tau\geqq 0$, $g\varepsilon H(f)$ and $\rho(f_\tau,g)\leqq\gamma(\varepsilon)$, then we have $|\tau-\sigma(\tau,g)|<\varepsilon$.

If f(t,x) is autonomous on K, that is, $f=f^*(x)$ on K, then for any $g\varepsilon H(f)$, $g=f^*(x)$. Therefore $\rho(f_\tau,g;K)=0$. Thus it is clear that $\phi(t)$ is stable under disturbances from H(f). Now we shall consider the case where f is not autonomous and we assume that ω is the smallest positive period of f. For t, $t'\varepsilon I$, we have $\|\phi(t)-\phi(t')\|<\frac{\delta(\varepsilon)}{2}$ if $|t-t'|<\frac{\delta(\varepsilon)}{2L}$, where $\delta(\varepsilon)$ is the one for the uniform stability of $\phi(t)$ and L is such that $\|f(t,x)\|\leqq L$ on R×K. For $\tau\geqq 0$, $\varepsilon>0$ and $g\varepsilon H(f)$, there is a $\sigma=\sigma(\tau,g)$ and there exists a $\gamma(\varepsilon)>0$ such that if $\rho(f_\tau,g)\leqq\gamma(\varepsilon)$, then we have $g(t,x)=f(t+\sigma, x)$ on R×K and

$$(4)\qquad |\tau-\sigma| < \frac{\delta(\varepsilon)}{2L},$$

where we can assume that $\gamma(\varepsilon)<\frac{\delta(\varepsilon)}{2}$ and $0<\varepsilon<\frac{B_1-B}{2}$. Set $\psi(t)=\phi(t+\tau)$. Then $\psi(t)$ is a solution through $(0,\phi(\tau))$ of the system

$$(5)\qquad x' = f(t+\tau, x).$$

Letting $\|\phi(\tau)-y_0\|\leqq\gamma(\varepsilon)$ and $g\varepsilon H(f)$ such that $\rho(f_\tau,g)\leqq\gamma(\varepsilon)$, consider a solution x(t) of (3) through $(0,y_0)$. As long as x(t) remains in K, x(t) is a solution of

$$(6)\qquad x' = f(t+\sigma, x)$$

through $(0,y_0)$.

Set $y(t)=\phi(t+\sigma)$. First of all, we assume that $\sigma\geqq 0$. Then y(t) is a solution of (6) through $(0,\phi(\sigma))$ and is uniformly stable with the same pair $(\varepsilon,\delta(\varepsilon))$ as the one for $\phi(t)$. We have $\|\phi(\sigma)-\phi(\tau)\|<\frac{\delta(\varepsilon)}{2}$, because of (4). Since we have

$$\|y(0)-y_0\| \leqq \|\phi(\sigma)-\phi(\tau)\| + \|\phi(\tau)-y_0\| < \delta(\varepsilon),$$

the uniform stability of y(t) implies that

$$(7)\qquad \|y(t) - x(t)\| < \varepsilon \quad \text{for all} \quad t \geqq 0.$$

Moreover $\|y(t)-\phi(t+\tau)\|=\|\phi(t+\sigma)-\phi(t+\tau)\|<\frac{\delta(\varepsilon)}{2}$ for all $t\geqq 0$ by (4), and hence, from this and (7) it follows that $\|\phi(t+\tau)-x(t)\|<2\varepsilon$ for all $t\geqq 0$. Next we shall consider the case where $\sigma<0$, and consequently $\tau-\sigma>0$. $z(t)=x(t+\tau-\sigma)$ is a solution of (5) through $(0, x(\tau-\sigma))$. Since (4) implies $\|y_0-x(\tau-\sigma)\|<\frac{\delta(\varepsilon)}{2}$, we have

$$\|\phi(\tau)-z(0)\| \leqq \|\phi(\tau)-y_0\| + \|y_0-x(\tau-\sigma)\| < \delta(\varepsilon).$$

Thus we have $\|\phi(t+\tau)-z(t)\|<\varepsilon$ for all $t\geqq 0$ since $\phi(t+\tau)$ is uniformly stable. Moreover, $\|z(t)-x(t)\|=\|x(t+\tau-\sigma)-x(t)\|<\frac{\delta(\varepsilon)}{2}<\varepsilon$ for $t\geqq 0$. Thus we have $\|\phi(t+\tau)-x(t)\|<2\varepsilon$ for all $t\geqq 0$. This completes the proof.

Proposition 4. For the almost periodic system (AP), if $\phi(t)$ is totally stable, then it is stable under disturbances from $H(f)$ with respect to K.

For the proof, see [6] and [23]. It is clear that the zero solution of $x'=0$ is stable under disturbances from the hull, but the zero solution is not totally stable as was stated before. For the almost periodic system (AP), we have the following proposition.

Proposition 5. Let $\{\tau_k\}$ be a sequence such that $\tau_k>0$, $f(t+\tau_k, x)\to g(t,x)$ uniformly for all $t\in R$ and $x\in K$ and $\phi(\tau_k)\to x_0$ as $k\to\infty$. If $\phi(t)$ is stable under disturbances from $H(f)$ with respect to K, then the solution $\psi(t)$ through $(0,x_0)$ of (3) is uniformly stable.

Proof. Since $\phi(t)$ is stable under disturbances from $H(f)$ with respect to K, for any $\varepsilon>0$ there is a $\delta(\varepsilon)>0$ such that $\|\phi(t+\tau)-x(t;x_0,0,g)\|\leqq\varepsilon$ for $t\geqq 0$, whenever $g\in H(f)$, $\|\phi(\tau)-x_0\|\leqq\delta(\varepsilon)$ and $\rho(f_\tau,g;K)\leqq\delta(\varepsilon)$ for some $\tau\geqq 0$. If we set $\phi_k(t)=\phi(t+\tau_k)$, $\phi_k(t)$ is a solution through $(0,\phi(\tau_k))$ of the system

$$(8)\qquad x' = f(t+\tau_k, x)$$

and clearly $\phi_k(t)$ also is stable under disturbances from $H(f_\tau)=H(f)$ with respect to K for the same pair $(\varepsilon,\delta(\varepsilon))$. Let $\psi(t)$ be a solution through $(0,x_0)$ of (3). For given $\varepsilon>0$, if k is sufficiently large, $\|\phi(\tau_k)-x_0\|\leqq\delta(\frac{\delta(\varepsilon)}{2})$ and

$$\|f(t+\tau_k, x) - g(t,x)\| \leqq \delta(\frac{\delta(\varepsilon)}{2}) \quad \text{on} \quad R\times K.$$

Therefore we have $\rho(f_{\tau_k},g;K)\leqq\delta(\frac{\delta(\varepsilon)}{2})$. Since $\phi_k(t)$ is stable under disturbances from $H(f_{\tau_k})$ and $\psi(t)$ is a solution of (3), we have

$$(9)\qquad \|\phi_k(t) - \psi(t)\| \leqq \frac{\delta(\varepsilon)}{2} \quad \text{for all } t \geqq 0.$$

For any $t_0\in I$, we have $\|\phi_k(t_0)-\psi(t_0)\|\leqq\frac{\delta(\varepsilon)}{2}$. Let $y(t;y_0,t_0)$ be a solution of (3) such that $\|\psi(t_0)-y_0\|<\frac{\delta(\varepsilon)}{2}$. Then $\|\phi_k(t_0)-y_0\|<\delta(\varepsilon)$. If we set $h(t,x)=g(t+t_0, x)$, clearly $h\in H(f)$ and $\|f(t+t_0+\tau_k, x)-h(t,x)\|\leqq\delta(\varepsilon)$ on $R\times K$. $z(t)=y(t+t_0;y_0,t_0)$ is a solution of

$$(10)\qquad x' = h(t,x)$$

and $z(0)=y_0$. Therefore the stability under disturbances of $\phi_k(t)$ implies that

$$(11) \qquad \|\phi_k(t+t_0)-z(t)\| \leqq \varepsilon \quad \text{for} \quad t \geqq 0.$$

Replacing t by $t-t_0$, we have $\|\phi_k(t)-z(t-t_0)\| \leqq \varepsilon$ for $t \geqq t_0$ or $\|\phi_k(t)-y(t;y_0,t_0)\| \leqq \varepsilon$ for $t \geqq t_0$. From this and (9), it follows that if $\|\psi(t_0)-y_0\| < \frac{\delta(\varepsilon)}{2}$,

$$\|\psi(t)-y(t;y_0,t_0)\| < 2\varepsilon \quad \text{for all} \quad t \geqq t_0,$$

which proves the uniform stability of $\psi(t)$.

For a more general result, see [6].

The following example due to Kato [6] is very interesting and is very important. This example shows that even for a scalar almost periodic equation, uniform stability and uniformly asymptotic stability also do not necessarily imply total stability and consequently stability under disturbances from the hull.

Let $a_0(t) \equiv 1$ and let $a_k(t)$ be a periodic function with period 2^k such that

$$a_k(t) = \begin{cases} 0 & (0 \leqq t < 2^{k-1}) \\ -\frac{1}{2^k} & (2^{k-1} \leqq t < 2^k). \end{cases}$$

Making $a_k(t)$ smooth, $a(t) = \sum_{k=0}^{\infty} a_k(t)$ is a continuous almost periodic function. Clearly $a(t)>0$. Let $h(x)$ be such that

$$h(x) = \begin{cases} 0 & (x = 0) \\ 2\sqrt{|nx-1|} & (\frac{2}{2n+1} \leqq x \leqq \frac{2}{2n-1}) \end{cases}$$

and define $f(t,x)$ for $x \geqq 0$ by $f(t,x)=h(x)-ca(t)\sqrt{x}$, $c>2\sqrt{2}$. Consider an almost periodic equation

$$(12) \qquad x' = \begin{cases} f(t,x) & (-\infty < t < \infty,\ x \geqq 0) \\ -f(t,-x) & (-\infty < t < \infty,\ x < 0). \end{cases}$$

It is not difficult to see that the zero solution of (12) is uniformly asymptotically stable. However the zero solution of an equation in the hull is not necessarily unique, and hence it is not necessarily stable. To see this, consider a sequence $\{2^k-1\}$. Then

$$f(t+2^k-1,\ x) \to h(x) \quad \text{on} \quad 0 \leqq t \leqq 1.$$

Let $g(t,x)$ be a function in $H(f)$ for the sequences $\{2^k-1\}$. Then the zero solution of $x'=g(t,x)$ is not unique to the right. Therefore the zero solution is not stable, and hence the zero solution of (12) is uniformly asymptotically stable, but it is neither totally stable nor stable under disturbances from the hull, because of Proposition 5.

Now we shall observe the case where uniformly asymptotic stability implies total stability. Following the paper of Kato [6], we shall consider a more general system. Let $F(t,x)$ be an R^n-valued continuous function defined on $I\times S_{B*}$ and let $T(F)$ be the set of continuous functions $F(t+s, x)$ for all $s\geqq 0$, which is a subset of $C(I\times S_{B*}, R^n)$. Here $C(\Omega, R^n)$ denote the set of all continuous R^n-valued functions defined on Ω and it is a topological space by compact-open topology. Let $H(F)$ be the closure of $T(F)$. If $F(t,x)$ is defined on $R\times S_{B*}$ and is almost periodic in t uniformly for $x\varepsilon S_{B*}$, then $H(F)$ consists of the restriction to $[0,\infty)$ of elements of the hull in the usual sense. Assume that $F(t,x)$ is uniformly continuous in (t,x) and bounded on $I\times S$ for any compact subset S of S_{B*}. Then $H(F)$ is compact.

Consider the systems

$$x' = F(t,x) \tag{13}$$

and

$$x' = G(t,x), \tag{14}$$

where $G\varepsilon H(F)$. Kato and the author [7] have shown that if the system (13) has a solution $\phi(t)$ defined on I which satisfies $\|\phi(t)\|\leqq B$, $B<B*$, for all $t\geqq 0$ and is uniformly asymptotically stable, and if for each $G\varepsilon H(F)$ the solution of (14) is unique to the right for the initial condition, then $\phi(t)$ is totally stable. Kato [6] has extended this result. Let $\phi(t)$ be a solution of (13) such that $\|\phi(t)\|\leqq B$, $B<B*$, for all $t\geqq 0$. Since $\|F(t,x)\|\leqq L$ for all $(t,x)\varepsilon I\times K$ and for some constant $L>0$, where $K=\{x;\ x\leqq \frac{B+B*}{2}\}$, we have $\|\phi(t)-\phi(s)\|\leqq L|t-s|$ for all $t,s\geqq 0$. Therefore the hulls $H(\phi)$ and $H(\phi,F)$ are compact. Let $(\psi(t),G(t,x))$ be an element of $H(\phi,F)$. Then $\psi(t)$ is a solution of (14). We shall simply call $\psi(t)$ a solution in $H(\phi)$. If for any $(\psi(t),G(t,x))\varepsilon H(\phi,F)$, $\psi(t)$ is a unique solution of (14) for the initial condition, we shall say that the solutions in $H(\phi)$ are unique for initial conditions. We shall consider the following condition (C): The solutions in $H(\phi)$ are uniformly asymptotically stable with a common triple $(\delta(\cdot),\delta_0,T(\cdot))$, that is, for any $\varepsilon>0$, any $t_0\geqq 0$ and any $(\psi,G)\varepsilon H(\phi,F)$, $\|\psi(t_0)-x_0\|<\delta(\varepsilon)$ implies

$$\|\psi(t) - x(t;x_0,t_0)\| < \varepsilon \quad \text{for all} \quad t \geqq t_0$$

and $\|\psi(t_0)-x_0\|\leqq\delta_0$ implies

$$\|\psi(t) - x(t;x_0,t_0)\| < \varepsilon \quad \text{for all} \quad t \geqq t_0+T(\varepsilon),$$

where $x(t;x_0,t_0)$ is a solution of (14) through (t_0,x_0).

Proposition 6. If the condition (C) holds good, then $\phi(t)$ is totally stable.

For the proof, see [6]. For the almost periodic system (AP), we know that if for each $g\varepsilon H(f)$ the solution of (3) is unique to the right for the initial condition and if $\phi(t)$ is uniformly asymptotically stable, then the condition (C) is satisfied (see [23]). For the periodic system (P), if $\phi(t)$ is uniformly asymptotically stable, the condition (C) always is satisfied as is shown in [23]. Therefore we have the following result.

Proposition 7. A bounded solution of the periodic system is totally stable, if it is uniformly asymptotically stable.

For the almost periodic system (AP), the following result is known [22].

Proposition 8. Suppose that the almost periodic system (AP) has a solution $\phi(t)$ defined on I such that $\|\phi(t)\| \leqq B$, $B < B^*$, for $t \geqq 0$. If the solution $\phi(t)$ is asymptotically almost periodic, then the system (AP) has an almost periodic solution.

A sufficient condition that $\phi(t)$ is asymptotically almost periodic is that $\phi(t)$ is stable under disturbances from H(f) with respect to K. Therefore, if for the periodic system, $\phi(t)$ is uniformly stable, then $\phi(t)$ is asymptotically almost periodic. For the almost periodic system, if $\phi(t)$ is totally stable, then $\phi(t)$ is asymptotically almost periodic, and also if the condition (C) holds good, then $\phi(t)$ is asymptotically almost periodic. Therefore, in these case, the system (AP) has an almost periodic solution.

For the existence of an almost periodic solution of the almost periodic system (AP), Seifert [13] has used the fact that, letting S be a compact set in S_{B^*}, if for each $g\varepsilon H(f)$ the system (3) has only one solution which remains in S on R and if $\phi(t)$ is a solution of (AP) such that $\phi(t)\varepsilon S$ on R, then $\phi(t)$ is an almost periodic solution and its module is contained in the module of f. Seifert has assumed that $\phi(t)$ is uniformly quasi asymptotically stable in the large to show that the system (3) has only one solution which remains in a compact set. Recently, by applying the fact above, Nakajima [10] has obtained a result which contains Seifert's and Yoshizawa's result [18].

Proposition 9. The following assumptions will be made:

(i) For each $g\varepsilon H(f)$, the solution of (3) is unique for the initial condition.

(ii) The system (AP) has a solution $\phi(t)$ defined on I such that $\|\phi(t)\| \leqq B$, $B < B^*$, for $t \geqq 0$.

(iii) There is a $T_0 \geqq 0$ such that if a solution $x(t)$ of (AP) satisfies $x(t) \varepsilon S_{B*}$ for all $t \varepsilon [t_0, t_0+T_0]$, $t_0 \geqq 0$, then $x(t) \varepsilon S_{B*}$ for all $t \geqq t_0$.

(iv) For any $\varepsilon > 0$ and any $r > 0$, $r < B^*$, there is a $T(\varepsilon, r) > 0$ such that if $\|x_0\| \leqq r$, $\|\phi(t) - x(t; x_0, t_0)\| < \varepsilon$ for $t \geqq t_0 + T(\varepsilon, r)$, where $x(t; x_0, t_0)$ is a solution of (AP) such that $x(t; x_0, t_0) \varepsilon S_{B*}$ for $t \geqq t_0$.

Then the system (AP) has a unique almost periodic solution in S_{B*} and its module is contained in the module of f.

References

[1] C.C. Conley and R.K. Miller, Asymptotic stability without uniform stability: almost periodic coeffients, J. Differential Eqs., 1 (1965), 333-336.

[2] W.A. Coppel, Almost periodic properties of ordinary differential equations, Ann. Mat. Pura Appl., 76(1967), 27-49.

[3] L.G. Deysach and G.R. Sell, On the existence of almost periodic motions, Michigan Math. J., 12(1965), 87-95.

[4] A.M. Fink and P.O. Frederickson, Ultimate boundedness does not imply almost periodicity, J. Differential Eqs., 9(1971), 280-284.

[5] J.K. Hale, J.P. LaSalle and Marshall Slemrod, Theory of a general class of dissipative processes, J. Math. Anal. Appl., (to appear).

[6] J. Kato, Uniformly asymptotic stability and total stability, Tohoku Math. J., 22(1970), 254-269.

[7] J. Kato and T. Yoshizawa, A relationship between uniformly asymptotic stability and total stability, Funkcialaj Ekvacioj, 12(1970), 233-238.

[8] J.L. Massera, Erratum: Contributions to stability theory, Ann. of Math., 68(1958), 202.

[9] R.K. Miller, Almost periodic differential equations as dynamical systems with applications to the existence of a. p. solutions, J. Differential Eqs., 1(1965), 337-345.

[10] F. Nakajima, Existence of quasi-periodic solutions of quasi-periodic systems, to appear.

[11] Z. Opial, Sur une équation différentielle presque-périodique sans solution presque-périodique, Bull. Acad. Polon. Sci. Ser. Sci. Math. Astron. Phys., 9(1961), 673-676.

[12] N. Paval, On dissipative systems, Boll. Unione Mat. Italiana, (to appear).

[13] G. Şeifert, Stability conditions for the existence of almost-periodic solutions of almost-periodic systems, J. Math. Anal. Appl., 10(1965), 409-418.

[14] G. Seifert, Almost periodic solutions for almost periodic systems of ordinary differential equations, J. Differential Eqs., 2(1966), 305-319.

[15] G. Seifert, Almost periodic solutions and asymptotic stability, J. Math. Anal. Appl., 21(1968), 136-149.

[16] G.R. Sell, Periodic solutions and asymptotic stability, J. Differential Eqs., 2(1966), 143-157.

[17] G.R. Sell, Nonautonomous differential equations and topological dynamics. I, II, Trans. Amer. Math. Soc., 127(1967), 241-262, 263-283.

[18] T. Yoshizawa, Extreme stability and almost periodic solutions of functional-differential equations, Arch. Rational Mech. Anal., 17(1964), 148-170.

[19] T. Yoshizawa, "Stability Theory by Liapunov's Second Method." The Mathematical Society of Japan, Tokyo, 1966.

[20] T. Yoshizawa, Stability and existence of periodic and almost periodic solutions, Proc. of United States-Japan Seminar on Differential and Functional Equations, 1967, 411-427.

[21] T. Yoshizawa, Stability and existence of a periodic solution, J. Differential Eqs., 4(1968), 121-129.

[22] T. Yoshizawa, Some remarks on the existence and the stability of almost periodic solutions, Studies in Applied Mathematics, 5 (1969), 166-172.

[23] T. Yoshizawa, Asymptotically almost periodic solutions of an almost periodic system, Funkcialaj Ekvacioj, 12(1969), 23-40.

PERTURBATIONS OF VOLTERRA EQUATIONS AND ADMISSIBILITY

John A. Nohel*

1. Introduction. Consider the system of Volterra equations

$$x(t) = f(t) + \int_0^t a(t,s)\,[x(s) + g(s, x(s))]\,ds \tag{1}$$

on the interval $0 \leq t < \infty$. In (1) x, f, g are vector functions with n components, a is a n by n matrix, and g is regarded as small in various senses. Our purpose is to summarize and also to extend research of several authors, particularly during the period since the first U.S.-Japan Seminar [21], in which solutions of (1) are compared in different ways to those of the unperturbed linear system

$$y(t) = f(t) + \int_0^t a(t,s)y(s)\,ds \;. \tag{2}$$

Also of interest are solutions of the initial value problem for the perturbed system of integrodifferential equations on $0 \leq t < \infty$

$$\begin{cases} x'(t) = F(t) + A(t)x(t) + \int_0^t B(t,s)x(s)\,ds + G(t, x(t)) \\ x(0) = x_0 \,, \end{cases} \tag{1d}$$

in relation to those of the unperturbed linear system

$$\begin{cases} y'(t) = F(t) + A(t)y(t) + \int_0^t B(t,s)y(s)\,ds \\ y(0) = y_0 \;. \end{cases} \tag{2d}$$

Here F, G, x_0, x are vectors with n components and A, B are n by n matrices; the perturbations G are again regarded as small.

We shall give criteria in terms of the resolvent kernel for the stability and asymptotic stability of continuous solutions of (1) and (1d), assuming a corresponding stability property of solutions of (2) and (2d). We shall also investigate the asymptotic equivalence of (1), (2) and of (1d), (2d). Satisfying these criteria in terms of the original kernels is discussed for a number of important subclasses of

*Supported by AROD - Durham.

problems. Systems (1) and (1d) arise in a variety of applications such as control theory, nuclear reactor dynamics, superfluidity, heat transfer, viscoelasticity, biology, population genetics and others. While (1d) can be reduced to the form (1) by an integration, doing so is undesirable for certain purposes, and separate results for (1) and (1d) are useful. These may be regarded as extensions to (1) and (1d) of classical theory of Poincaré, Perron, Weyl and others for ordinary differential equations. A further generalization is possible in the general framework of operators on Fréchet spaces utilizing the concept of admissibility of subspaces with respect to an operator.

We remark that perturbation problems in which the unperturbed system is a nonlinear Volterra equation also arise in applications. These cannot in general be handled by techniques discussed here, but the reader may consult [11], [9], [7], [13] for such problems. For the study of the asymptotic behavior of bounded solutions of general functional equations (including Volterra equations) which may be regarded as perturbations of an entirely different sort (namely of certain limiting equations) and which also cannot be handled by these methods we refer the reader to [12].

2. Some "Variation of Constants" formulae. With the kernel $a(t,s)$ in (1) we associate the resolvent kernel $r(t,s)$ which is defined to be the (unique) solution of the linear system (see [20], [22])

$$r(t,s) = a(t,s) + \int_s^t a(t,u)r(u,s)\,du \qquad (0 \le s \le t < \infty) \; . \tag{3}$$

With the kernel $A(t,s)$ in (1d) we associate the kernel $R(t,s)$ which is defined to be the solution of the initial value problem (see also [4] where a different definition is taken)

$$\begin{cases} \dfrac{\partial R}{\partial t}(t,s) = A(t)R(t,s) + \displaystyle\int_s^t B(t,u)R(u,s)\,du \\ R(s,s) = I \; , \quad t \ge s \ge 0 \; , \end{cases} \tag{3d}$$

where I is the identity matrix. If $s > t \ge 0$ we define $r(t,s) \equiv 0$, $R(t,s) \equiv 0$.

Let $LL'(S)$ denote the set of all measurable functions on S such that the seminorms

$$\|f\|_\Sigma = \int_\Sigma |f(t)|\,dt$$

are finite for all compact subsets Σ of S. We shall require throughout that at least $r(t,s) \in LL'(R^+ \times R^+)$, where $R^+ = \{t : 0 \le t < \infty\}$. It can be verified by direct substitution that under a variety of hypotheses (see remarks following Theorem 1)

(4) $$y(t) = f(t) + \int_0^t r(t,s)f(s)\,ds$$

is a solution of (2). As far as $R(t,s)$ is concerned it is reasonable to assume (see remarks following Theorem 1d) that $R(t,s)$ is continuous; one also readily verifies that

(4d) $$Y(t) = R(t,0)x_0 + \int_0^t R(t,s)F(s)\,ds$$

is in a variety of senses a solution for the system (2d). Another straight forward calculation shows that the perturbed system (1) is equivalent to the system

(5) $$x(t) = y(t) + \int_0^t r(t,s)g(s,x(s))\,ds\ ,$$

where $y(t)$ is given by (4), while the system (1d) is equivalent to the system

(5d) $$x(t) = Y(t) + \int_0^t R(t,s)G(s,x(s))\,ds$$

where $Y(t)$ is given by (4d).

Taking the special case $B(t) \equiv 0$ it follows from (3d) that $R(t,s) = \Phi(t)\Phi^{-1}(s)$, where Φ is the fundamental matrix of $y' = A(t)y$ for which $\Phi(0) = I$, and (4d) and (5d) are classical variations of constants formulae for ordinary differential equations.

3. <u>Stability</u>. We shall be interested in stability type results of <u>continuous solutions</u> of (1) and (1d) on R^+. Let $BC = \{\varphi \text{ continuous on } R^+ : \|\varphi\| = \sup_{t\in R^+} |\varphi(t)| < \infty\}$ where $|\ |$ is any convenient norm in $\mathbb{R}^n$. We make the following assumptions

(H_1) $r(t,s) \in LL'(R^+ \times R^+)$

(H_2) $g(t,x)$ is continuous in (t,x) for $t \in R^+$, $|x| < \infty$ and $g(t,0) \equiv 0$

(H_3) For every $\alpha > 0$ there exists a $\beta > 0$ such that $|g(t,x)| \le \alpha|x|$, uniformly in t, whenever $|x| \le \beta$;

the following result is established in [22, Theorem 1]:

<u>Theorem 1</u>. <u>Suppose</u> $y \in BC$ <u>is a solution of</u> (2). <u>Let there exist a constant</u> $K > 0$ <u>such that</u>

$$(6) \qquad \int_0^t |r(t,s)|ds \le K \qquad (t \in R^+) ,$$

and let

$$(7) \qquad \lim_{h \to 0} \left[\int_t^{t+h} |r(t+h,s)|ds + \int_0^t |r(t+h,s) - r(t,s)|ds \right] = 0$$

for any t on R^+. Then given λ, $0 < \lambda < 1$, there exists a number $\varepsilon_0 > 0$ such that $0 < \varepsilon \le \varepsilon_0$ and $\|y\| \le \lambda\varepsilon$ together imply that (1) has at least one solution $x \in BC$ and $\|x\| \le \varepsilon$. If in addition $\lim_{t \to \infty} y(t) = 0$ and if for each $T > 0$

$$(8) \qquad \lim_{t \to \infty} \int_0^t |r(t,s)|ds = 0 ,$$

then also $\lim_{t \to \infty} x(t) = 0$.

The proof of Theorem 1 makes use of the Schauder-Tychonoff fixed point theorem applied to the operator Ω defined by the right hand side of (5) acting on the subspace $S_\varepsilon = \{\varphi \in BC : \|\varphi\| \le \varepsilon\}$.

We observe that if $f \in BC$ and if r satisfies (6), (7) then by (4) $y \in BC$ and $\|y\| \le (1 + K)\|f\|$. Thus if $\|f\| \le \dfrac{\lambda\varepsilon}{1 + K}$, Theorem 1 implies that (1) has a solution $x \in BC$ such that $\|x\| \le \varepsilon$. If in addition $f \to 0$ as $t \to \infty$ and if r satisfies (8) then from (4) $y \to 0$ and by Theorem 1 the solution $x \to 0$. Thus Theorem 1 is a stability type result with the forcing term f playing the role of initial conditions in ordinary differential equations. A somewhat different result and proof is obtained in [24, Theorem 1*]

If assumption (H_3) is strengthened as follows:

($\tilde{H}_3$) For every $\alpha > 0$ there exists $\beta > 0$ such that $|g(t,u) - g(t,v)| \le \alpha|u-v|$, uniformly in t, whenever $|u| \le \beta$, $|v| \le \beta$,

then the above mentioned mapping Ω is a contraction on S_ε and the solution x of (1) is unique (see [20, Theorem 3 and Corollary 3.1]; for earlier versions of this latter result in the convolution case see [16], [21]. This technique has also been used in [20] to establish the existence of asymptotically periodic and almost periodic solutions of (1) assuming that (2) has a solution with this property; see also [8].

The applicability of results like Theorem 1 depends mainly on satisfying assumptions (6) and (8). The following are criteria which do this in terms of the original kernel $a(t,s)$.

(i) $a(t,s) = a(t-s)$

Paley-Wiener Theorem [23]. Let $a(t) \in L'(R^+)$; then the resolvent kernel $r(t) \in L'(R^+)$ if and only if

$$\det(I - \hat{a}(s)) \neq 0 \quad (\operatorname{Re} s \geq 0)\ ,$$

where $\hat{a}(s)$ is the Laplace transform of $a(t)$.

(ii) if $a(t) \notin L'(R^+)$ (e.g. one may have $a(t) = t^{-\sigma}(0 < \sigma < 1)$) one can often apply Laplace transforms and Tauberian theorems to (3) and satisfy conditions (6) and (8). Examples of this, though in a different context, may be found in [10], [1], [6].

(iii) If $n = 1$, $a(t,s) = A(t-s)B(s)$ then it is shown in [20, Theorem 6], see also Miller [17] that if $A(t)$ is completely monotonic on $0 < t < \infty$ and if $B(t) \in BC$ and $B(t) \geq 0$ (e.g. $a(t,s) = \cos^2 ws/(t-s)^\sigma$) then $r(t,s)$ satisfies (6) and (7). If in addition $\lim_{t\to\infty} A(t) = 0$, then $r(t,s)$ satisfies (8).

We now turn to the differential system (1d). Similar results are established in [4, Theorems 4 and 5].

Theorem 1d. Let $A \in LL'(R^+)$, $B \in LL'(R^+ \times R^+)$. Let $G(t,x)$ satisfy (H_2) and (H_3) and let $R(t,s)$ satisfy (6). Let $Y(t) \in BC$ be a solution of (2d) satisfying the initial condition $Y(0) = x_0$. Then given $0 < \lambda < 1$ there exists an $\varepsilon_0 > 0$ such that $0 < \varepsilon \leq \varepsilon_0$ and $\|Y\| \leq \lambda\varepsilon$ together imply that the system (1d) has at least one solution $x \in BC$, $x(0) = x_0$, and $\|x\| \leq \varepsilon$. If in addition $\lim_{t\to\infty} Y(t) = 0$ and if $R(t,s)$ satisfied (8), then $\lim_{t\to\infty} x(t) = 0$. (As was the case for (1) if $G(t,x)$ satisfies $\tilde{H}_3$, the solution x is unique).

The proof of Theorem 1d is essentially the same as that of Theorem 1 making use of (5d) (equivalent to 1d). Observe that the hypothesis concerning A, B implies from (3d) and from (integrating (3d))

$$R(t,s) = I + \int_s^t [A(u) + \int_u^t B(\sigma,u)\,d\sigma] R(u,s)\,du$$

that $R(t,s)$ is continuous in (t,s) for $0 \leq s \leq t$; thus (7) is satisfied. Also observe that if $R(t,0) \in BC$, $F \in BC$ and if (6) is satisfied, then (from (4d)) $Y \in BC$; moreover, if $\|F\|$ and $|x_0|$ are sufficiently small one can in particular make $\|Y\| \leq \lambda\varepsilon$. Thus Theorem 1d is a stability theorem. The method of proof of

Theorem 1d can easily be modified to establish the following result which extends well known results in ordinary differential equations [2, p. 327].

Corollary 1d. *Consider the system*

$$(*)\qquad x'(t) = A(t)x + \int_0^t B(t,s)x(s)ds + G(t,x(t)) + H(t,x(t)) \quad .$$

Let A, B, G *satisfy the hypothesis of Theorem* 1d. *For small* $|x|$ $H(t,x) \to 0$ *as* $t \to \infty$ *uniformly in* x. *Let* $R(t,s)$ *satisfy* (6), (8). *Then there exists a* $T > 0$ *such that any solution* $\varphi(t)$ *of* (*) *approaches zero provided* $|\varphi(T)|$ *is sufficiently small.*

For the proof one notes that (*) is equivalent to

$$x(t) = Y(t) + \int_T^t R(t,s)G(s,x(s))ds + \int_T^t R(t,s)H(s,x(s))ds$$

where $Y(t) = R(t,T)\varphi(T)$ and proceeds much as in the proof of Theorem 1d for $t \geq T$.

The applicability of Theorem 1d or the corollary again depends on whether $R(t,s)$ satisfies (6) and (8). This is a more difficult problem than in Theorem 1. In the special case $A(t) \equiv A$ (a constant matrix), $B(t,s) = B(t-s)$, the resolvent equation (3d) becomes

$$(R)\qquad R'(t) = AR + \int_0^t B(t-s)R(s)ds \quad , \qquad R(0) = I \quad .$$

Clearly, if $B \equiv 0$ and if all the eigenvalues of A have negative real parts, we are in the O.D.E. case and (6), (8) are satisfied. More generally, one can take Laplace transforms of (R) and often use Tauberian theorems to verify the hypotheses needed. For different illustrations of this technique see [10] where $n = 1$, $A = 0$, $B(t) = \int_0^\infty h(x)\exp(-x^2t)dx$, $h(x) \in L'(R^+)$, $h(x) \geq 0$ (this does not imply $B(t) \in L'(R^+)$), [4, Theorem 6], and [6]. An important general result for (R) recently has been obtained by Grossman and Miller [5][#]:

Theorem (Generalized Paley-Wiener). *Let* $B(t) \in L'(R^+)$. *Then* $R(t) \in L'(R^+)$ *if and only if*

$$(PW)\qquad \det(sI - A - \hat{B}(s)) \neq 0 \qquad (\mathrm{Re}\, s \geq 0) \quad ,$$

where $\hat{B}$ *is the Laplace transform of* B.

[#]This result has recently been established by D. F. Shea by a completely different technique using methods of harmonic analysis.

To obtain this result they make crucial use of the work of the second author [19] concerning the initial value problem

$$(9) \qquad \begin{cases} y'(t) = Ay(t) + \int_0^t B(t-s)y(s)ds \\ y(t) = h(t) \quad (0 \leq t \leq \tau) \end{cases}$$

where $B \in LL'(R^+)$ and where h is a given continuous function on $0 \leq t \leq \tau$ (the usual problem has $\tau = 0$); $y \equiv 0$ is a solution of (9) and one defines its stability, uniform stability and uniform asymptotic stability (with respect to τ) in the way it is done for delay equations. (e.g. $y \equiv 0$ is uniformly stable if given $\varepsilon > 0$ and $\tau \geq 0$ there exists a $\delta = \delta(\varepsilon)$ independent of τ such that whenever h is a given continuous function for which $\sup_{0 \leq t \leq \tau} |h(t)| \leq \delta$, then (9) has a solution $y(t, \tau, h)$ for $t \geq \tau$ satisfying $y(\tau, \tau, h) = h(\tau)$ and $|y(t, \tau, h)| \leq \varepsilon$). In rough outline these concepts are related to $R(t) \in L^1(0, \infty)$ and the Paley-Wiener Theorem for (R) is as follows:

(i) If $B \in L'(0,\infty)$ and if the solution $y \equiv 0$ of (9) is uniformly asymptotically stable, then condition (PW) is satisfied.

(ii) If $B, R \in L'(0, \infty)$ then

(a) the solution $y \equiv 0$ of (9) is uniformly asymptotically stable and

(b) for any τ, h the solution $y(t, \tau, h)$ of (9) is in $L^p(0, \infty)$ $(1 \leq p \leq \infty)$.

(iii) If $B \in L'(0, \infty)$ and if the solution $y \equiv 0$ is uniformly asymptotically stable then $R(t) \in L'(0, \infty)$.

To prove the sufficiency of (PW) it suffices to show that condition (PW) implies that the solution $y(t) \equiv 0$ of (9) is uniformly asymptotically stable. The necessity follows from (ii)a and (i). Statements (i), (ii), (iii) are established in [5] and [19]. The proof of (iii) involves establishing a converse theorem showing that (9) possesses a Liapunov functional of a certain type; this is done by adapting to (9) the ingenious construction of Massera for ordinary differential equations [14].

4. Asymptotic Equivalence. We now generalize Theorems 1, 1d in a different way. We say that systems (1) and (2) are asymptotically equivalent if given a bounded solution $y(t)$ of (2) on R^+, there exists a bounded solution $x(t)$ of (1) on R^+ such that $\lim_{t\to\infty} (x(t) - y(t)) = 0$ and conversely. The same definition holds for systems (1d) and (2d). It is clear that this is a more general concept than asymptotic

stability. In place of assumptions (H_2), (H_3) in Section 3 we now assume

$$
(H_4) \quad \begin{cases} g(t,x) \text{ is continuous in } (t,x) \text{ for } t \in R^+, \ |x| < \infty \\ |g(t,x)| \le \begin{cases} \lambda(t)|x| & \text{if } |x| \ge 1 \\ \lambda(t) & \text{if } |x| < 1 \end{cases} \\ \text{where } \lambda(t) \ge 0, \ \lambda \in BC, \ \lim_{t\to\infty} \lambda(t) = 0 \\ \text{and } \|\lambda\| K \le 1/2 \text{ where } K \text{ is the a priori constant in (6)} . \end{cases}
$$

One can then establish the following result [22, Theorem 3].

<u>Theorem 3.</u> <u>Let</u> g <u>satisfy</u> (H_4) <u>and let the resolvent</u> r <u>satisfy</u> (6), (7), (8). <u>Then the systems</u> (1) <u>and</u> (2) <u>are asymptotically equivalent.</u>

A part of this result was established in [24, Theorem 2] under a different and more stringent hypothesis concerning the original kernel a and by different methods without, however, requiring the condition $\|\lambda\|K \le 1/2$. It is also shown in [24, Ex. 3] that the requirement $\lim_{t\to\infty} \lambda(t) = 0$ cannot be improved to $\lambda \in L^1(R^+)$. For the asymptotic equivalence of systems (1d) and (2d) are

<u>Theorem 3d.</u> <u>Let</u> G <u>satisfy</u> (H_4) <u>and let the differential resolvent</u> R <u>satisfy</u> (6) <u>and</u> (8). <u>Then the systems</u> (1d) <u>and</u> (2d) <u>are asymptotically equivalent.</u>

In the proof of Theorem 3 (similarly 3d) one uses the Schauder-Tychonoff fixed point theorem to establish the existence of a bounded solution of (1) given a bounded solution of (2) (it is here where one needs the condition $\|\lambda\| K \le 1/2$) and then a direct limiting argument to show that the difference between $x(t)$ and $y(t)$ tends to zero. For the converse given a bounded solution $u(t)$ of (1) one defines

$$
v(t) = u(t) - \int_0^t r(t,s)g(s,u(s))\,ds
$$

and shows (using the resolvent equation) that v is a (bounded) solution of (2); then one verifies directly that $\lim_{t\to\infty} (u(t) - v(t)) = 0$.

The condition $\|\lambda\|K \le 1/2$ in (H_4) can be dropped if one strengthens the other condition on $g(t,x)$ slightly, (see [22, Theorem 4]); we have:

Theorem 4. Let $g(t,x)$ be continuous in (t,x) for $t \in R^+$, $|x| < \infty$, and let $|g(t,x)| \le \lambda(t)\,|x|^\sigma$ $(0 \le \sigma < 1)$, where $\lambda(t) \ge 0$, $\lambda \in BC$, $\lim_{t\to\infty} \lambda(t) = 0$ and let r satisfy (6), (7), (8). Then (1) and (2) are asymptotically equivalent.

A similar result holds for systems (1d) and (2d). Theorem 4 is proved by a different agrument in that, instead of using the Schauder-Tychonoff theorem, one obtains an a priori estimate for the solution of (1) from which one can deduce the existence of a bounded solution of (1) given a bounded solution of (2). The remainder of the argument is the same as in the proof of Theorem 3.

5. Admissibility. Let $\mathfrak{F}$ be a Fréchet space. Let $T : \mathfrak{F} \to \mathfrak{F}$ be a continuous linear map and let $g : \mathfrak{F} \to \mathfrak{F}$ be a continuous nonlinear map. We now wish to compare the perturbed nonlinear system

$$\text{(N)} \qquad x = f + T(x + g(x)) \ ,$$

where $f \in \mathfrak{F}$ is given, with the linear unperturbed system

$$\text{(L)} \qquad y = f + Ty \ .$$

Let X be a linear subspace of $\mathfrak{F}$ with a topology stronger than the topology induced from $\mathfrak{F}$, and assume that X is admissible with respect to (L). This means that for each $f \in X$, the system (L) has a unique solution $y \in X$. The concept of admissibility was introduced in ordinary differential equations by Massera and Schäffer [15], and for more general equations by Corduneanu [3] and Miller [18]. In the analysis to follow an important role is played by the resolvent operator ρ which is a linear map : $\mathfrak{F} \to \mathfrak{F}$ given by

$$(\rho) \qquad \rho = -I + (I - T)^{-1} \ .$$

It is easily verified that

$$y = (I - T)^{-1}(f) = (I + \rho)(f) = f + \rho(f)$$

is the unique solution of (L). For the linear Volterra equation (2), the map T is defined by

$$\text{(10)} \qquad (T\phi)(t) = \int_0^t a(t,s)\phi(s)\,ds \ ,$$

and the map ρ by

$$(11) \qquad (\rho\phi)(t) = \int_0^t r(t,s)\phi(s)\,ds \ ,$$

where r is the solution of (3). Taking for example $\mathfrak{F} = C(R^+)$, $X = BC$ it follows from the earlier analysis that X is admissible with respect to (L) if r satisfies (6) and (7).

If X is a linear subspace of $\mathfrak{F}$ admissible with respect to (L), and if ρ is the resolvent corresponding to T, then it is easy to verify (see [19]) that (N) is equivalent to the system

$$(V) \qquad x = y + \rho\, g(x) \ ,$$

where $y = f + \rho f$ is the solution of (L).

Now let X_1, X_2 be linear subspaces of $\mathfrak{F}$ with norms $\| \ \|_1$, $\| \ \|_2$ respectively such that X_i are Banach subspaces each with a stronger topology than the topology induced by $\mathfrak{F}$. Let $I - T : \mathfrak{F} \to \mathfrak{F}$ be one-to-one and onto. We require the following generalization of admissibility due to Corduneanu [3]; see also Miller [18].

<u>Definition</u>. <u>The pair of Banach subspaces</u> (X_1, X_2) <u>of</u> $\mathfrak{F}$ <u>is admissible with respect to the resolvent operator</u> ρ <u>if and only if</u> $\rho f \in X_2$ <u>for every</u> $f \in X_1$.

As an example take $\mathfrak{F} = C(R^+)$, $X_1 = X_2 = BC$ and define ρ by (11). If r satisfies (6), (7) then the pair (X_1, X_2) is admissible. If instead we take $X_1 = X_2 = BC_0 = \{x \in BC \mid \lim_{t\to\infty} x(t) = 0\}$ and if we require that $r(t,s)$ also satisfies (8), then (X_1, X_2) is another admissible pair. As another illustration take $\mathfrak{F} = LL^2(R^+)$ with topology of L^2 convergence on compact subsets, $X_1 = X_2 = L^2(R^+)$; if we define ρ by (10) where $r(t,s) = r(t-s)$ and if $r(t) \in L'(R^+)$, then (X_1, X_2) is admissible. It is clear that in all cases where $X_1 = X_2$ the above definition of admissiblity reduces to the previous one.

Since $\rho : \mathfrak{F} \to \mathfrak{F}$ is continuous and linear, it is clear that whenever (X_1, X_2) is admissible with respect to ρ, then ρ is a continuous linear map : $X_1 \to X_2$ and

$$\|\rho\| = \sup\{ \|\rho f\|_2 \ \big| \ \|f\|_1 = 1\} < \infty \ .$$

We may now state the following generalization of Theorem 1 to convex Fréchet spaces (every neighborhood of the origin has a convex subneighborhood - the Schauder fixed point theorem holds for such spaces), see Miller [18, Theorem 3] where a very slightly more general result is given.

Theorem 5. Let $\mathfrak{F}$ be a convex Fréchet space; let (X_1, X_2) be Banach subspaces of $\mathfrak{F}$ admissible with respect to the resolvent operator ρ associated with the operator T and let ρ be compact. Let y (the solution of (L)) $\in X_2$. Let $g : X_2 \to X_1$ be continuous, $g(0) = 0$ and let g be of higher order (in the sense that for every $\alpha > 0$ there exists a $\beta > 0$ such that $\|g(x)\|_1 \leq \alpha \|x\|_2$ whenever $\|x\|_2 \leq \beta$). Then given λ, $0 < \lambda < 1$ there exists an ε_0 such that $0 < \varepsilon \leq \varepsilon_0$ and $\|y\|_2 \leq \lambda\varepsilon$ together imply that (N) has at least one solution $x \in X_2$ such that $\|x\|_2 \leq \varepsilon$.

The assumptions on $r(t, s)$ in Theorem 1 insure both the admissibility of (BC, BC) (also of (BC_0, BC_0)) with respect to ρ defined by (11) and the compactness of ρ on $\mathfrak{F} = C(R^+)$. Remarks (i) and (ii) following Theorem 1 show that if ρ is the integral operator (11), then rather mild conditions on the original kernel $a(t, s)$ imply the compactness of ρ on $\mathfrak{F} = C(R^+)$. For most subspaces X_1, X_2, however, the compactness of ρ as a map from X_1 to X_2 is difficult to check.

We also remark that if in Theorem 5 the nonlinear map g satisfies the stronger higher order condition that given $\alpha > 0$ there exists a $\beta > 0$ such that $\|g(u) - g(v)\|_1 \leq \alpha \|u - v\|_2$ whenever $u, v \in X_2$ and $\|u\|_2 \leq \beta$, $\|v\|_2 \leq \beta$, then the solution x of (N) in Theorem 2 is unique (the right hand side of N defines a contraction on X_2).

We now turn briefly to (1d) and a generalization of Theorem 1d. Consider the (more general) system

$$\text{(Nd)} \qquad x'(t) = F(t) + A(t)x(t) + \int_0^t B(t, s)x(s)ds + (Gx)(t) \ , \quad x(0) = x_0 \ .$$

Here F, A, B are as in Theorem 1d. Let $\mathfrak{F}$ be a Fréchet subspace of $LL'(R^+)$ with norm $\| \ \|$ and with a stronger topology than the topology induced by $LL'(R^+)$. Let $G : \mathfrak{F} \to \mathfrak{F}$ be a continuous map. Define the map $\mathfrak{D}$ by

$$(\mathfrak{D}) \qquad (\mathfrak{D}F)(t) = \int_0^t R(t, s)F(s)ds \qquad (0 \leq t < \infty)$$

for any $F \in LL'(R^+)$, where $R(t, s)$ is the differential resolvent defined in Section 2. It is easily verified that if $\mathfrak{D}(\mathfrak{F}) \subset \mathfrak{F}$, then $\mathfrak{D}$ is a continuous linear map from $\mathfrak{F} \to \mathfrak{F}$($R(t, s)$ is continuous). It is also easily verified that if $F \in \mathfrak{F}$, if $\mathfrak{D} : \mathfrak{F} \to \mathfrak{F}$

and $G : \mathfrak{F} \to \mathfrak{F}$ then (Nd) is equivalent to

$$\text{(Vd)} \qquad x = Y + \mathfrak{K}x ,$$

where Y is the solution of the linear unperturbed system (2d) given by $Y(t) = R(t,0)x_0 + (\mathfrak{K}\mathfrak{F})(t)$ (see also (4d) and (5d)). Note that if also $R(t,0) \in \mathfrak{F}$, then $Y \in \mathfrak{F}$.

Theorem 5d. *Let* $A \in LL'(R^+)$, $B \in LL'(R^+ \times R^+)$; *let* $Y \in \mathfrak{F}$ *be a solution of* (2d), $Y(0) = x_0$. *Let* $G : \mathfrak{F} \to \mathfrak{F}$, $G(0) = 0$ *and let* G *be of higher order with respect to* $\mathfrak{F}$ (in the sense that for every $\alpha > 0$ there exists a $\beta > 0$ such that $\|Gx\| \le \alpha \|x\|$ whenever $x \in \mathfrak{F}$ and $\|x\| \le \beta$). *Let* $\mathfrak{K} : \mathfrak{F} \to \mathfrak{F}$ *be defined by* $(\mathfrak{K})$ *and let* $\mathfrak{K}$ *be a compact operator. Then given* λ, $0 < \lambda < 1$, *there exists a* $\varepsilon_0 > 0$ *such that* $0 < \varepsilon \le \varepsilon_0$ *and* $\|Y\| \le \lambda\varepsilon$ *together imply that* (Nd) *has a solution* $x \in \mathfrak{F}$ *and* $\|x\| \le \varepsilon$.

The case when $\mathfrak{F} = BC$ and subsequently $\mathfrak{F} = BC_0$ is treated in Theorem 1d. Theorem 5d is a small generalization of the result of Grossman and Miller [4, Theorem 4] in which G satisfies the more stringent order condition that for every $\alpha > 0$ there exists a $\beta > 0$ such that $\|G(u) - G(v)\| \le \alpha \|u - v\|$ whenever, $u, v \in \mathfrak{F}$ and $\|u\|$, $\|v\| \le \beta$; in this case the right hand side of (Vd) defines a contraction on $\mathfrak{F}$. Theorem 5d is important because one sometimes wishes to consider perturbations G which are nonlinear functionals such as

$$(Gx)(t) = \int_0^t c(t,s)x^2(s)\,ds \quad \text{or} \quad x(t)\int_0^t c(t,s)x(s)\,ds ,$$

rather than merely perturbations which are continuous functions. For such applications and others in which $\mathfrak{F} = L^2(R^+) \cap BC_0$ with $\|\ \| = \|\ \|_{L^2} + \|\ \|_{BC}$, see [4, Theorems 6-10].

BIBLIOGRAPHY

1. Bronikowski, T. A., An integrodifferential system which occurs in reactor dynamics, Arch. Rational Mech. Anal. 37 (1970), 363-380.

2. Coddington, E. A. and Levinson N., Theory of Ordinary Differential Equations, McGraw Hill Book Co., Inc., 1955.

3. Corduneanu, C., Problème globaux dans la théorie des équations intégrals de Volterra, Ann.Math. Pura Appl. 67 (1965), 349-363.

4. Grossman, S. I and Miller R. K., Perturbation theory for Volterra integro-differential systems, J. Diff. Eq. 8 (1970), 457-474.

5. Grossman, S. I. and Miller, R. K., Nonlinear Volterra integrodifferential systems with L' kernels, (to appear).

6. Hannsgen, K. B., Indirect abelian theorems and a linear Volterra equation, Trans. Amer. Math. Soc. 142 (1969), 539-555.

7. Hannsgen, K. B., On a nonlinear Volterra equation, Mich. Math. J. 16 (1969), 365-376.

8. Kaplan, J. L., On the asymptotic behavior of Volterra integral equations, J. Math. Anal. Appl.

9. Levin, J. J., A nonlinear Volterra equation, not of convolution type, J. Differential Eq. 4 (1968), 176-186.

10. Levin, J. J. and Nohel J. A., A system of integrodifferential equations occuring in reactor dynamics, J. Math. Mech. 9 (1960), 347-368.

11. Levin, J. J. and Nohel, J. A., Perturbations of a nonlinear Volterra equation, Mich. Math. J. 12 (1965) 431-446.

12. Levin, J. J. and Shea, D. F., On the asymptotic behavior of the bounded solutions of some integral equations, I, II, III, J. Math. Anal. Appl. (to appear).

13. MacCamy, R. C. and Wong, J. S. W., Stability theorems for some functional differential equations, Trans, AMS (to appear).

14. Massera, J. L., On Liapunov's conditions for stability, Ann. Math. 50 (1949), 705-721.

15. Massera, J. L. and Schäffer J. J., Linear Differential Equations and Function Spaces, Academic Press, N.Y. & London, 1966.

16. Miller, R. K., On the linearization of Volterra integral equations, J. Math. Anal. Appl. 23 (1968), 198-208.

17. Miller, R. K., On Volterra integral equations with nonnegative integrable resolvents, J. Math. Anal. Appl. 22 (1968), 319-340.

18. Miller, R. K., Admissibility and nonlinear Volterra integral equations, Proc. Amer. Math. Soc. 25 (1970), 45-71.

19. Miller, R. K., Asymptotic stability properties of linear Volterra integro-differential equations, J. Differential Equations (to appear).

20. Miller, R. K., Nohel, J. A., and Wong, J. S. W., Perturbations of Volterra integral equations, J. Math. Anal. Appl. 25 (1969), 676-691.

21. Nohel, J. A., Remarks on nonlinear Volterra equations, Proc. U.S.-Japan Seminar Differential and Functional Equations, W. A. Benjamin, Inc., 1967, pp. 249-266.

22. Nohel, J. A., Asymptotic relationships between systems of Volterra equations, Annali di Mat. Pura ed Applic. (to appear).

23. Paley, R.E.A.C. and Wiener, N., Fourier Transforms in the Complex Domain, Amer. Math. Soc., Providence 1934.

24. Strauss, A., On a perturbed Volterra integral equation, J. Math. Anal. Appl. 30 (1970), 564-575.

ON LIAPUNOV-RAZUMIKHIN TYPE THEOREMS

Junji KATO

Introduction. Liapunov's direct method is a powerful tool in the stability theory of ordinary differential equations. This method has also been extended for functional differential equations by many authors.

A natural generalization in functional differential equations is done by using Liapunov functionals instead of Liapunov functions, and we can find almost same numbers of results as for ordinary differential equations (cf. [4],[6],[7],[13]). On the other hand, Razumikhin [10],[11] has obtained stability theorems for functional differential equations by utilizing Liapunov function (not functional). Our aim in this paper is to unify and generalize these results.

Recently, Yorke [12] and Grossman [3] gave new results in stability theory for functional differential equations. In this paper it will be shown that under the same assumptions as given by them we can reproduce their results by applying our results, though we need fundamental lemmas used in their proofs.

§ 1. Liapunov-Razumikhin type theorems. In Liapunov's direct method, a Liapunov function and a related differential inequality play essential roles. Here we shall deal with a Liapunov function merely as a non-negative continuous function $v(t;\xi)$ with a parameter ξ in Ξ, and throughout this paper, $\dot{v}(t;\xi)$ means the upper time-derivative of $v(t;\xi)$ for fixed ξ. For example, if Ξ is a family of solutions $\xi = x(\cdot)$ of a system (E) of differential equations and if $V(t,x)$ is a Liapunov function, then

$$v(t;\xi) = V(t,x(t)) ,$$

and $\dot{v}(t;\xi)$ is the derivative $\dot{V}_{(E)}(t,x)$ of $V(t,x)$ along the solutions of the system.

The famous theorem due to Razumikhin can be stated in the following form. Here and henceforth, for constant $\tau \geqq 0$ we set

$$V(t,\tau;\xi) = \sup \{v(s;\xi) \ ; \ t-\tau \leqq s \leqq t\}.$$

Theorem A (see [11]). *Let $f(u)$ be a continuous and non-decreasing function such that $f(u) > u$ for $u > 0$, and let $c(u)$ be continuous and positive definite.*

Suppose that

$$\dot{v}(t;\xi) \leqq -c(v(t;\xi)) \quad \text{if} \quad f(v(t;\xi)) \geqq V(t,\tau;\xi).$$

Then $v(t;\xi)$ is uniformly asymptotically stable, that is, $v(t;\xi)$ remains small for $t \geqq a$ if $V(a,\tau;\xi)$ is small, and for any $\varepsilon > 0$ and $\alpha > 0$ there exists a $T(\alpha,\varepsilon)$ such that

$$v(t;\xi) < \varepsilon \quad \text{if} \quad t-s \geqq T(\alpha,\varepsilon) \quad \text{and} \quad V(s,\tau;\xi) \leqq \alpha.$$

This theorem is cited also in [7] and [8] with different proofs, but neither of them apply the comparison theorem in apparent form. My motivation is to obtain a kind of the comparison theorem which includes Theorem A as a corollary.

Consider an ordinary differential equation

$$\dot{u}(t) = U(t,u(t)) \ , \tag{1.1}$$

where $U(t,u)$ is a continuous scalar function on $[0,\infty) \times R^1$, and denote by $u(t;s,\alpha)$ and $r(t;s,\alpha)$ the maximal solution of (1.1) through (s, α) to the right and to the left, respectively.

Then we have the following theorem.

Theorem 1. Suppose that

$$\dot{v}(t;\xi) \leqq U(t,v(t;\xi)) \quad \text{if} \quad v(s;\xi) \leqq r(s;t,v(t;\xi)) \, (s \in [t-\tau,t]). \tag{1.2}$$

Then we have

$$v(t;\xi) \leqq u(t;a,\alpha) \quad \text{for all} \quad t \geqq a \ ,$$

where (a,α) satisfies $v(s;\xi) \leqq r(s;a,\alpha)$ for all $s \in [a-\tau, a]$.

Thus Theorem A is a simple consequence of Theorem 1, since we have the following lemma.

Lemma 1. Let $c(t,u)$ and $f(t,u)$ be continuous functions such that $f(t,u) \geqq u$ and $c(t,u) \geqq 0$ for any t and any $u \geqq 0$ and that $f(t,u) - u$ and $c(t,u)$ are non-decreasing in u for each fixed t. Then, if $v(t;\xi)$ satisfies the condition

$$\dot{v}(t;\xi) \leqq -c(t,v(t;\xi)) \quad \text{when} \quad f(t,v(t;\xi)) \geqq V(t,\tau;\xi), \tag{1.3}$$

then $v(t;\xi)$ satisfies the condition (1.2) with respect to $U(t,u)$, where $U(t,u)$ is given by

$$U(t,u) = -\min \left\{\frac{1}{3\tau} u, \ \frac{1}{\tau} \min_{t \leqq s \leqq t+\tau} [f(s,\tfrac{2}{3} u) - \tfrac{2}{3} u], \ c(t,u)\right\}. \tag{1.4}$$

Moreover, Theorem 1 and Lemma 1 suggest that we can generalize Theorem A.

Theorem 2. *Let* $c(t,u)$ *and* $f(t,u)$ *satisfies the conditions in Lemma 1. Suppose that*

(1.5) $\quad f(t,u) = f(u)$ *is independent of* t, $f(u) > u$ *for* $u > 0$

and

(1.6) $c(t,u) \leqq K(u)$ *and* $\int^{\infty} c(t,u)dt = \infty$ *for any constant* $u > 0$,

where $K(u)$ *is a continuous function.*

If $v(t;\xi)$ *satisfies the condition* (1.3), *then* $v(t;\xi)$ *is asymptotically stable* (T *in Theorem A may depend on* s).

The proof of this theorem can be done by showing that the function $U(t,u)$ given by (1.4) satisfies

(1.7) $$\int^{\infty} U(t,u)dt = -\infty \quad \text{for any constant } u > 0.$$

Remark. It is clear from the form (1.4) that in Theorem 2 the conditions (1.5) and (1.6) can be replaced by the conditions ;

$$f(t,u) \leqq K(u) \quad \text{for some continuous function } K(u),$$

$$\int^{\infty} \min_{t \leqq s \leqq t+\tau} [f(s,u) - u]dt = \infty \quad \text{for any constant } u > 0$$

and

$$c(t,u) = c(u) \text{ is independent of } t, \; c(u) > 0 \text{ for } u > 0.$$

As was shown in the above, Theorem A can be proved by a comparison theorem. However, recently, we have recognized that if there exists a Liapunov function satisfying the conditions in Theorem A, then it is possible to construct a Liapunov functional of the usual type. More generally, we have the following theorem.

Theorem 3. *If a Liapunov function* $v(t;\xi)$ *satisfies the condition* (1.3) *for continuous functions* $c(t,u) \geqq 0$ *and* $f(t,u) \geqq u$ *such that* $f(t,0) = 0$ and $c(t,u)$, $f(t,u)/u$ *are non-decreasing in* u *for each fixed* t, *then there exists a Liapunov function* $w(t;\xi)$ *such that* $v(t;\xi) \leqq w(t;\xi) \leqq V(t,\tau;\xi)$ *and that*

$$\dot{w}(t;\xi) \leqq U(t,w(t;\xi))$$

for a continuous function $U(t,u)$ *given by*

$$U(t,u) = -\min\{c(t,u),\ u\alpha(t,u)\},$$

where $\alpha(t,u) = \{\log[u/f^{-1}(t,u)]\}/\tau$ *and* $f^{-1}(t,u)$ *is the inverse function of* $f(t,u)$ *with respect to* u for each fixed t.

This theorem can be proved by putting

$$w(t;\xi) = \sup_{-\tau \leqq \theta \leqq 0} v(t+\theta;\xi)\exp[\theta\alpha(t+\theta, v(t+\theta;\xi))].$$

As is clear from the definition of $w(t;\xi)$, if $v(t;\xi)$, for $\xi = x(\cdot)$, is given by

$$v(t;\xi) = V(t,x(t))$$

for a Liapunov function $V(t,x)$ defined on $[0,\infty)\times R^n$, then

$$w(t;\xi) = W(t,x_{t,\tau})$$

for a Liapunov functional $W(t,\varphi)$ defined on $[0,\infty)\times C([-\tau,0], R^n)$, where for any continuous R^n-valued function $x(s)$ and for given constant $\tau \geqq 0$, $x_{t,\tau}$ denotes the element of $C([-\tau,0], R^n)$ defined by

$$x_{t,\tau}(\theta) = x(t+\theta) \quad \text{for} \quad \theta \in [-\tau,0].$$

When a constant $h > 0$ is the time-lag for functional differential equations under the consideration and $\tau = kh$ for an integer $k > 0$, a type of the Liapunov functional $W(t,\varphi)$ in the above has been considered by Barnea [1] and by Halanay and Yorke [5].

If $f(t,u) \equiv u$ and $c(t,u) \equiv 0$, then we have

(1.8) $\quad V(t,\tau;\xi)$ is non-increasing for each fixed ξ,

which is a most expectable conclusion. For the case where $f(t,u) \equiv u$, we have the following theorem.

Theorem 4. Suppose that $v(t;\xi)$ *satisfies the conditions* ;

(1.9) $\quad \dot{v}(t;\xi)$ *is continuous in* ξ *uniformly for* t

and

(1.10) *for any* $\varepsilon > 0$ *and any* (t,ξ) *with* $v(t;\xi) \geqq V(t,\tau;\xi) - \varepsilon > 0$ *there exists a* ζ *such that* $d(\xi,\zeta;t) \leqq \varepsilon$ *and*

$$v(t;\zeta) = V(t,\tau;\zeta) \geqq v(t;\xi),$$

where $d(\cdot,\cdot;t)$ is a distance properly introduced in Ξ. For a positive-definite continuous function $c^*(u)$, if we have

$$\dot{v}(t;\xi) \leqq -c^*(v(t;\xi)) \quad \text{when} \quad v(t;\xi) = V(t,\tau;\xi) ,$$

then there exist continuous functions $f(u)$ and $c(u)$ for which the conditions in Theorem A hold good.

Thus, under the conditions (1.9) and (1.10), we can omit the condition $f(u) > u$ in Theorem A, as was shown in [10] (in [10] the condition (1.10) is implicitly assumed).

On the other hand, for the case where $c(t,u) \equiv 0$ but $f(t,u) = f(u) > u$, we can give the following results.

Theorem 5. Let $f(u)$ be a continuous, increasing function such that $f(u) > u$ for $u > 0$ and $f(0) = 0$. Suppose that

$$\dot{v}(t;\xi) \leqq 0 \quad \text{if} \quad f(v(t;\xi)) \geqq V(t,\tau;\xi) . \tag{1.11}$$

Then we have the following conclusions :

(i) $v(t;\xi) \leqq \max\{v(s;\xi) , f^{-1}(V(s,\tau;\xi))\}$ for all $t \geqq s$.

(ii) There exists $\lim_{t\to\infty} v(t;\xi)$.

Theorem 6. In addition to the condition (1.11), we assume the condition

(1.12) $\dot{v}(t_m;\xi_m)$ does not converge to zero for any sequences $\{t_m\}$, with $t_m \to \infty$, and $\{\xi_m\}$ such that

$$f(v(t_m;\xi_m)) \geqq V(t_m,\tau;\xi_m)$$

and that $v(t_m+\theta;\xi_m)$ converges to a non-zero constant independent of $\theta \in [-\tau,0]$.

Then $v(t;\xi)$ is uniformly asymptotically stable.

Remark. As a corollary of Theorem 5, Theorem 2 can be proved in the following way : Under the assumptions in Theorem 2, there exists the limit

$$2\alpha = \lim_{t\to\infty} v(t;\xi)$$

by Theorem 5. Suppose $\alpha > 0$. Then we can find a T such that

$$v(t;\xi) \geqq \alpha \quad \text{and} \quad f(v(t;\xi)) \geqq V(t,\tau;\xi) \quad \text{for all} \quad t \geqq T.$$

Hence we have

$$\dot{v}(t;\xi) \leqq -c(t,\alpha) \quad \text{for all} \quad t \geqq T.$$

Thus there arises a contradiction, because

$$\alpha \leqq v(t;\xi) \leqq v(T;\xi) - \int_T^t c(s,\alpha)ds \quad \text{and} \int^\infty c(s,\alpha)ds = \infty .$$

§ 2. Application I. In the study of stability in functional differential equations, the norm $\|\cdot\|_1$ has been utilized by several authors [2], [3],[5]. For a $\varphi \in C([-h,0], R^n)$, the norm $\|\cdot\|_1$ is defined by

$$\|\varphi\|_1 = \max \{ \|\varphi\| , \sup_{-h \leqq \theta' < \theta \leqq 0} \frac{|\varphi(\theta)-\varphi(\theta')|}{\theta - \theta'} \}$$

for a given constant $h > 0$.

Consider a system of functional differential equations

(2.1) $$\dot{x}(t) = F(t,x_t)$$

and its perturbed system

(2.2) $$\dot{x}(t) = F(t,x_t) + X(t,x_t) ,$$

where $x_t = x_{t,h}$ and $F(t,\varphi)$ and $X(t,\varphi)$ are continuous R^n-valued functions defined on $[0,\infty) \times C([-h,0], R^n)$. Suppose that $F(t,\varphi)$ and $X(t,\varphi)$ satisfy the conditions

(2.3) $$|F(t,\varphi) - F(t,\psi)| \leqq L_1\|\varphi - \psi\|$$

and

(2.4) $$|F(t,\varphi) + X(t,\varphi)| \leqq L_2\|\varphi\|$$

for all t, φ, ψ and for some constants L_1 and L_2.

Then we have the following theorem.

Theorem 7. Suppose that the zero solution of the system (2.1) is uniformly asymptotically stable. Then there exists a positive -definite continuous function $\varepsilon(u)$, for which under the condition

(2.5) $$|X(t,\varphi)| \leqq \varepsilon(\|\varphi\|_1)$$

the zero solution of the system (2.2) is also uniformly asymptotically stable.

In the above, if the zero solution of (2.1) is exponentially stable, then we can choose $\varepsilon(u)$ in a linear form $\varepsilon(u) = \varepsilon u$

<u>for a constant</u> $\hat{\varepsilon} > 0$, <u>and under the condition</u> (2.5) <u>the zero solution of</u> (2.2) <u>is also exponentially stable</u>.

<u>Proof</u>. By a theorem due to Krasovskii [7; Theorem 30.2], we can construct a Liapunov functional $V(t,\varphi)$ satisfies conditions ;

$$a(\|\varphi\|) \leqq V(t,\varphi) \leqq b(\|\varphi\|) ,$$
$$|V(t,\varphi) - V(t,\psi)| \leqq L\|\varphi - \psi\| ,$$
$$\dot{V}_{(2.1)}(t,\varphi) \leqq -c(V(t,\varphi)) ,$$

where $a(u)$, $b(u)$ and $c(u)$ are continuous increasing and positive -definite functions for $u \geqq 0$, L is a constant and $\dot{V}_{(2.1)}(t,\varphi)$ is the upper derivative of $V(t,\varphi)$ along the solutions of the system (2.1).

Let Ξ be the family of solutions $\xi = y(\cdot)$ of the system (2.2), and put

$$v(t;\xi) = V(t,y_t) .$$

Then, clearly

$$\dot{v}(t;\xi) \leqq -c(v(t;\xi)) + L\varepsilon(\|y_t\|_1) .$$

On the other hand,

$$|\dot{y}(s)| \leqq |F(s,y_s) + X(s,y_s)| \leqq L_2\|y_s\| ,$$

and hence

$$\|y_t\|_1 \leqq M\,\|y_{t,2h}\| , \quad M = \max\{L_2, 1\} .$$

Therefore, we have

$$\dot{v}(t;\xi) \leqq -c(v(t;\xi)) + L\varepsilon(Ma^{-1}(V(t,h;\xi))) ,$$

since clearly

$$\|y_{t,2h}\| \leqq a^{-1}(V(t,h;\xi)).$$

Let $f(t,u) = qu$ for a constant $q > 1$, and let $c(t,u) = pc(u)$ for a constant $p \in (0,1)$. Putting

$$\varepsilon(u) = \frac{1-p}{L}\, c\left(\frac{1}{q}\, a\left(\frac{u}{M}\right)\right) , \tag{2.6}$$

we can verify the condition (1.3).

Thus we can show the uniformly asymptotic stability of the zero solution of the system (2.2) by applying Theorem 2 and the standard arguments. Here we should note that for a solution starting from $t = t_o$, we may not be allowed to apply the Liapunov function for $t < t_o + h$. Therefore, to complete the proof, it needs to be shown that

$$\|y_{t_o+h,2h}\| \leqq K\,\|y_{t_o}\| \quad \text{for a constant } K > 0 .$$

However, letting $K = e^{L_2 h}$, this follows immediately from the condition (2.4).

For the second part, it is sufficient to recall that for the exponential stability there exists a Liapunov function $V(t,\varphi)$ with

$$a(u) = u \quad \text{and} \quad c(u) = cu$$

for a positive constant c (see lemma 33.1 in [7]). Therefore, $\varepsilon(u)$ in (2.6) becomes a linear function

$$\varepsilon(u) = \varepsilon u \ , \quad \varepsilon = \frac{c(1-p)}{LqM} > 0 \ .$$

As a simple consequence, we have the following corollary.

Corollary. Consider the system (2.1) under the condition (2.3). Put

$$F_o(t,x) = F(t,\varphi) \quad \text{for} \quad \varphi(\theta) \equiv x \ ,$$

and suppose that the zero solution of the system of ordinary differential equations

$$\dot{x}(t) = F_o(t,x(t))$$

is exponentially stable. Then there exists a constant $h_o > 0$ such that the zero solution of (2.1) is exponentially stable whenever $0 \leqq h < h_o$.

Since we have

$$|X(t,\varphi)| = |F(t,\varphi) - F_o(t,\varphi(0))| \leqq L_1 h \|\varphi\|_1 \ ,$$

the proof is immediate.

In the case where

$$F(t,\varphi) = A\varphi(0) + B\varphi(-h)$$

for constant matrices A and B, Corollary has been proved in [9] by the same way.

§ 3. Application II. We shall apply our results to the theorems presented in [3] and [12].

For the case where (2.1) is a scalar functional differential equation, Yorke [12] has stated the following theorem.

Theorem B. Suppose that $F(t,\varphi)$ in (2.1) satisfies

$$(3.1) \qquad -\alpha M(\varphi) \leqq F(t,\varphi) \leqq \alpha M(-\varphi) \quad \text{for a constant} \quad \alpha > 0 \ ,$$

where $M(\varphi) = \max\{0, \max_{-h\leqq\theta\leqq 0} \varphi(\theta)\}$.

(i) *If* $\alpha h \leqq 3/2$, *the zero solution of the equation* (2.1) *is uniformly stable*.

(ii) *If* $\alpha h < 3/2$ *and if*

(3.2) *for any divergent sequence* $\{t_m\}$ *and any sequence* $\{\varphi_m\}$ *converging to a non-zero constant function*, $F(t_m, \varphi_m)$ *does not tend to zero*,

then the zero solution of (2.1) *is uniformly asymptotically stable*.

The essential part of his proof is the following.

Lemma 2. *Under the condition* (3.1), *a solution* $x(t)$ *of the equation* (2.1) *satisfies the estimation*

$$|x(t)| \leqq (\alpha h - \frac{1}{2})\|x_{s,2h}\| \quad \text{for all} \quad t \geqq s$$

if $x(s) = 0$ (*here we may assume* $\alpha h > 1$), *where the equality holds only when* $\dot{x}(t) = 0$.

Theorem B will be proved in the following way : For a solution $\xi = x(\cdot)$ of (2.1), set

$$v(t;\xi) = \frac{1}{2} x(t)^2 .$$

Then we have

$$\dot{v}(t;\xi) = x(t)F(t,x_t) .$$

If $x(t)x(s) \geqq 0$ for all $s \in [t-h,t]$, then we have $\dot{v}(t;\xi) \leqq 0$. In the case where $x(t)x(s) < 0$ for some $s \in [t-h,t]$, there exists an $r \in [t-h,t]$ such that $x(r) = 0$, and hence by Lemma 2 we have

$$|x(t)| < (\alpha h - \frac{1}{2})\|x_{r,2h}\| \leqq (\alpha h - \frac{1}{2})\|x_{t,3h}\| \quad \text{or} \quad \dot{x}(t) = 0 ,$$

that is,

$$\frac{1}{(\alpha h - \frac{1}{2})^2} v(t;\xi) < V(t,3h;\xi) \quad \text{or} \quad \dot{v}(t;\xi) = 0 .$$

Therefore, if

$$qv(t;\xi) \geqq V(t,3h;\xi) ,$$

there can not exist an $r \in [t-h,t]$ such that $x(r) = 0$ (otherwise $\dot{x}(t) = 0$), and hence

$$\dot{v}(t;\xi) \leqq 0 ,$$

where $q = 1/(\alpha h - \frac{1}{2})^2$.

Thus the conclusion (i) of Theorem B is an immediate consequence of Theorem 1 with $U(t,u) \equiv 0$, while the conclusion (ii) follows from Theorem 6, because the condition (3.2) implies the condition (1.12).

For an n-dimensional functional differential equations (2.1), Grossman [3] has shown the following result.

Theorem C. *Suppose that*

$$|F(t,\varphi)| \leqq M \|\varphi\| \quad \text{for a constant} \quad M > 0$$

and that for positive constants ε *and* K, *we have*

$$\inf_{-h \leqq \theta \leqq 0} \frac{{}^T\varphi(\theta) F(t,\varphi)}{|\varphi(\theta)| \cdot |F(t,\varphi)|} \leqq -\varepsilon, \tag{3.3}$$

whenever $F(t,\varphi) \neq 0$ and $|\varphi(\theta) - \varphi(0)| \leqq K|\varphi(0)|$ *for all* $\theta \in [-h,0]$. *Then we have the following conclusions* :

(i) *If* $\varepsilon \geqq K \geqq Mh$, *then the zero solution of* (2.1) *is uniformly stable*.

(ii) *If* $\varepsilon > K > Mh$ *and if*

(3.4) *there exists a continuous function* $m(t) \geqq 0$ *such that*

$$\int^{\infty} m(t)dt = \infty \ , \quad |F(t,\varphi)| \geqq m(t) \inf_{-h \leqq \theta \leqq 0} |\varphi(\theta)| \ ,$$

then the zero solution of (2.1) *is asymptotically stable*.

The essense of the proof in [3] is based on the following algebraic fact.

Lemma 3. *Let* a, b *and* c *be any points of* R^n, *and let* α *be a constant such that* $0 \leqq \alpha \leqq 1$.

(i) *If* $|a-b| \leqq \alpha|b|$, *then*

$${}^Tab \geqq |a| \cdot |b| \sqrt{1 - \alpha^2} \quad \text{and} \quad |a| \geqq (1-\alpha)|b| \ .$$

(ii) *If* $|a| = |b| = |c| = 1$, ${}^Tab > 0$, ${}^Tac < 0$ *and if*

$$({}^Tab)^2 + ({}^Tac)^2 \geqq 1 + \beta \ ,$$

then ${}^Tbc \leqq -\beta$.

By applying our results, Theorem C will be proved in the following way : First of all, we shall note that under the assumptions in Theorem C, we have

$${}^T\varphi(0)F(t,\varphi) \leqq -\beta|\varphi(0)| \cdot |F(t,\varphi)| \tag{3.5}$$

for $\beta = \varepsilon^2 - K^2$, whenever $\varepsilon \geqq K$ and φ satisfies

$$|\varphi(\Theta) - \varphi(0)| \leqq K|\varphi(0)| \quad \text{for all} \quad \Theta \in [-h,0] .$$

For a solution $\xi = x(\cdot)$ of (2.1), put

$$v(t;\xi) = \frac{1}{2}{}^{T}x(t)x(t) .$$

Then we have

$$\dot{v}(t;\xi) = {}^{T}x(t)F(t,x_t) .$$

Therefore, if we can show that

(3.6) $$|x(s) - x(t)| \leqq K|x(t)| \quad \text{for all} \quad s \in [t-h,t] ,$$

the relation (3.5) means

$$\dot{v}(t;\xi) \leqq 0 .$$

On the other hand, since we have

$$|x(s) - x(t)| \leqq Mh \, \|x_{t,2h}\| \quad \text{for all} \quad s \in [t-h,t] ,$$

the relation

(3.7) $$qv(t;\xi) \geqq V(t,2h;\xi)$$

implies (3.6), where $q = \{K/(Mh)\}^2$.

Therefore, $K \geqq Mh$ implies $q \geqq 1$, from which the conclusion (i) follows.

In the case where $\varepsilon > K > Mh$, we have $\beta > 0$ and $q > 1$.

Moreover, from the condition (3.4) we can easily show that

$$\dot{v}(t;\xi) \leqq -2m(t)\beta(1-K)v(t;\xi)$$

under the condition (3.7). Therefore, letting

$$f(t,u) = qu \quad \text{and} \quad c(t,u) = 2m(t)\beta(1-K)u ,$$

all the conditions in Theorem 2 are satisfied. Thus the conclusion (ii) follows from Theorem 2.

Furthermore, from (3.3) we have

$${}^{T}xF(t,\varphi) \leqq -\varepsilon|x|\cdot|F(t,\varphi)| \quad \text{for a constant function} \quad \varphi(\Theta) \equiv x .$$

Hence, it is easy to see that the condition (3.2) with the condition (3.3) implies the condition (1.12) for the Liapunov function $v(t;\xi)$ considered in the proof of Theorem C.

Thus the following theorem can be proved by applying Theorem 6.

Theorem 8. In the conclusion (ii) of Theorem C, if the condition (3.4) is replaced by the condition (3.2), then the zero solution of the system (2.1) is uniformly asymptotically stable.

REFERENCES

[1] D.I. Barnea, A method and new results for stability and instability of autonomous functional differential equations, SIAM J. Appl. Math., 17(1969), 681-697.

[2] K.L. Cooke, Linear functional differential equations of asymptotically autonomous type, J. Differential Eqs., 7(1970), 154-174.

[3] S.E. Grossman, Stability and asymptotic behavior of differential-delay equations, Tech. Rep. BN 611, IFDAM, Univ. of Md., 1969.

[4] A. Halanay, Differential Equations, Acad. Press, 1966.

[5] A. Halanay and J.A. Yorke, Some new results and problems in the theory of differential-delay equations, SIAM Review, 13(1971), 55-80.

[6] J.K. Hale, Sufficient conditions for stability and instability of autonomous functional-differential equations, J. Differential Eqs., 1(1965), 452-482.

[7] N.N. Krasovskii, Stability of Motion, Stanford Univ. Press, 1963.

[8] V. Lakshmikantham and S. Leela, Differential and Integral Inequalities, Vol.2, Acad. Press, 1969.

[9] Y.-X. Qin, I.-Q. Liou and L. Wang, Effect of time-lags on stability of dynamical systems, Scientia Sinica, 9(1960), 719-747.

[10] B.S. Razumikhin, On the stability of a system with deviations, PMM, 20(1956), 500-512.

[11] B.S. Razumikhin, An application of Liapunov method to a problem on the stability of systems with a lag, Autom. and Remote Cont., 21(1960), 740-748.

[12] J.A. Yorke, Asymptotic stability for one dimensional differential delay equations, J. Differential Eqs., 7(1970), 189-202.

[13] T. Yoshizawa, Stability Theory by Liapunov's Second Method, Math. Soc. Japan, 1966.

ON A NONLINEAR VOLTERRA EQUATION

J. J. Levin*

1. INTRODUCTION

We are concerned here with the asymptotic behavior as $t \to \infty$ of the solutions of the Volterra equation

$$(1.1) \quad x(t) + \int_0^t g(x(\xi))\, a(t-\xi)\, d\xi = f(t) \qquad (0 \leq t < \infty),$$

where x is the unknown and g, a, and f are prescribed real functions. This talk is mainly a summary of our paper [5], to which we refer for additional details and complete proofs.

The behavior of the solutions of (1.1) naturally depend on the assumptions made on the prescribed functions. We shall assume that

(1.2) $g \in C(-\infty, \infty), \quad xg(x) \geq 0 \quad (-\infty < x < \infty)$

(1.3) $a \geq 0$, a is non-increasing on $[0, \infty)$

(1.4) $f \in C[0,\infty) \cap BV[0,\infty)$.

Hypothesis (1.4) implies, of course, that $\lim_{t\to\infty} f(t) = f(\infty)$ exists.

Under assumptions (1.2)-(1.4) one has the following intuitive picture of the behavior of the solutions of (1.1). If t denotes a fixed moment in time, say the present moment, then the variable ξ can be described as being at the present moment if $\xi = t$ and as being in the recent past if $\xi < t$ but near t. Because of (1.3), the function $a(t-\xi)$ of ξ assumes its largest value, possibly ∞, at the present moment and, roughly speaking, its largest values in the recent past. Therefore, in the integral term the values of $g(x(\xi))$ taken in the recent past are weighted most heavily. To simplify matters, let us suppose for the moment (this assumption is not made in the theorems) that g, in addition to (1.2), is also monotone increasing. Then if, for example, x were to get fairly large for a long time, it would follow that the integral term would get large. But then the sum of the two terms on the left-hand side of (1.1) would be large, which is impossible since f(t) must be fairly close to its limiting value for large t. Thus one expects a bounded and possibly oscillatory behavior for the solutions of (1.1) under our assumptions - at least if g is monotone. This, in fact, is the case even without g being monotone.

*This research, mainly performed while the author was an NSF Senior Postdoctoral Fellow at UCLA, was partially supported by the US Army Research Office, Durham.

It is interesting to compare the behavior of (1.1) under assumptions (1.2)-(1.4) with that of the renewal equation of probability theory, which is (1.1) under the hypothesis

$$(1.5)\quad \begin{cases} g(x) = x \\ a(t) \leq 0, \quad -\int_0^\infty a(t)\,dt = 1, \quad -\int_0^\infty t\,a(t)\,dt = m < \infty \\ f(t) \equiv 1. \end{cases}$$

Here (1.1) may be written as

$$(1.6)\quad x(t) = \int_0^t x(\xi)\,[-a(t-\xi)]\,d\xi + 1.$$

Hence $x(0) = 1$, from which it is obvious that $x(t) > 0$ $(t \geq 0)$. But then (1.6) implies $x(t) > 1$ $(t > 0)$. From (1.5) one has, for any $\varepsilon > 0$, that

$$-\int_0^t a(t-\xi)\,d\xi = -\int_0^t a(\xi)\,d\xi > 1 - \varepsilon \qquad (t \geq T_1),$$

for some $T_1 = T_1(\varepsilon) < \infty$. Hence (1.6) now implies $x(t) > 2 - \varepsilon$ $(t \geq T_1)$ Similarly one shows that $x(t) > 3 - \varepsilon$ $(t \geq T_2)$, for some $T_2(\varepsilon) \geq T_1(\varepsilon)$. Thus x increases without bound as $t \to \infty$, which is quite different from the bounded behavior we expect under (1.2)-(1.4). (It is well known, see, e.g., Bellman and Cooke [1], that (1.5) implies

$$\lim_{t\to\infty} \frac{x(t)}{t} = \frac{1}{m}\,.)$$

Many studies of (1.1) under hypotheses similar to (1.2)-(1.4) have been motivated by applications. Thus, e.g., stemming from heat conduction problems are the papers of Mann and Wolf [9] and Padmavally [14], and from superfluidity the paper of Levinson [7]. Later investigations, not necessarily tied to specific applications, include Friedman [2], [3]; Levin [4], [5]; Miller [11]; and Londen [8]. There are studies of (1.1) and other equations which assume existence and boundedness of solutions and then attempt to get asymptotic information on the solutions under less stringent assumptions on f than (1.4). In this direction are the papers, e.g., of Miller [10], Miller and Sell [12], and Levin and Shea [6].

A distinctive feature of the present analysis is that $a \in C(0,\infty)$ is not assumed, whereas in the earlier literature $a \in AC(t_1, t_2)$, for every $0 < t_1 < t_2 < \infty$, has generally been assumed. Also, for existence and boundedness results $f \in AC[0,\infty)$ rather than the weaker hypothesis of (1.4) on f has often been assumed. The reason for these assumptions of absolute continuity is that the equation which one obtains by formally differentiating (1.1) has often played an important role. Under our assumptions (1.2)-(1.4) one is not able to differentiate (1.1).

The first result is concerned with the questions of existence and boundedness of solutions of (1.1) on $[0,\infty)$.

Theorem 1. Let (1.2), (1.3), (1.4), and

(1.7) $a(0) < \infty$

hold. Then

(i) there exists a continuous solution of (1.1) on $[0,\infty)$.

(ii) if $x \in C[0,\infty)$ is a solution of (1.1) it satisfies

$$(1.8) \qquad \sup_{0 \le t < \infty} |x(t)| \le V(f) + \sup_{0 \le t < \infty} |f(t)| \quad ,$$

where $V(f)$ is the total variation of f on $[0,\infty)$.

It should be observed that the bound (1.8) does not depend on g or a. The finiteness assumption (1.7) can be relaxed if more is assumed of g. (In (1.9) it is understood that the Lipschitz constant is allowed to depend on the interval.)

Corollary. Let (1.2), (1.3), (1.4), and

(1.9) g is locally Lipschitzian

$$(1.10) \qquad \int_0^1 a(t)\, dt < \infty$$

hold. Then (1.1) has a unique solution and it satifies (1.8).

The Corollary follows rather easily from Theorem 1 and well known approximation arguments. One observes that because (1.8) is independent of a, Theoreml yields a bound independent of $\varepsilon > 0$ for the unique solution, $x_\varepsilon(t)$, of

$$x_\varepsilon(t) + \int_0^t g(x_\varepsilon(\xi))\, a(t - \xi + \varepsilon)\, d\xi = f(t) \qquad (0 \le t < \infty).$$

In another corollary of Theorem 1 the condition $xg(x)) \ge 0$ of (1.2) is replaced by: there exists a constant c such that $x\,[g(x + c) - g(c)] \ge 0$ on $(-\infty,\infty)$. Here, however, $a \in L^1(0,\infty)$ is assumed and (1.8) takes a somewhat different form.

Having established the existence and boundedness of solutions of (1.1), we wish to look more closely at their asymptotic behavior.

Theorem 2. Let (1.2), (1.3), (1.4), (1.7), and

$$(1.11) \qquad a \in L^1(0,\infty), \quad \limsup_{x \to 0} \frac{|g(x)|}{|x|} < \infty$$

hold and let $x \in C[0,\infty)$ be a solution of (1.1). Then

(i) $f(\infty) \ge 0$ implies $0 \le \liminf_{t\to\infty} x(t) \le \limsup_{t\to\infty} x(t) \le f(\infty)$

(ii) $f(\infty) \le 0$ implies $f(\infty) \le \liminf_{t\to\infty} x(t) \le \limsup_{t\to\infty} x(t) \le 0.$

The oscillatory behavior as $t \to \infty$ allowed for by this result really can happen. In particular, an example is constructed of the following type: For each $\varepsilon \in (0,1/2)$ functions $g_\varepsilon(x)$, $a(t)$, $f_\varepsilon(t)$ are defined which satisfy the hypothesis of Theorem 2, for which $f_\varepsilon(\infty) = 1$, and for which the unique solution of (1.1) is

$$x_\varepsilon(t) = \frac{1}{2} + \left(\frac{1}{2} - \varepsilon\right) \sin(\log(1 + t^2)) \qquad (0 \leq t < \infty).$$

Thus, for this example, one obviously has

$$0 < \varepsilon = \liminf_{t \to \infty} x_\varepsilon(t) < \limsup_{t \to \infty} x_\varepsilon(t) = 1 - \varepsilon < 1 = f_\varepsilon(\infty).$$

In corollaries of Theorem 2 it is shown that fairly natural hypotheses, in addition to those of Theorem 2, can be made to insure the existence of $\lim_{t \to \infty} x(t) = x(\infty)$, which, of course, satisfies

$$x(\infty) + g(x(\infty)) \int_0^\infty a(\xi)\, d\xi = f(\infty).$$

2. PROOFS

In the proofs we need the

__Lemma__. __Let__ a __satisfy__ (1.3) __and__ (1.7) __and let__ $h(t) \in C[0,T]$ __for some__ $0 < T < \infty$. __Define__

$$z(t) = \int_0^t h(\xi)\, a(t - \xi)\, d\xi .$$

__Then__ $z \in AC[0,T]$.

The proof of the lemma is a fairly straightforward real variable exercise, see [5].

__Proof of Theorem 1__. Here we restrict ourselves to establishing the a priori bound (1.8), i.e., to (ii); it is a routine matter (see [5] and [13]) to show that existence, (i), follows from (ii).

Define

$$\begin{cases} P = \{\xi \mid x(\xi) > 0\}, \quad Q = \{\xi \mid x(\xi) < 0\}, \quad R = \{\xi \mid x(\xi) = 0\} \\ P_t = P \cap [0,t], \quad Q_t = Q \cap [0,t], \quad R_t = R \cap [0,t] \\ p(t) = \int_{P_t} g(x(\xi))\, a(t - \xi)\, d\xi \geq 0, \\ q(t) = -\int_{Q_t} g(x(\xi))\, a(t - \xi)\, d\xi \geq 0. \end{cases}$$

Then (1.2), (1.3) readily imply that (1.1) may be written as

(2.1) $\quad x(t) + p(t) - q(t) = f(t) \qquad (0 \leq t < \infty).$

Let

$$x^+(t) = \max\,(x(t),0), \qquad x^-(t) = \max\,(-x(t),0).$$

Thus, $x(t) = x^+(t) - x^-(t)$. It follows from (1.2), (1.3) that p,q can also be written as

$$p(t) = \int_0^t g(x^+(\xi))\, a(t - \xi)\, d\xi \;,\quad q(t) = -\int_0^t g(-x^-(\xi))\, a(t - \xi)\, d\xi \;.$$

From the Lemma it now follows that

$$p, q \in LAC\ [0,\infty),$$

i. e. , that p and q are absolutely continuous on every compact subinterval of $[0,\infty)$.

We now show that

(2.2) $p'(t) \leq 0$ a.e. on $Q \cup R$.

Let $t \geq 0$ and $h > 0$. Then

$$\begin{aligned} p(t + h) - p(t) &= \int_t^{t+h} g(x^+(\xi))\, a(t + h - \xi)\, d\xi \\ &\quad + \int_0^t g(x^+(\xi))\, [a(t + h - \xi) - a(t - \xi)]\, d\xi \\ &\leq \int_t^{t+h} g(x^+(\xi))\, a(t + h - \xi)\, d\xi \;. \end{aligned}$$

Thus

$$\frac{p(t + h) - p(t)}{h} \leq \frac{1}{h} \int_t^{t+h} g(x^+(\xi))\, a(t + h - \xi)\, d\xi \;.$$

Letting $h \to 0+$ yields

$$p'(t) \leq g(x^+(t))\, a(0) \qquad \text{a.e. on } [0,\infty),$$

from which (2.2) is obvious. In the same way one shows that

(2.3) $q'(t) \leq 0$ a.e. on $P \cup R$.

We assume, in the remainder of this proof, that $f(0) \leq 0$. This involves no loss of generality, for if $f(0) > 0$ then $\tilde{x}(t) = -x(t)$ satisfies an equation of type (1.1) with (1.2)-(1.4) satisfied and with $\tilde{f}(t) = -f(t)$ so that $\tilde{f}(0) < 0$. Thus (1.1) now implies

$$x(0) = f(0) \leq 0.$$

Hence

(2.4) $$P = \bigcup_{k=1}^{\infty} (\alpha_k', \alpha_k''), \qquad \alpha_k', \alpha_k'' \in R,$$

a disjoint union which might be finite or empty. Since it is possible that $x(t) > 0$ on (T,∞) for some $T \geq 0$, it is allowed in (2.4) that $\alpha_{k_0}'' = \infty$ for some k_0 rather than $\alpha_{k_0}'' \in R$. Similarly

(2.5) $$Q = \bigcup_{k=1}^{\infty} (\beta_k', \beta_k'') \;;\quad \beta_k', \beta_k'' \in R \qquad (\text{if } f(0) = 0)$$

(2.6) $$Q = [0, \beta_0'') \cup \bigcup_{k=1}^{\infty} (\beta_k', \beta_k'') \;;\quad \beta_k', \beta_k'' \in R \qquad (\text{if } f(0) < 0),$$

where again these disjoint unions might be finite or empty and the possibility of $\beta''_{k_0} = \infty$ for some k_0 is allowed. Somewhat surprisingly, perhaps, the arguments for (2.5) and (2.6) are not exactly the same. Consider the simpler case (2.5), in which $f(0) = 0$.

Clearly (1.8) is an assertion about both $\sup_{0 \le t < \infty} x(t)$ and $\inf_{0 \le t < \infty} x(t)$. As the argument for each is similar, we consider only the one for $\sup_{0 \le t < \infty} x(t)$. For this it clearly suffices to only consider $t \in P$, for when $t \in Q \cup R$ one trivially has $x(t) \le 0$. Thus let $t \in P$ be fixed. Then there exists a unique positive integer, j, such that $\alpha'_j < t < \alpha''_j$. Let $I = \{k \mid \beta'_k < \alpha'_j\}$. Thus

$$s < t \quad \text{and} \quad x(s) < 0 \quad \text{if and only if} \quad \beta'_k < s < \beta''_k \quad \text{for some} \quad k \in I.$$

From (2.1) it is evident that

$$x(t) \le q(t) + f(t).$$

Clearly $q(0) = 0$. Since q is absolutely continuous one has, using (2.2), (2.3), (2.4), and (2.5), that

$$\begin{aligned} q(t) = \int_0^t q'(\xi)\,d\xi &\le \int_{Q_t} q'(\xi)\,d\xi = \sum_{k\in I} \int_{\beta'_k}^{\beta''_k} q'(\xi)\,d\xi \\ &= \sum_{k\in I} [q(\beta''_k) - q(\beta'_k)] \\ &= \sum_{k\in I} [p(\beta''_k) - p(\beta'_k)] + \sum_{k\in I} [f(\beta'_k) - f(\beta''_k)] \\ &\le \sum_{k\in I} [f(\beta'_k) - f(\beta''_k)] \le V(f). \end{aligned}$$

Hence

$$x(t) \le V(f) + f(t) \le V(f) + \sup_{0 \le t < \infty} |f(t)|,$$

from which the inequality for $\sup_{0 \le t < \infty} x(t)$ follows immediately. This completes the argument for $\sup_{0 \le t < \infty} x(t)$ in the case (2.5). In the case (2.6) the argument is slightly harder; see [5].

<u>Proof of Theorem 2</u>. We need only consider the case (i) ($f(\infty) \ge 0$), since the case (ii) is transformed into it by the change of variable $\tilde{x}(t) = -x(t)$ already mentioned. Of course, we can no longer make a sign assumption on $f(0)$. However, formulas for P and Q as the union of disjoint intervals, as in Theorem 1 but with the added possibility

$$P = [0, \alpha''_0) \cup \bigcup_{k=1}^{\infty} (\alpha'_k, \alpha''_k) \qquad (\text{if } f(0) > 0)$$

are also available here.

The result is almost obvious if either P or Q are bounded sets. To see this suppose, for example, that Q is bounded. Then

$$0 \leq x(t) \qquad (T \leq t < \infty)$$

for some finite T. Then

$$x(t) + \int_0^T g(x(\xi))\, a(t - \xi)\, d\xi \leq f(t) \qquad (T \leq t < \infty).$$

Hence, obviously,

$$0 \leq \liminf_{t\to\infty} x(t) \leq \limsup_{t\to\infty} x(t) \leq f(\infty),$$

as asserted. Thus suppose P, Q, and, hence, also R are unbounded.

We sketch the proof of

$$(2.7) \quad \limsup_{t\to\infty} x(t) \leq f(\infty)$$

only, since the one for

$$0 \leq \liminf_{t\to\infty} x(t)$$

is similar.

One first shows that

$$(2.8) \qquad \lim_{t\to\infty,\, t\in P\cup R} q(t) = q^*, \qquad \lim_{t\to\infty,\, t\in Q\cup R} p(t) = p^*$$

exist. The first assertion of (2.8), e.g., means that there exists $q^* \geq 0$ such that for each $\varepsilon > 0$ there exists $T(\varepsilon) < \infty$ with the property that

$$t \geq T(\varepsilon) \text{ and } t \in P \cup R \quad \text{imply} \quad |q(t) - q^*| < \varepsilon .$$

Let $\varepsilon > 0$. Choose $T = T(\varepsilon) < \infty$ so that $V(f,[T,\infty)) < \varepsilon$, where $V(f,I)$ denotes the total variation of f on the interval I. Let $T \leq t_1 < t_2$ and $t_1, t_2 \in P \cup R$. Define $J = \{k \mid t_1 \leq \beta_k' < t_2\}$. Then (2.2) and (2.3) imply, as in the proof of Theorem 1,

$$q(t_2) - q(t_1) \leq \int_{Q\cap[t_1,t_2]} q'(\xi)\, d\xi = \sum_{k\in J} \int_{\beta_k'}^{\beta_k''} q'(\xi)\, d\xi$$

$$\leq \sum_{k\in J} [f(\beta_k') - f(\beta_k'')] \leq V(f,[t_1,t_2]) < \varepsilon .$$

First letting $t_2 \to \infty$ and then letting $t_1 \to \infty$ yields

$$\limsup_{t\to\infty,\, t\in P\cup R} q(t) \leq q(t_1) + \varepsilon$$

$$\limsup_{t\to\infty,\, t\in P\cup R} q(t) \leq \liminf_{t\to\infty,\, t\in P\cup R} q(t) + \varepsilon,$$

which implies the first half of (2.8). The second half of (2.8) is similarly established.

Suppose (2.7) is false. Then there exists $\hat{x}$ and a subsequence

$\{\alpha'_{k_j}\}$ of $\{\alpha'_k\}$ such that

$$(2.9)\quad \begin{cases} \max\limits_{\alpha'_{k_j} \le t \le \alpha''_{k_j}} x(t) \ge \hat{x} > f(\infty) \ge 0 \quad (j = 1,2,\dots) \\ \lim\limits_{j\to\infty} \alpha'_{k_j} = \infty . \end{cases}$$

We now study the behavior of $x(t)$ on the intervals $[\alpha'_{k_j}, \alpha''_{k_j}]$ $(j = 1,2,\dots)$.

These intervals are of bounded length, that is,

$$(2.10)\qquad \sup_j (\alpha''_{k_j} - \alpha'_{k_j}) < \infty .$$

For one has

$$\begin{aligned} q(\alpha''_{k_j}) &= -\int_0^{\alpha''_{k_j}} g(-x^-(\xi))\, a(\alpha''_{k_j} - \xi)\, d\xi \\ &= -\int_0^{\alpha'_{k_j}} g(-x^-(\xi))\, a(\alpha''_{k_j} - \xi)\, d\xi \\ &\le \sup_{0 \le \xi < \infty} |g(x(\xi))| \int_0^{\alpha'_{k_j}} a(\alpha''_{k_j} - \xi)\, d\xi \\ &\le \sup_{0 \le \xi < \infty} |g(x(\xi))| \int_{\alpha''_{k_j} - \alpha'_{k_j}}^{\infty} a(\xi)\, d\xi , \end{aligned}$$

so that if (2.10) is false, then $q^* = 0$. However, since

$$x(t) \le q(t) + f(t),$$

$q^* = 0$ implies

$$\limsup_{t\to\infty} x(t) \le f(\infty),$$

which contradicts (2.9). Thus (2.10) is established.

On $[\alpha'_{k_j}, \alpha''_{k_j}]$ we rewrite (1.1) as

$$(2.11)\quad x(t) + \int_0^{\alpha'_{k_j}} g(x^+(\xi))\, a(t - \xi)\, d\xi + \int_{\alpha'_{k_j}}^{t} g(x^+(\xi)) a(t - \xi)\, d\xi = q(t) + f(t).$$

Notice that for large j the right hand side is near its limiting value $q^* + f(\infty)$. From (2.11) One has

$$(2.12)\quad \int_0^{\alpha'_{k_j}} g(x^+(\xi))\, a(t-\xi)\, d\xi \le q(t) + f(t) - x(t) \quad (\alpha'_{k_j} \le t \le \alpha''_{k_j}).$$

Since the left-hand side of (2.12) is monotone non-increasing, it now follows from (2.8), (2.9), (2.12), and (1.4) that for some J

$$(2.13)\quad \int_0^{\alpha'_{k_j}} g(x^+(\xi))\, a(\alpha''_{k_j} - \xi)\, d\xi \le q(\alpha''_{k_j}) + f(\alpha''_{k_j}) - \frac{1}{2}\left(f(\infty) + \hat{x} \right) \quad (j \ge J).$$

Theorem 1 and the hypothesis imply

$$(2.14)\quad |g(x(t))| < \rho |x(t)| \qquad (t \in P \cup Q)$$

for some constant $0 < \rho < \infty$.

Define $w_j(t)$ on $[\alpha'_{k_j}, \alpha''_{k_j}]$ as the unique solution of the linear equation

$$(2.15)\quad w_j(t) + \int_0^{\alpha'_{k_j}} g(x^+(\xi))\, a(t-\xi)\, d\xi + \rho\, a(0) \int_{\alpha'_{k_j}}^{t} w_j(\xi)\, d\xi = q(t) + f(t),$$

which is intimately related to (2.11). It is a simple matter to solve (2.15) and with (2.9) and (2.13) to then show that

$$(2.16)\quad \liminf_{j \to \infty} w_j(\alpha''_{k_j}) > 0.$$

One next shows that

$$(2.17)\quad x(t) \ge w_j(t) \qquad (\alpha'_{k_j} \le t \le \alpha''_{k_j}).$$

This is done (see [5]) with the aid of (2.14) and the linear equation

$$z_{j,\varepsilon}(t) + \int_0^{\alpha'_{k_j}} g(x^+(\xi))\, a(t-\xi)\, d\xi + \rho\, a(0) \int_{\alpha'_{k_j}}^{t} z_{j,\varepsilon}(\xi)\, d\xi = q(t) + f(t) - \varepsilon,$$

where $\varepsilon > 0$. However, (2.16) and (2.17) imply

$$\liminf_{j\to\infty} x(\alpha''_{k_j}) > 0 \quad ,$$

which, of course, contradicts $x(\alpha''_{k_j}) = 0$. The contradiction establishes (2.7) and completes the proof.

REFERENCES

1. Bellman, R. and Cooke, K. L. - Differential-Difference Equations, Academic Press Inc., New York, 1963.
2. Friedman, A. - On integral equations of Volterra type, J. d'Analyse Math. 11(1963),381-413.
3. Friedman, A. - Periodic behavior of solutions of Volterra integral equations, J. d'Analyse Math. 15(1965), 287-303.
4. Levin, J. J. - The qualitative behavior of a nonlinear Volterra equation, Proc. Amer. Math. Soc. 16(1965), 711-718.
5. Levin, J. J. - On a nonlinear Volterra equation, J. Math. Anal. Appl., to appear.
6. Levin, J. J. and Shea, D. F. - On the asymptotic behavior of the bounded solutions of some integral equations, I, II, III, J. Math. Anal. Appl., to appear.
7. Levinson, N. - A nonlinear Volterra equation arising in the theory of superfluidity, J. Math. Anal. Appl. 1(1960), 1-11.
8. Londen, S-O. - On the solutions of a nonlinear Volterra equation, J. Math. Anal. Appl., to appear.
9. Mann, W. R. and Wolf, F. - Heat transfer between solids and gases under nonlinear boundary conditions, Quart. Appl. Math. 9 (1951), 163-184.
10. Miller, R. K. - Asymptotic behavior of solutions of nonlinear Volterra equations, Bull. Amer. Math. Soc. 72(1966), 153-156.
11. Miller, R. K. - On Volterra integral equations with nonnegative integrable resolvents, J. Math. Anal. Appl. 22(1968),319-340.
12. Miller, R. K. and Sell, G. R. - Volterra Integral Equations and Topological Dynamics, Memoir 102 Amer. Math. Soc.(1970).
13. Nohel, J. A. - Some problems in nonlinear Volterra integral equations, Bull. Amer. Math. Soc. 68(1962), 323-329.
14. Padmavally, K. - On a nonlinear integral equation, J. Math. Mech. 7(1958), 533-555.

LOCAL DYNAMICAL SYSTEMS AND THEIR ISOMORPHISMS

Taro URA

Introduction

A system of axioms which is a generalization of that of a dynamical system was constructed by T. Ura in 1959 [20] and later an essentially equivalent system was introduced and the term " local dynamical system " was given by O. Hajek in 1965 [11] and by G. R. Sell in 1967 [19], independently. Recently S. Ahmad clarified in what sense they are equivalent [7]. Applications of the theory of local systems have been amazingly developed during the last five years (cf. [3], [19]).

On the other hand, O. Hajek proposed the categorical treatment of dynamical systems, at an International Symposium on Dynamical Systems, Fort Collins, Colo. 1967 [12]. As we know well, the Category Theory started as a systematization of known results for some branches of mathematics. Of course this is the same for dynamical systems and in fact there had been many outstanding results belonging to the categorical theory of them. However, as far as I know, my paper [21] contributed to the same symposium was the first one giving some concrete results explicitly mentioning this standpoint.

The purpose of this lecture is to exhibit some results on local dynamical systems from the categorical point of view. The topics are selected from the results obtained by my colleagues. Here my colleagues include Otomar Hajek, Roger C. McCann in Cleveland, Ohio, David H. Carlson formerly in Cleveland at present in Colombus, Mo., Ikuo Kimura, Jiro Egawa in Kobe and Giko Ikegami formerly in Kobe at present in Nagoya.

The lecture consists of :

(1) an explanation of local dynamical systems,

(2) an introduction of two kinds of morphisms, the first called NS-isomorphisms and the second GH-isomorphisms, with the further classification of the second,

(3) I. Kimura - T. Ura's Theorems showing the relation between these two kinds of morphisms, more precisely, sufficient conditions for existence, uniqueness and continuity of reparametrizations for a given NS-isomorphism in order to make this a GH-isomorphism,

(4) a classification of centers by R. G. McCann by types of morphisms, together with an example of pairs of local systems to

explain the need of introducing the various types of GH-isomorphisms,

(5) the problem of restrictions of isomorphisms to subsystems treated by I. Kimura and T. Ura,

(6) the works on Local parallel flows by T. Ura and J. Egawa, i.e., relations between local and global parallelizability of local systems and the considered morphisms, together with examples showing the limit of extendability of Vinograd's Theorem to general phase spaces (cf. (7) below),

(7) G. Ikegami's extension of Vinograd's Theorem to C^r differentiable local systems on C^∞ n-manifolds, with emphasis on C^r differentiability of the reparametrization mappings, and D. H. Carlson's extension of the same theorem to local systems on metrizable phase spaces,

(8) the problem of universal dynamical systems treated by O.Hajek and D. H. Carlson, i.e., two generalizations of Beboutov - Kakutani's Embedding Theorem of dynamical systems of a special character on compact separable metric spaces to those of a more general property.

Local Dynamical Systems

The notion of a local dynamical system has been getting more and more popular. I think at present everyone is familiar with it. However to make the notations fixed, I would like to begin this lecture with an explanation of it.

Definition. (Cf. [22], [7]) Let X be a topological space. Assume that to each point $x \in X$ is assigned an interval I_x containing 0. Put

$$D = \bigcup_{x \in X} \{x\} \times I_x .$$

Then D is a subset of $X \times R$, where R denotes the set of real numbers. (This will be our standing notation.) We consider the subspace topology of D in $X \times R$. Let π be a mapping of D into X. If π satisfies the following axioms, then the triple (X, D, π) is called a *local dynamical system* (abbreviated as *local system*) or a *local flow*, or π is called a *local system* or a *local flow on the phase space* X *with domain* D.

(LD-1) (*Open Interval Axiom*) $\forall x \in X$, I_x is an open interval (say (a_x, b_x)).

(LD 0) (*Openness or Kamke Axiom*) D is open in $X \times R$.

(LD 1) (*Identity Axiom*) $\forall x \in X$, $\pi(x, 0) = x$.

(LD 2) (*Homomorphism Axiom*) $(x, t) \in D$, $(x, t + s) \in D$ and $(\pi(x,$

$t), s) \in D$

$\Longrightarrow \quad \pi(\pi(x, t), s) = \pi(x, t + s).$

(LD 3) (*Continuity Axiom*) π is continuous in D.

(LD 4) (*Non-extendability Axiom*) If b_x is finite, then $L^+(x) = \phi$; if a_x is finite, then $L^-(x) = \phi$.

Here, for example, $L^+(x)$ denotes the cluster set of $\pi(x, t)$ as $t \uparrow b_x$.

It is known that condition (LD 0) is equivalent to : *the mapping* $x \mapsto a_x$ *and the mapping* $x \mapsto b_x$ *are upper and lower semi-continuous respectively.*

If in particular $D = X \times R$, then (LD-1), (LD 0), (LD 4) and the assumptions in (LD 2) are superfluous and hence we have the well known classical system of axioms of a dynamical system. Therefore the notion of a local system is a generalization of that of a dynamical system in the classical sense. If $D = X \times R$, we shall say π is *global*. For example, if the phase space X is compact, then (LD 4) implies $\forall x \in X$, $b_x = \infty$, $a_x = -\infty$ or equivalently $D = X \times R$, thus every local system on a compact phase space is global. In contrast to this, if we want to put emphasis on the fact $D \neq X \times R$ for a given local system π, we shall say π is *strictly local*. The same convention of terminology will be pursued in the sequel.

Applications

A local system is originally an abstraction of an autonomous system of differential equations. However recently, many applications were found. In particular Richard K. Miller and George R. Sell succeeded in treating *Volterra integral equations*

$$x(t) = f(t) + \int_0^t a(t, s)g(x(s), s)ds$$

as a local system [3]. One may notice that in this case the phase space is non-metrizable. Thus one should not be satisfied with treating local systems on metrizable phase spaces only.

Restriction of Local Systems.

Let $M \subset X$. It is easily shown that for every $x \in M$ there exists the maximum interval I_x containing 0 such that $\pi(x, I_x) \subset M$. Put

$$D_M = \bigcup_{x \in M} \{x\} \times I_x.$$

The restriction $\pi|_{D_M}$ will be denoted by $\pi||_M$. In general $(M, D_M, \pi||_M)$ is not a local system.

Definition. If $\pi||_M$ is a local system on M, then M is said to be *admissible*, and $\pi||_M$ is called *the restriction* or *the subsystem of* π *in* M.

Definition. Let $x \in X$. The set

$$C_\pi(x) \equiv C(x) \equiv \pi(x, I_x)$$

is called the orbit *through* (or *of*) x.

Definition. Let $\phi \neq M \subset X$. If $C(M) = M$, then M is said to be *quasi-invariant*. A quasi-invariant set M is said to be *invariant* if for every $x \in M$ we have $I_x = R$, in accordance with the usual terminology in the Dynamical Systems Theory.

Theorem. *Every quasi-invariant set is admissible and every open set is admissible.*

(This theorem for open sets could be considered as a motivation for introducing local systems.)

The definition of a singular point and a periodic point remain the same as for a global system. The set of singular points will be denoted by S or S_π, and the set of periodic points will be denoted by P or P_π.

Categories of Local Systems

We shall consider categories whose objects are local systems. The first problem is how to determine the morphisms or to answer the question what are the most appropriate morphisms for our research of local systems. There should be many ways to do it. Here we shall introduce two kinds of morphisms, called NS and GH. The introduction will be done in a very formal way. The formal explanation and the obtained results would suggest, I hope, the concrete meaning of the isomorphisms considered.

Let (X, D, π), (Y, E, ρ) denote two local systems in the sequel, unless otherwise stated.

NS-Isomorphisms.

Definition. Let h be a homeomorphism $X \to Y$. If the diagram

$$\begin{array}{ccc} X & \xrightarrow{h} & Y \\ \downarrow C_\pi & & \downarrow C_\rho \\ 2^X & \longrightarrow & 2^Y \end{array}$$

commutes, then h is called an *NS-isomorphism* $\pi \to \rho$.

In short, an NS-isomorphism is a *homeomorphism* of the phase spaces *preserving the orbits*.

We said " NS-isomorphism ". It is quite easy to prove that NS-isomorphisms are actually isomorphisms in the sense of the Category Theory (cf. [6]).

GH-Isomorphisms for Global Systems

We shall explain various types of GH-isomorphisms. To make the motivation clear, we begin with such morphisms for global systems.

Let π, ρ be two global systems, so that $\pi: X \times R \to X$, $\rho: Y \times R \to Y$ are continuous and surjective.

Let $h: X \to Y$ be a homeomorphism, $\phi: X \times R \to R$ or $R \to R$ a mapping. (h, ϕ) will denote the mapping defined by

$$(x, t) \mapsto (h(x), \phi(x, t))$$

or

$$(x, t) \mapsto (h(x), \phi(t)).$$

Our standing algebraic postulate is that the diagram

$$\begin{array}{ccc} X \times R & \xrightarrow{(h,\ \phi)} & Y \times R \\ \downarrow\pi & & \downarrow\rho \\ X & \xrightarrow{h} & Y \end{array}$$

commutes, (in contrast to the definition of an NS-isomorphism).

Type 0: $\phi: R \to R$ is a **continuous field** isomorphism
$\iff \phi(t) = t \iff \phi: R \to R$ is the identity.

Type 1: $\phi: R \to R$ is a **continuous group** (cf.[2]) isomorphism
$\iff \exists c \neq 0\ (\in R)$, $\phi(t) = ct$.

Type 2: (h, ϕ) is a **vector bundle** isomorphism
$\iff \exists c: X \to L_t(R, R) = R - \{0\}$ continuous,
such that $\phi(x, t) = c(x)t$.
(Here, $L_t(R, R)$ denotes the space of toplinear maps: $R \to R$.)

Type 3: (h, ϕ) is a **product bundle** isomorphism and
$\forall x \in X$, $\phi(x, 0) = 0$
$\iff \phi: X \times R \to R$ is continuous and
$\forall x \in X$, $\phi(X, \cdot): R \to R$ is a homeomorphism, such that
$\phi(x, 0) = 0$.

(The condition $\phi(x, 0) = 0$ will be used only for $x \in P_\pi$.)

In the conditions for type 2 or type 3 isomorphisms, the continuity condition may be dropped. Then we shall say that (h, ϕ) is of type 2.5 or type 3.5. We may only assume the continuity in $(X - S_\pi) \times R$; then we shall say that (h, ϕ) is of type 2.25 or type 3.25.

The homeomorphism h is called the *phase mapping* and ϕ the *reparametrization mapping* (or simply the *reparametrization*).

We may consider another important structure of R, namely the total order of R. We shall say that the reparametrization ϕ is *increasing* or *decreasing* according as for every $x \in X$, $\phi(x, \cdot)$ is increasing or decreasing.

GH-Isomorphisms for Local Systems

Now let us extend various types of GH-isomorphisms to local systems. Only one ambiguity arises in the fact that π and ρ may not be defined in $X \times R$ and in $Y \times R$. Our diagram can be written as

$$\begin{array}{ccc} X \times R & \xrightarrow{(h,\ \phi)} & Y \times R \\ \cup & & \cup \\ D & & E \\ \downarrow \pi & & \downarrow \rho \\ X & \xrightarrow{\quad h \quad} & Y \end{array}$$

Firstly, we would tentatively understand that our diagram commutes as far as it has meaning. One may notice that this understanding means:

$$(x, t) \in D \text{ and } (h(x), \phi(x, t)) \in E$$
$$\Longrightarrow \ h \circ \pi(x, t) = \rho(h(x), \phi(x, t)).$$

Secondly, continuity of the reparametrization ϕ should be considered only in D or $D - (S_\pi \times R)$.

Contemplating the diagram written as above, I think that it is quite natural to extend type 3,... isomorphisms as follows.

Let I_x^* and J_y^* be intervals assigned to each $x \in X$ and to each $y \in Y$ respectively. Put

$$D^* = \bigcup_{x \in X} \{x\} \times I_x^*, \qquad E^* = \bigcup_{y \in Y} \{y\} \times J_y^*$$

and assume that

$$D^* \supset D \qquad \text{and} \qquad E^* \supset E.$$

Let h be a homeomorphism $X \to Y$ and ϕ a mapping $D^* \to R$ such that for every $x \in X$, $\phi(x, \cdot): I_x^* \to J_{h(x)}^*$ is a homeomorphism and $\phi(x, 0) = 0$.

If the diagram

$$\begin{array}{ccc} D^* & \xrightarrow{(h,\ \phi)} & E^* \\ \cup & & \cup \\ D & & E \\ \downarrow\pi & & \downarrow\rho \\ X & \xrightarrow{\quad h \quad} & Y \end{array}$$

commutes in the same sense as explained above, then (h, ϕ) is called a *type* 4, 4.25, 4.5 (= *GH*-)*isomorphism* $\pi \to \rho$, according as ϕ is continuous in D or in $D - (S_\pi \times R)$ or may not be continuous. Of course a type 4, 4.25, or 4.5 isomorphism is of type 3, 3.25, or 3.5, if and only if $D^* = X \times R$, $E^* = Y \times R$ respectively, in particular if both π and ρ are global. The following is obvious:

$$\begin{array}{ccccccccc} (0) & \Longrightarrow & (1) & \Longrightarrow & (2) & \Longrightarrow & (3) & \Longrightarrow & (4) \\ & & & & \Downarrow & & \Downarrow & & \Downarrow \\ & & & & (2.25) & \Longrightarrow & (3.25) & \Longrightarrow & (4.25) \\ & & & & \Downarrow & & \Downarrow & & \Downarrow \\ & & & & (2.5) & \Longrightarrow & (3.5) & \Longrightarrow & (4.5)\ (= \mathrm{GH}). \end{array}$$

Theorem. [22] *If* (h, ϕ) *is a GH-isomorphism, then for every* $x \in X$, $\phi(x, \cdot)$ *maps* I_x *homeomorphically onto* $J_{h(x)}$, *or in other words,* $(h, \phi)(D) = E$.

Corollary. *The condition*

$$(x, t) \in D \ \text{ and } \ (h(x), \phi(x, t)) \in E \implies h \circ \pi(x, t) = \rho(h(x), \phi(x, t))$$

is equivalent to

$$(x, t) \in D \implies (h(x), \phi(x, t)) \in E \ \text{ and } \ h \circ \pi(x, t) = \rho(h(x), \phi(x, t)).$$

In contrast to NS-isomorphisms, it is not easy, but one can prove that each type of isomorphisms are isomorphisms in the sense of the Category Theory. The essential points to prove this are (see [22], [23]):

(1) the preceding theorem,

(2) invariance of S_π, i.e. $h(S_\pi) = S_\rho$,

(3) existence of the inverse $(:E^* \to R)$ of the reparametrization,

(4) corresponding continuity of this inverse mapping.

Embeddings

Once the notions of isomorphisms are introduced, it is not hard to extend them to those of embeddings, namely;

Let h be an embedding $X \to Y$ and ϕ a mapping $D^* \to R$.

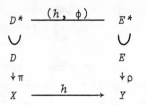

Assume that $h(X)$ is admissible with respect to ρ. If h or (h,ϕ) is an NS-isomorphism or a type n isomorphism $\pi \to \rho||_{h(X)}$, then h or (h, ϕ) is called a *NS* or a *type n embedding* $\pi \to \rho$.

Therefore, as far as intrinsic properties are concerned, the study of a single embedding reduces to that of a single isomorphism.

Kimura - Ura's Theorems

Theorem. *If* (h, ϕ) *is a GH-isomorphism* $\pi \to \rho$, *then* h *is an NS-isomorphism* $\pi \to \rho$.

This theorem is easy to prove and is expressed in terms of the Category Theory as follows.

Theorem. *Let* C, C' *be the categories defined by*

Obj C = Obj C' = {*local system*},

Mor C = $\{(h,\phi) \mid \pi,\rho \in$ Obj C, $(h, \phi):\pi \to \rho$: *GH-iso.*},

Mor C' = $\{h \mid \pi,\rho \in$ Obj C', $h:\pi \to \rho$: *NS-iso.*}.

Then T *defined by*

$$T: \pi \mapsto \pi$$

$$T: (h, \phi) \mapsto h$$

is a functor $C \to C'$.

The converse problem was solved by Ura under the assumption that the phase space is completely regular [23] and this result was improved by Kimura as follows.

Theorem. [17] *Assume that the phase spaces* X *and* Y *are* T_1-*spaces. If* h *is an NS-isomorphism* $\pi \to \rho$, *then there exists a reparametrization* ϕ *such that* (h, ϕ) *is a GH-isomorphism* $\pi \to \rho$. *In addition for every* $x \in X-S$, $\phi(x, \cdot)$ *is uniquely determined.*

Theorem. [23] *Assume that* X *and* Y *are Hausdorff. If* (h, ϕ) *is a type* $n.5$ *isomorphism* $\pi \to \rho$, *then* (h, ϕ) *is of type* $n.25$. *In other words, if* X *and* Y *are Hausdorff, then for every GH-isomorphism* (h, ϕ), ϕ *is continuous in* $D - S_\pi \times R$.

These two theorems are stated in terms of the Category Theory as follows.

Theorem. *Let* C_1, C_1' *be two categories defined by*

Obj C_1 = Obj C_1' = {*local system on a* T_1-*space*},

Mor C_1 = $\{(h, \phi) \mid \pi,\rho \in$ Obj C_1, $(h, \phi):\pi \to \rho$:*GH-iso.*},

Mor C_1' = $\{h \mid \pi, \rho \in$ Obj C_1', $h:\pi \to \rho$:*NS-iso.*}.

Then there exists a functor $T':C_1' \to C_1$ *such that* $T \circ T'$ *is the identity* $C_1' \to C_1'$, *for the functor* T *introduced above.*

Theorem. *Let* C_2 *be the category defined by*

Obj C_2 = {*local system on a Hausdorff space*},

Mor C_2 = $\{(h, \phi) \mid \pi, \rho \in \text{Obj } C_2,\ (h, \phi): \pi \to \rho: \textit{type } 4.25 \textit{ iso.}\}$.

Then C_2 *is a full subcategory of the category* C.

From these theorems, many corollaries can be derived, e.g.

Corollary. *Let* $C_{2,R}$, $C'_{2,R}$ *be two subcategories of* C_1, C'_1 *defined by*

Obj $C_{2,R}$ = Obj $C'_{2,R}$ = $\{\pi \mid S_\pi = \phi,\ \pi$ *on a Hausdorff space*$\}$,

Mor $C_{2,R}$ = $\{(h, \phi) \mid \pi, \rho \in \text{Obj } C_{2,R},\ (h, \phi): \pi \to \rho: \textit{type } 4 \textit{ iso.}\}$,

Mor $C'_{2,R}$ = $\{h \mid \pi, \rho \in \text{Obj } C'_{2,R},\ h: \pi \to \rho: \textit{NS-iso.}\}$.

Then $C_{2,R}$ *is full in* C *and*

$$T'|_{C'_{2,R}} = (T|_{C_{2,R}})^{-1}.$$

These theorems were stated for isomorphisms. However, as we remarked above, they hold for the corresponding embeddings.

The author's theorem recalled above tells us that in usual applications, we need not type $n.5$ isomorphisms. On the other hand the distinction of types 0, 1, 2, 2.25, 3, 3.25, 4, 4.25 is important. The final verification will be given in the next paragraph. Here we shall show a partial verification.

Firstly, as is easily seen, if π is strictly local and ρ is global, then π and ρ can not be type n isomorphic with $n < 4$.

Secondly, could we have type $n.25$ =>type $n.0$ under suitable condition on the phase spaces ? This question is negatively answered. It is quite easy to show two simple global systems on R^1 which are type 2.25 isomorphic but not type 2.0 isomorphic. One also finds another pair of examples on R^2 having the same equivalence properties (cf. [23]).

The following example shows us there is clear distinction between type 2,... isomorphisms and type 3,... isomorphisms [unpublished].

Let X be the closed ring domain $1 \leq r \leq 2$. ((r, θ) is the polar coordinate in the plane). Let π and ρ be two global systems on X defined by

$$\begin{cases} \dfrac{d\theta}{dt} = 2\pi \\ \dfrac{dr}{dt} = (r - 1)(2 - r) \end{cases} \quad \text{and} \quad \begin{cases} \dfrac{d\theta}{dt} = 2\pi(2r - 1) \\ \dfrac{dr}{dt} = (r - 1)(2 - r)(2r - 1) \end{cases}$$

respectively. Then π and ρ are type 3 isomorphic but not type 2.25

(*a fortiori* 2) isomorphic.

Centers

Localization of the notions of our isomorphisms is easily done in a natural way. Thus Poincaré centers in the plane could be classified modulo type n local isomorphisms for each n. This research was done by McCann in [18]. By Hajek's Theorem concerning the continuity of local sections in the plane, it is shown that all Poincaré centers are locally NS-isomorphic and hence, by Kimura - Ura's Theorems, they are type 3.25 isomorphic. McCann proved that they are type 2.25 isomorphic and obtained a necessary and sufficient condition that two Poincaré centers are type 2 isomorphic. It is easy to show examples of pairs of Poincare centers each of which satisfies one of the following conditions:

(1) $\neg$ type 0 iso. $\wedge$ type 1 iso.
(2) $\neg$ type 1 iso. $\wedge$ type 2 iso.
(3) $\neg$ type 2 iso. $\wedge$ type 2.25 iso.

Thus, combining with the remark at the end of the last paragraph, every type number except $n.5$ has its own meaning.

Local Categories

A remarkable categorical property of GH-isomorphisms is:

Theorem. *If* (h, ϕ) *is a GH-isomorphism* $\pi \to \rho$, *and* U *is any open subset of the phase space* X *of* π, *then* $(h|_U, \phi|_{D_U})$ *is also a GH-isomorphism* $\pi||_U \to \rho||_{h(U)}$.

In terms of S. Eilenberg and H. Cartan (cf. [6]), this is stated as :

The category whose objects are local systems and whose morphisms are GH-embeddings is a local category.

By Kimura's Theorem, *the same is true for the category whose objects are local sytems on T_1-spaces and whose morphisms are NS-embeddings.* However we do not know we can omit the T_1-assumption.

Parallelizability

Parallelizability of a global system has been studied for a long time and many elegant results have been obtained (cf. [13]). As is easily understood, the notion of parallelizability is strongly related to what kind of isomorphisms we consider [21], [22].

As local systems are introduced, it is natural to introduce the notions such as a local parallel flow and local parallelizability [21], [10].

Definition. [10] Let H be a topological space. Assume that for each $\eta \in H$ is assinged an open interval J_η containing 0 and that the set

$$Y = \bigcup_{\eta \in H} \{\eta\} \times J_\eta \quad (\subset H \times R)$$

is open (cf. (LD-1) and (LD 0)). Put

$$E = \bigcup_{(\eta, \sigma) \in Y} \{(\eta, \sigma)\} \times (J_\eta - \sigma).$$

The local system $\rho_p : E \to Y$ defined by

$$\rho_p((\eta, \sigma), t) = (\eta, \sigma + t)$$

is called a *local parallel flow on* Y.

If $\forall \eta \in H$, $J_\eta = R$ and so $Y = H \times R$, then ρ_p is global and is called a *global parallel flow*, in accordance with the classical definition of it.

Definition. [10] Let π be a given local system. If there exists a local (or global) parallel flow ρ_p which is type n (or NS) isomorphic to π, then π is said to be *locally* (or *globally*) *type* n (or *NS*) *parallelizable* (abbreviated as $l.$ (or $g.$) n (or *NS*) $p.$).

The notion of global type 0 parallelizability of a global system is not other than that of parallelizability in the classical sense.

One may notice that a local parallel flow is type 0 embeddable in a global parallel flow, with the inclusion map as the phase map.

Fortunately or unfortunately, local parallelizability does not depend on the type of GH-isomorphism considered and there is no need of distinction between NS and GH [24]: schematically we have
$l.$ 0 $p.$ <=> $l.$ 3.5 $p.$ <=> $l.$ 4 $p.$ <=> $l.$ 4.5 $p.$ <=> $l.$ NS $p.$
However, for global parallelizability, there is a distinction.
In general, we have
$g.$ 0 $p.$ <=> $g.$ 3.5 $p.$ => $g.$ NS $p.$ <=> $g.$ 4.5 $p.$ <=> $g.$ 4 $p.$;
and for global systems

$g.$ 3.5 $p.$ <=> $g.$ NS $p.$

Egawa proved in [10] that dispersivity is a necessary and sufficient condition for local parallelizability of a local system, as it is so for global parallelizability of a global system, if the phase space is locally compact and paracompact [13]. However, we know only that the same condition is necessary and sufficient for global parallelizability of a local system, if the phase space is metrizable.

Egawa also constructed a local system on a completely regular, separable and first countable space and another on a locally compact space which are locally parallelizable but not globally. These examples, in turn, tell us limits of extension of Vinograd's theorem, which we shall explain in the next paragraph.

Vinograd's Theorem

In the celebrated book by Nemytskii-Stepanov, Vinograd's theorem is explained. The essential part of the explanation can be stated in our terms as follows [9], [10], [15].

Let X be a domain in R^n and π a local system defined by an autonomous system of differential equations

$$\frac{dx}{dt} = f(x),$$

where f is locally lipschitzian in X. Then there exists an autonomous system of differential equations

$$\frac{dx}{dt} = g(x),$$

where g is locally lipschitzian, and which defines a *global system* ρ *on* X *NS-isomorphic to* π.

However, by careful reading of the proof, we see the global system ρ is type 4 isomorphic to π with the identity as phase map and with increasing reparametrization.

By understanding Vinograd's theorem in this way, G. Ikegami succeeded in extending it to C^r differentiable local systems on C^∞ n-manifolds. Moreover he showed the reparametrization is of C^r-class and increasing [15].

Recently Carlson extended Vinograd's theorem to local systems on metric phase spaces; more precisely

Theorem. [9] *Let* π *be a (strictly) local system on a metric space* X. *Then there exists a global system* ρ *on* X *which is type 4 isomorphic to* π *with the identity as phase map and with increasing reparametrization.*

Universal Systems

Definition. Let C be a category whose Obj C consists of local systems and let $\pi_0 \in$ Obj C. If

$$\forall \pi \in \text{Obj } C, \ \exists f \in \text{Mor } C, \quad \text{such that } f: \pi \to \pi_0,$$

then π_0 is said to be *universal for the category* C.

Our problem is to find a category C which admits a universal system π_0. The larger Obj C, the smaller Mor C and the simpler π_0, the better the result.

Theorem. *Every global parallel flow on a separable metric space is type* 0 *embeddable in a certain global parallel flow* π_0 *on a Hilbert space.*

This is a well known result (cf. [4]). Since every local parallel flow is type 0 embeddable in a global parallel flow, we have the following:

Corollary. *Let* π *be a local parallel flow on a separable metric space. Then there exists a type* 0 *embedding* $\pi \to \pi_0$, *where* π_0 *is the same as in the preceding theorem.*

Another classical result concerning universal systems is due to Beboutov (cf. [5]). His theorem was refined by S. Kakutani in [16] and extended by Hajek as follows in [14].

Theorem. *Let* N *be a positive integer and* π_0 *the shift transformation group on* $C(R, R^N)$.

If π *is a global system on a locally compact metric space such that* S_π *is homeomorphic to a closed subset of* R^N, *then there exists a type* 0 *embedding* $\pi \to \pi_0$.

Since, by Carlson's extension of Vinograd's theorem, every local system on a metric space is type 4 isomorphic to a global system on the same space with increasing reparametrization, in the preceding theorem we can change " global " to " local " in the assumption and " type 0 embedding " to " type 4 embedding with increasing reparametrization " in the conclusion.

The most recent and elegant result on this subject is due to Carlson [8]. Let C_v be the set defined by the following:

$$C_v \ni \phi \overset{d}{\iff}$$

(1) $\phi \in C(R^2, R)$,

(2) $|\phi(x_1, x_2)| \leq e^{-\frac{1}{2}(x_1^2 + x_2^2)}$,

(3) $\phi(x_1, x_2)\, e^{\frac{1}{2}(x_1^2 + x_2^2)} \to 0$ as $(x_1^2 + x_2^2) \to \infty$.

We consider in C_v the metric d defined by

$$d(\phi_1, \phi_2) = \sup |e^{\frac{1}{2}(x_1^2 + x_2^2)}(\phi_1(x_1, x_2) - \phi_2(x_1, x_2))|$$

and the mapping $\pi_0 \colon C_v \times R \to C_v$ defined by

$$(\phi, t) \mapsto \psi$$

where ψ is a mapping $R^2 \to R$ defined by

$$(x_1, x_2) \mapsto e^{(x_1 + x_2)t + t^2}\phi(x_1 + t, x_2 + t).$$

Then π_0 is a global system on C_v and

Theorem. *Let* π *be a global system on a separable metric space. Then there exists a type* 0 *embedding* $\pi \to \pi_0$.

The same consideration as above leads us to:

Corollary. *Let* π *be a local system on a separable metric space. Then there exists a type* 4 *embedding* $\pi \to \pi_0$ *with increasing reparametrization.*

References

Books

[1] Bhatia, N. P. and Szegö, G. P.; Stability Theory of Dynamical Systems, Springer-Verlag, Berlin, 1970

[2] Gottschalk, W. H. and Hedlund, G. A.; Topological Dynamics, Amer. Math. Soc. Coll. Publ., XXXVI, Amer. Math. Soc., Providence, R. I., 1955

[3] Miller, R. K. and Sell, G. R.; Volterra Integral Equations and Topological Dynamics, Memoirs of the Amer. Math. Soc., **102**, 1970

[4] Nemytskii, v. v. and Stepanov, v. v.; Qualitative Theory of Differential Equations, (English tr.), Princeton Univ. Press., Princeton N. J., 1960

[5] Zubov, V. I.; Methods of A. M. Lyapunov and Their Application, P. Noordhoff Ltd., The Netherlands, 1964

[6] Eilenberg, S.; Foundations of Fiber Bundles; Lecture Notes, University of Chicago, 1957

Papers

[7] Ahmad, S.; *On Ura's Axioms and Local Dynamical Systems*, Funkc. Ekvac., **12** (1969), 181 - 191

[8] Carlson, D. H.; *Universal Dynamical Systems*, (to appear)

[9] ________; *A Generalization of Vinograd's Theorem for Dynamical Systems*, (to appear)

[10] Egawa, J.; *Global Parallelizability of Local Dynamical Systems*, Math. Syst. Theory (to appear)

[11] Hajek, O.; *Structure of Dynamical Systems*, Commentat. Math. Univ. Carol., **6**, **1**, (1965), 53 - 72

[12] ________; *Categorial Concepts in Dynamical Systems Theory*, Topological Dynamics, An International Symposium, Benjamin, New York 1968, 243 - 258

[13] ________; *Parallelizability Revisited*, Proc. of Amer. Math. Soc., **27** (1971), 77 - 84

[14] Hajek, O.; *Representation of Dynamical Systems*, Funkc. Ekvac. **14** (1971), 25 - 34

[15] Ikegami, G.; *Vinograd's Theorem on Manifolds*, Proc. of Japan Academy (to appear)

[16] Kakutani, S.; *A Proof of Beboutov's Theorem*, Journ. of Diff. Equat., **4** (1968), 194 - 201

[17] Kimura, I.; *Isomorphisms of Local Dynamical Systems and Separation Axioms for Phase-Spaces*, Funkc. Ekvac., **13** (1970), 23 - 34

[18] McCann, R. C.; *A Classification of Centers*, Pacific Journ. of Math., **30** (1969), 733 - 746

[19] Sell, G. R.; *Nonautonomous Differential Equations and Topological Dynamics*, 1. The Basic Theory, Trans. Amer. Math. Soc., **127** (1967), 241 - 262

[20] Ura, T.; *Sur le courant extérieur à une région invariante, Prolongement d'une caractéristique et l'ordre de stabilité*, Funkc. Ekvac., **2** (1959), 143 - 200; 2e éd., 105 - 143

[21] _______; *Local Isomorphisms and Local Parallelizability of Dynamical Systems*, Dynamical Systems, An International Symposium, Benjamin, New York, 1968

[22] _______; *Local Isomorphism and Local Parallelizability in Dynamical Systems Theory*, Math. Syst. Theory, **3** (1969),1-16

[23] _______; *Isomorphisms and Local Characterization of Local Dynamical Systems*, Funkc. Ekvac., **12** (1969), 99 - 122

[24] _______ and Egawa, J.; *Isomorphism and Parallelizability in Dynamical Systems Theory*, (in preparation)

SOME RESULTS AND APPLICATIONS OF GENERALIZED DYNAMICAL SYSTEMS

E. F. Infante

The object of this paper is to present an overview of some recent results related to the stability theory of general dynamical systems. The results and the applications described briefly below, which are due to a number of different members of the staff of the Center, display a central viewpoint of trying to obtain for general dynamical systems results similar in nature to those known for ordinary differential equations and to develop methods and techniques which are of interest in applications.

In this paper, rather than to attempt to present the general theoretical results in detail, three applications which motivate the theoretical developments are presented together with appropriate specific theorems. In this manner, perhaps, it is possible to better understand the general trend of the theoretical results and their applicability. The reader interested in theoretical developments is referred to the papers quoted. As mentioned above, not all the work reported here is of the author who is indebted to N. Chafee, J. K. Hale, J. P. LaSalle and M. Slemrod.

1. A Problem of Nonexistence of Oscillations

Consider the network shown in Figure 1. In this circuit the section between 0 and 1 is a lossless transmission line with specific capacitance C_s and specific inductance L_s. The current i and the voltage v of this line are functions of ξ and t and satisfy the equations

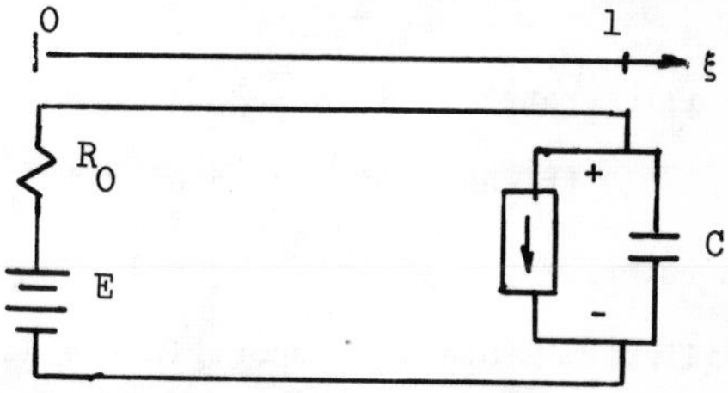

Figure 1

$$L_s \frac{\partial i}{\partial t} = - \frac{\partial v}{\partial \xi}, \qquad 0 < \xi < 1, \quad t > 0. \qquad (1.1)$$
$$-C_s \frac{\partial v}{\partial t} = \frac{\partial i}{\partial \xi},$$

The circuits at the ends of the line give rise to the boundary conditions

$$E = v_0 + R_0 i_0, \qquad (1.2)$$
$$C \frac{dv_1}{dt} + f(v_1) = i_1, \quad t > 0,$$

where $v_0(t) = v(0,t)$, $v_1(t) = v(1,t)$, $i_0(t) = i(0,t)$ and $i_1(t) = i(1,t)$, The function f which renders the problem nonlinear is pictured in Figure 2 and represents the general characteristic on an Esaki diode.

There has been considerable recent interest in circuits of this type, generally called flip-flops, particularly regarding the existence and nonexistence of oscillations. Moser [12], Brayton [2] and Brayton and Miranker [3] have considered increasingly sophisticated mathematical models for the study of such circuits, from lumped models to the present one. The equilibrium states of (1.1), (1.2) are given by

$$E = v_1 + R_0 i_1, \qquad (1.3)$$
$$i_1 = f(v_1),$$

and, as illustrated in Figure 2, we shall consider only the case of a unique equilibrium point, say (v^*, i^*). Translating the equilibrium state to the origin and de noting the new variables by the same notation yields

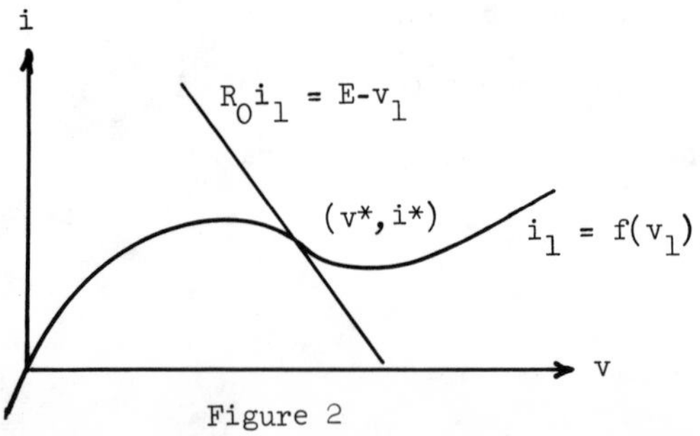

Figure 2

$$L_s \frac{\partial i}{\partial t} = - \frac{\partial v}{\partial \xi}, \quad 0 = v_0 + R_0 i_0,$$
$$-C_s \frac{\partial v}{\partial t} = \frac{\partial i}{\partial \xi}, \quad C\frac{dv_1}{dt} + g(v_1) = i_1, \tag{1.4}$$

with $g(v_1) = f(v_1+v^*) - f(v^*)$, which is assumed continuously differentiable and globally lipschitzian.

The behavior of the solutions of (1.4) is far from obvious. What is desired is to determine conditions on the parameters that guarantee the global asymptotic stability of the solution; because of the nature of the circuit, the lossless transmission line, it is suspected that periodic oscillations are possible.

To study this problem with some mathematical care it is necessary to have an existence theorem which suggests the appropriate space in which the problem should be viewed; for this purpose it is fairly simple to prove [13]:

<u>Theorem 1.1</u>: For the system (1.4), let the initial conditions $i(\xi,0) = \hat{i}(\xi)$ and $v(\xi,0) = \hat{v}(\xi)$ belong to $C^1[0,1]$ and satisfy the consistency conditions

i) $0 = -\hat{v}(0) - R_0\hat{i}(0)$

ii) $0 = L_s \hat{i}'(0) + R_0 C_s \hat{v}'(0)$,

iii) $\frac{C}{C_s} \hat{i}'(1) = -\hat{i}(1) + f(\hat{v}(1))$,

then there exists a unique solution $v(\xi,t)$, $i(\xi,t)$ in $C^1[0,1] \times C^1[0,\infty)$. Furthermore, this solution has the representation

$$v(\xi,t) = \frac{1}{2}[\phi(\xi-\sigma t) + \psi(\xi+\sigma t)],$$
$$i(\xi,t) = \frac{1}{2z}[\phi(\xi-\sigma t) - \psi(\xi+\sigma t)], \tag{1.5}$$

with $\sigma = \frac{1}{(L_s C_s)^{1/2}}$, $z = (\frac{L_s}{C_s})^{1/2}$.

This theorem yields a representation for the solutions which is very suggestive: through the use of this representation it is possible to reduce this problem to a more tractable one. Indeed, introducing (1.5) into (1.4), the wave equation is automatically satisfied and the boundary conditions become

$$v_1(t) + zi_1(t) = -\psi_1\left(t - \frac{2}{\sigma}\right)\left(\frac{z-R_0}{z+R_0}\right),$$

$$v_1(t) - zi_1(t) = \psi_1(t), \tag{1.6}$$

$$c\,\frac{dv_1}{dt} + g(v_1) = i_1.$$

Eliminating i_1 and ψ_1 then yields the neutral functional differential equation

$$C\,\frac{d}{dt}[v_1(t) + kv_1(t-r)] = -\frac{v_1(t)}{z} + \frac{k}{z}\,v_1(t-r) - g(v_1(t)) - kg(v_1(t-r)), \tag{1.7}$$

where $r = \frac{2}{\sigma}$ and $k = \frac{R_0-z}{R_0+z}$. It is also simple to see that the given initial data $\hat{i}(\xi)$, $\hat{v}(\xi)$ in $C^1[0,1]$ completely determines the initial data $v_1 \in C^1[-r,0]$ for (1.7). Furthermore, it is not difficult to see that since $|k| < 1$ if $\lim_{t\to\infty} v_1(t) = 0$, then $\lim_{t\to\infty} i(\xi,t) = 0$ and $\lim_{t\to\infty} v(\xi,t) = 0$ uniformly in ξ and that therefore oscillations will not exist.

The problem has then been reduced to the determination of conditions for the global asymptotic stability of (1.7), which is rewritten for convenience of later computations as

$$\frac{d}{dt}[Dv_{1t}] = -\left[\frac{1}{Cz} + \frac{g(v_1(t))}{Cv_1(t)}\right]v_1(t) + \left[\frac{k}{Cz} - \frac{k}{C}\,\frac{g(v_1(t-r))}{v_1(t-r)}\right]v_1(t-r), \tag{1.8}$$

where $D\varphi = \varphi(0) + k\varphi(-r)$, $x_t(\theta) = x(t+\theta)$ with $-r \le \theta \le 0$. Cruz and Hale [9] have developed existence, uniqueness and continuous dependence results for this type of neutral functional differential equation. In particular, the following results lead to a determination of conditions for global asymptotic stability.

Consider the equation

$$\frac{d}{dt}[Dx_t] = f(t,x_t),\ t \ge \sigma,\ x_\sigma = \varphi \in C[-r,0], \tag{1.9}$$

where D is a stable difference operator and $f: [\sigma,\infty) \times C \to R^n$. $|f(t,\psi)|$ is uniformly bounded in t for ψ in closed bounded sets of C and assume that f is such that existence, uniqueness and continuous dependence of the solutions $x(t,\varphi)$ on the initial data C are guaranteed. Define a Liapunov functional $V: [\sigma,\infty) \times C \to R$ which is continuous along the solutions of (1.9), and let

$$\dot{V}(t_0,\psi) = \lim_{h \to 0^+} \frac{1}{h}[V(t_0+h, x_{t_0+h}(t_0,\psi)) - V(t_0,\psi)].$$

Then [9]:

<u>Theorem 1.2</u>. Suppose $u(s)$, $v(s)$, $w(s)$ are continuous functions for s in $[0,\infty)$; $u(s)$, $v(s)$, positive nondecreasing with $u(0) = v(0) = w(0) = 0$ and $w(s) > 0$ for $s > 0$. Then, if $u(|D\psi|) \le V(t,\psi) \le v(\|\psi\|)$ and $\dot{V}(t,\psi) \le -w(|D\psi|)$. Then the solution $x = 0$ is uniformly asymptotically stable.

The result is precisely the one expected as a generalization of the usual theorems for ordinary differential equations. Now, through the use of this theorem it is not too difficult to obtain some stability results for our problem. Indeed, it is possible to prove [13]:

<u>Theorem 1.3</u>: If g satisfies the sector criterion

$$\sup_\sigma \left(\frac{g(\sigma)}{\sigma}\right) \le \left(\frac{1-|k|}{1+|k|}\right)\frac{1}{z} + \inf_\sigma \left(\frac{g(\sigma)}{\sigma}\right),$$

and

$$\inf_\sigma\left(\frac{g(\sigma)}{\sigma}\right) \ge -\frac{1}{z}\left(\frac{1-|k|}{1+|k|}\right),$$

then the equilibrium solution $v_1 = 0$ of Equation (1.8) is globally asymptotically uniformly stable.

The proof of this theorem is straightforward, although the detailed computations are involved. In essence, the Liapunov functional $V(t,\varphi) =$

$\frac{1}{2}[D\varphi]^2 + \alpha \int_{-r}^{0} \varphi^2(\theta)d\theta$ is used and conditions for the existence of a nonnegative α such that $\dot{V}(t,\varphi) \leq -\beta[D\varphi]^2$, $\beta > 0$, are determined. These conditions yield the sector criterion quoted in the theorem.

From what has been said above, these sector criteria naturally also imply the nonexistence of oscillations in the original problem. It is of interest to note that these criteria are sharp in the following sense. If the problem is linear, that is, $g(\sigma) = -\gamma\sigma$, then it is a simple exercise to determine that the condition $-\gamma \geq -\frac{1}{2}(\frac{1-|k|}{1+|k|})$ is a necessary and sufficient condition for the nonexistence of oscillations. But in the linear case, this is precisely the condition given by Theorem 1.3, which implies that a type of Aizerman conjecture is valid for this problem.

2. A Bifurcation Problem

A number of applications, especially those arising from chemical reactor stability problems [1] give rise to a problem of the following nature. Consider the partial differential equation

$$u_t = u_{xx} + \lambda f(u), \quad \lambda \geq 0, \quad 0 \leq x \leq \pi, \quad t > 0 \tag{2.1}$$

which satisfies the boundary and initial conditions

$$\begin{aligned} u(0,t) &= u(\pi,t) = 0, \quad t \geq 0, \\ u(x,0) &= \phi(x), \qquad 0 \leq x \leq \pi. \end{aligned} \tag{2.2}$$

where f is a given function defined on the real line, $f(0) = 0$, $uf(u) > 0$ for $u \neq 0$ and $f(u)u^{-1} \to 0$ as $|u| \to \infty$. Assume for simplicity that f is C^2 smooth, odd and $\operatorname{sgn} f''(u) = -\operatorname{sgn} u$. With the given hypotheses $u \equiv 0$ is an equilibrium solution of this problem. For $\lambda = 0$ it is well known that this solution of the heat equation is stable in any usual meaning of the word, and the qualitative behavior of the solutions of (2.1), (2.2) is clear. What is of interest here is to determine how this picture changes as λ is allowed to increase from the zero value; if the equilibrium solution $u \equiv 0$ loses its property of stability, do there appear any new equilibrium solutions which inherit this property? This problem has been investigated by Matkowsky [4] using formal asymptotic methods under hypothesis differing somewhat from these given here. The viewpoint here is to interpret (2.1), (2.2) as a dynamical system in an appropriate Banach space and to apply Liapunov methods of the type developed in [8,10, 11]. Again, the details are omitted for the sake of the brevity of exposition. This specific application is more fully described in [4].

Here, the notion of a dynamical system is heavily used. Recall [8 ,10] that given a Banach space β with norm $|\ |$ then by a dynamical system in this space is meant a function $w\colon R^+ \times \beta \to \beta$ such that w is i) continuous, ii) $w(0,\phi) = \phi$ for every $\phi \in \beta$ and iii) $w(t+\tau,\phi) = w(t,w(\tau,\phi))$ for every $t,\tau \geq 0$ and every $\phi \in \beta$. The positive orbit $O^+(\phi)$ through $\phi \in \beta$ is $O^+(\phi) = \bigcup_{t\geq 0} w(t,\phi)$ and a set M in β is positively invariant with respect to w if of every $\phi \in M$ $O^+(\phi) \subset M$ and the set M in β is invariant with respect to w if for each $\phi \in M$ there exists a function U continuously mapping $(-\infty,\infty)$ into M such that $U(0,\phi) = \phi$ and for $s \in (-\infty,\infty)$ $u(t,U(s,\phi)) = U(s+t,\phi)$ for every $\phi \in M$ and $t \in R^+$.

The first task here is to show that (2.1)-(2.2) defines a dynamical system. As a first step in this direction, consider the Banach space X of functions $\phi\colon [0,\pi] \to R$ continuously differentiable on $[0,\pi]$ with $\phi(0) = \phi(\pi) = 0$ and with norm $\|\phi\|_1 = \sup\{|\phi'(x)| : 0 \leq x \leq \pi\}$. Define also the norms $\|\phi\|_0 = \sup\{|\phi(x)| : 0 \leq x \leq \pi\}$ and $\|\phi\|_{W_2^1} = (\int_0^\pi \phi'(x)^2 dx)^{1/2}$, and note that $\|\phi\|_0 \leq \sqrt{\pi}\,\|\phi\|_{W_2^1} \leq \pi\|\phi\|_1$. Let $B_0(r)$ be open balls centered at zero with radius r in the $\|\ \|_0$ norm. Then it is possible to prove [4]:

Theorem 2.1: For any $\phi \in X$ and $\lambda \in [0,\infty)$ Equations (2.1), (2.2) have unique solutions $u(x,t;\phi,\lambda)$ denoted by $u(\phi,\lambda)(t) \in X$ defined for $0 < t < s(\phi,\lambda) \leq \infty$. Furthermore, if $u(\phi,\lambda)(t) \in B_0(r)$ for some r then $s(\phi,\lambda) = \infty$, the map

$(t,\phi) \to u(\phi,\lambda)(t)$ defined for all $\phi \in X$ is a dynamical system in X with $\| \|_1$ and furthermore the positive orbit $0^+(\phi,\lambda)$ of $u(\phi,\lambda)(t)$ is relatively compact in this space.

Note that, except for the hypothesis that the orbits are bounded in the $\| \|_0$ norm, the theorem states that we are dealing with a dynamical system; furthermore, that the dynamical system is self compactifying. This last property is precisely the expected result, given the smoothing properties of the heat equation which, this theorem states, are not affected by the nonlinearity.

Let us now define for every $\lambda \in [0,\infty)$ the Liapunov functional $V_\lambda(\phi) = \int_0^\pi \{\frac{1}{2} \phi^1(x)^2 - \lambda \int_0^{\phi(x)} f(\xi)d\xi\}dx$ for $\phi \in X$. Note that V_λ is continuous on X relative to $\| \|_1$ and $\| \|_{W_2^1}$ and that, given the assumptions it is not too difficult to see that $V_\lambda(\phi) \to \infty$ as $\|\phi\|_0 \to \infty$. Furthermore, it is of interest to see that $\frac{d}{dt} V_\lambda(u(\phi,\lambda)(t)) = -\int_0^\pi u_t^2(x,t;\phi,\lambda)^2 dx \leq 0$, for $0 < t < s(\phi,\lambda)$. These observations lead to

Theorem 2.2: For any $\phi \in X$ and $\lambda \in [0,\infty)$ the map $t,\phi \to u(\phi,\lambda)(t)$ is a dynamical system in X normed by $\| \|_1$. Furthermore the positive orbit $0^+(\phi,\lambda)$ is relatively compact in this space.

Note that the use of the Liapunov functional was essential in proving global existence. But now, since the Liapunov function has already been constructed it is possible to conclude much more. Let, with the notation of dynamical systems

Definition 2.1: Consider the Liapunov functional $V: \beta \to R$ be defined on a set $G \subset \beta$, the Banach space, be continuous on $\overline{G}$ and let $\dot{V}(\phi) \leq 0$ for every $\phi \in G$. Denote by M the largest invariant set in the set $S = \{\phi \in G \mid \dot{V}(\phi) = 0\}$.

On the basis of this devinition it is not difficult to prove [8,10,11] the invariance result:

Theorem 2.3: Let w be a dynamical system on β. If V has the properties described in Def. 2.1 for a subset $G \subset \beta$ and the positive orbit $0^+(\phi)$ belongs G and is a compact set of β, then $w(t,\phi) \to M$ as $t \to \infty$. Furthermore M is compact and connected.

This powerful result is tailor made for our purposes. For given what is known about our dynamical system and noting that the largest invariant set M within our context consists precisely the equilibrium solutions of our problem. Hence, the dynamical system represented by (2.1)-(2.2) is totally stable; i.e., every solu- of the problem approaches an equilibrium solution as time increases.

Actually, much more can be said about the qualitative picture by analyzing the equilibrium solutions, which are the solutions of the two point boundary value problem

$$u''(x) + \lambda f(u(x)) = 0, \quad u(0) = u(\pi) = 0, \quad 0 \leq \lambda < \infty. \tag{2.3}$$

Using methods inspired by the work of Urabe [15] it is possible to prove

Theorem 2.4: Let $\lambda_n = \frac{n^2}{f'(0)}$, $n = 1,2,\ldots$. Then, for any $\lambda \in [\lambda_n,\infty)$ Equation (2.3) has two solutions $u_n^{\pm}(\lambda) \in B_0(r_0)$ with the properties that i) $u_n^{\pm}(\lambda_n) = 0$, ii) $u_n^{\pm}(\lambda)$ have exactly $n+1$ zeros in $[0,\pi]$, iii) $u_n^{\pm}(\lambda)$ varies continuously in λ relative to the norm $\| \ \|_1$ with $\|u_n^{\pm}(\lambda)\|_1 \to \infty$ as $\lambda \to \infty$. Furthermore, for any $\lambda \in [0,\infty)$, (2.1) have no equilibrium points in X other than the origin $u_0 \equiv 0$ and those elements $u_n^{\pm}(\lambda)$, $n \geq 1$, for which $\lambda_n < \lambda$.

It is then quite clear that if $\lambda \leq \lambda_1$ then for every $\phi \in X$ the corresponding solution $u(\phi,\lambda)(t) \to 0$ as $t \to \infty$, the convergence naturally being in the norm $\| \ \|_1$. The question arises, given $\phi \in X$ and $\lambda \in (\lambda_1,\infty)$ to which equilibrium point $u(\phi,\lambda)(t)$ will converge. Again, it is possible to answer, at least partially, this query by an appropriate analysis of the Liapunov functional. Indeed, we have

Theorem 2.5: For each integer $n \geq 1$, let $u_n^{\pm}(\lambda)$, $\lambda_n \leq \lambda < \infty$ be as in Theorem 2.4. Then for any $\lambda \in (\lambda_1,\infty)$ the origin $u_0 = 0$ is unstable. For any $\lambda \in [\lambda_1,\infty)$, $u_1^{\pm}(\lambda)$ is asymptotically stable and for $\lambda \in [\lambda_n,\infty), n \geq 2$, $u_n^{\pm}(\lambda)$ is unstable. (All these assertions valid in X normed by $\| \ \|_1$).

These five theorems give a rather clear picture of the qualitative behavior of the solutions. All solutions will in general approach either u_0 or $u_1^{\pm}(\lambda)$.

3. The General Problem of Thermoelasticity

In the previous problem it was possible to find a Banach space in which the dynamical system was self compactifying. It was this property that was heavily exploited and which is essential in the application of invariance principles. It is to be suspected that such self compactifying properties can be expected in dynamical systems which arise from functional differential equations of the retarded type and partial differential equations of parabolic nature. For hyperbolic partial differential equations clearly this property would be very surprising. The example presented now is of hyperbolic nature, yet it is possible, through a little more work to still apply the principle.

Elastic stability is usually discussed from strictly mechanical considerations; here the concern is with thermodynamic properties of elastic materials. More specifically, one may ask what effects the second law of thermodynamics has on the asymptotic stability of equilibrium of otherwise non-dissipative materials [5, 6].

A material point is identified by $x = (x_1, x_2, x_3)$ in its state of equilibrium (no stresses, constant temperature = γ_0). The displacement field at some time t following an initial disturbance at time $t = 0$ is given by $u(x,t)$ and the temperature deviation by $T(x,t)$; $\rho(x)$ denotes the density at x in the equilibrium state. Let Ω be an open, bounded, connected set in E^3 which is properly regular [7]; let $\partial\Omega$ denote the boundary of Ω. The constitutive equations of thermoelasticity are taken then in the form

$$\rho \ddot{u}_i = (C_{ijk\ell} u_{k,\ell})_{,j} - (m_{ij} T)_{,j}, \tag{3.1}$$

$$\rho C_D \dot{T} + m_{ij} \gamma_0 \dot{u}_{i,j} = (K_{ij} T_{,j})_{,i}; \tag{3.2}$$

where body forces and heat sources have been excluded. In these equations $C_{ijk\ell} = C_{jik\ell} = C_{k\ell ij}$, $m_{ij} = m_{ji}$, $K_{ij} = K_{ji}$ and C_D, ρ, $C_{ijk\ell}$, m_{ij} and K_{ij} are assumed to be smooth functions of x.

Let now $t_0 > 0$. By a classical solution of the mixed initial-boundary value problem in $\Omega \times (0,t_0)$ we mean a pair (u,T) satisfying equations (3.1) and (3.2) together with the boundary conditions

$$u = 0 \quad \text{on} \quad \partial\Omega \times (0,t_0) \quad \text{(clamped boundary)}, \tag{3.3}$$

$$T = 0 \quad \text{on} \quad \partial\Omega \times (0,t_0) \quad \text{(constant temperature)}; \tag{3.4}$$

and with initial conditions

$$(u(x,0), \dot{u}(x,0), T(x,0)) = (u_0(x), \dot{u}_0(x), T_0(x)), \tag{3.5}$$

where $u_0(x)$, $\dot{u}_0(x)$ and $T_0(x)$ are given functions on Ω.

The generalized solutions of the mixed initial boundary value problem described above can be viewed on an appropriate Banach space as a dynamical system. Once this is done, the application of Theorem 2.3 permits us to draw immediate conclusions on the asymptotic behavior of the solutions of our problem.

Consider the Sobolev spaces $W_2^{(k)}(\Omega)$ and $W_{20}^{(k)}(\Omega)$, $k = 1,2,\ldots$. Assume that

$$\text{ess inf } \rho(x) > 0, \ \text{ess inf } C_D(x) > 0, \tag{3.6}$$

$$K_{ij}\xi_i\xi_j \geq C_1\xi_i\xi_i, \ C_1 > 0 \ \text{constant}, \tag{3.7}$$

(the second law of thermodynamics requires K_{ij} positive semidefinite at $x \in \Omega$; we make the stronger assumption of positive definiteness). Also for all $v_i \in W_{20}^{(1)}(\Omega)$

$$\int_\Omega C_{ijk\ell} v_{i,j} v_{k,\ell} dx \geq C_2 \int_\Omega v_{i,j} v_{i,j} dx, \ C_2 > 0 \ \text{constant} \tag{3.8}$$

Define now the spaces $H_0(\Omega) \approx W_{20}^{(1)}(\Omega) \times L_2(\Omega) \times L_2(\Omega)$ with norm $|(v_i, w_i, R)|_0^2 = \int_\Omega [\rho w_i w_i + C_{ijk\ell} v_{i,j} v_{k,\ell} + \frac{\rho C_D}{\gamma_0} R^2] dx$ and $H(\Omega) = W_{20}^{(1)}(\Omega) \times W_{20}^{(1)}(\Omega) \times W_{20}^{(1)}(\Omega)$. Define the map $P: H_0(\Omega) \overset{\text{onto}}{\to} H_1(\Omega)$ sending $(v_{i,} w_{i,} R) \in H_0(\Omega)$ onto $(u_{i,} v_{i,} T) \in H(\Omega) \subset H(\Omega)$ where $(u_i, T) \in W_{20}^{(1)}(\Omega) \times W_{20}^{(1)}(\Omega)$ is defined by the solution of the system

$$\int_\Omega C_{ijk\ell} u_{k,\ell} \theta_{i,j} dx = -\int_\Omega [\rho w_i \theta_i - m_{ij} T \theta_{i,j}] dx$$

$$\int_\Omega K_{ij} T_{,j} D_{,i} dx = -\int_\Omega [\rho C_D R + m_{ij} \gamma_0 v_{i,j}] D\, dx.$$

for every $D_, \theta_i \in W_{20}^{(1)}(\Omega)$. The mapping P is linear, well defined on $H_0(\Omega)$ and one to one. Hence, defining $P_m = \overbrace{P^\circ P^\circ \ldots {}^\circ P}^{m}$ let $H_m(\Omega)$ denote the range of the map P_m. It is clear that P_m^{-1} exists and maps $H_m(\Omega)$ onto $H_0(\Omega)$. Let $\psi \in H_m(\Omega)$ and define $|\psi|_m = |P_m^{-1}\psi|_0$. Then [5],

<u>Lemma 3.1.</u> H_m is a Banach space with norm $|\cdot|_m$. $H_0(\Omega) \supset H(\Omega) \supset \ldots \supset H_m(\Omega)$ algebraically and topologically. Furthermore, $H_m(\Omega)$ is dense in $H_\ell(\Omega)$ for $m > \ell$ and the imbedding $I: H_m(\Omega) \to H_\ell(\Omega)$ is compact.

Let us now define appropriately a <u>generalized solution</u> of our problem:

<u>Definition 3.1.</u> $(u_{i;} \dot{u}_{i,} T)$ will be called a <u>generalized solution</u> of (3.1) - (3.5) on $\Omega \times (0, t_0)$ if for all smooth test functions $(v_{i,} R)$ with compact support on Ω and v_i vanishing on $\Omega \times 0$,

$$\begin{aligned} &\int_0^{t_0} \int_\Omega \{(t-t_0)[\rho \dot{u}_i \dot{v}_i - C_{ijk\ell} u_{k,\ell} \dot{v}_{i,j} + m_{ij} T \dot{v}_{i,j} + \\ &+ \frac{\rho C_D}{\gamma_0} T\dot{R} + m_{ij} u_{i,j} \dot{R}] + \rho \dot{u}_i \dot{v}_i + \rho \frac{C_D}{\gamma_0} TR + \\ &+ m_{ij} u_{i,j} R - \frac{1}{\gamma_0} \int_0^t (K_{ij} R_{,i})_{,j} T\, dt\} dx dt \qquad (3.9) \\ &= -t_0 \int_\Omega [\rho \dot{u}_{0_i} \dot{v}_i|_{t=0} + \frac{\rho C_D}{\gamma_0} T_0 R|_{t=0} + m_{ij} u_{0_{i,j}} R|_{t=0}] dx \end{aligned}$$

With this definition it follows that [5]:

Theorem 3.1. Under assumptions (3.6) - (3.8) the triple $(u_i, \dot{u}_i, T)$ describes a dynamical system on $H_m(\Omega)$, $m = 0,1,2,\ldots$, where $(u_i, \dot{u}_i, T)$ is the generalized solution to the equations of linear thermoelasticity satisfying equation (3.9). Furthermore, for t in $(0,t_0)$

$$
\begin{aligned}
&|(u_i, \dot{u}_i, T)(t)|_m^2 + \frac{1}{\gamma_0} \int_0^t \int_\Omega K_{ij} T^{(m)}_{,i} T^{(m)}_{,j} \, dx\, d\tau \\
&\quad = |(u_{i_0}, \dot{u}_{i_0}, T_0)|_m^2 ,
\end{aligned}
\tag{3.10}
$$

where $T^{(m)}(x,t)$ denotes the generalized m^{th} derivative in time of $T(x,t)$.

The problem of thermoelastic stability has now been put in a setting appropriate for the application of Theorem 2.3, which allows us to obtain stability results in a simple and direct manner.

For the trajectory $(u_i, \dot{u}_i, T)$ in $H_m(\Omega)$ define $P \circ (u_i, \dot{u}_i, T) \equiv (\bar{u}_i, \dot{\bar{u}}_i, \bar{T})$. It follows from the definition of the map P that $(\bar{u}_i, \dot{\bar{u}}_i, \bar{T})$ is a dynamical system on $H_{m+1}(\Omega)$ with initial data $P \circ (u_{0_i}, \dot{u}_{0_i}, T_0)$ in $H_{m+1}(\Omega)$ satisfying (4.4) and Theorem 4.1. Therefore, Theorem 3.1 and (3.10) imply that for any initial data $(u_{0_i}, \dot{u}_{0_i}, T_0)$ in $H_m(\Omega)$ the trajectory $(\bar{u}_i, \dot{\bar{u}}_i, \bar{T})(t)$ will lie in a bounded set of $H_m(\Omega)$ for all $t \geq 0$. Hence by Lemma 3.1 the trajectory $(\bar{u}_i, \dot{\bar{u}}_i, \bar{T})$ will lie in a compact set G of $H_\ell(\Omega)$, $\ell \leq m$. But then all the hypotheses of Theorem 2.3 are met with $\beta = H_\ell(\Omega)$. For simplicity let $\ell = 1$ and $V = |(\bar{u}_i, \dot{\bar{u}}_i, \bar{T})|_1^2$. From (3.7) and (3.10) it immediately follows that $\dot{V} = \frac{-1}{\gamma_0} \int_\Omega K_{ij} \overset{(1)}{T}_{,i} \overset{(1)}{T}_{,j} \, dx \leq -c_3 |(0,0,\bar{T})|_1^2$. The set S is then $S = \{(\bar{u}_i, \dot{\bar{u}}_i, \bar{T}) \in H_1(\Omega) \mid \bar{T} = 0\}$. The determination of M, the largest invariant set in S, which is not trivial, then leads to [14]:

Theorem 3.2. For any initial data $(u_{0_i}, \dot{u}_{0_i}, T_0)$ in $H_m(\Omega)$, $m \geq 1$, and under assumptions (3.6) - (3.7), $(u_i, \dot{u}_i, T)(t)$ approaches the set $M = \{(w_i, \dot{w}_i, Y)$ in

$H_0(\Omega) | m_{ij} w_{i,j} = 0,\ Y = 0,\ \int_0^{t_0} \int_\Omega \{(t-t_0)[\rho \dot{w}_i \ddot{v}_i - C_{ijk\ell} w_{k,\ell} \dot{v}_{i,j}] + \rho \dot{w}_i \dot{v}_i\} dxdt =$

$-t_0 \int_\Omega \rho \dot{w}_{0_i} \dot{v}_i |_{t=0} dx$ for all v_i test functions with compact support on Ω and vanishing on $\Omega \times 0\}$ in the norm of the space $H_0(\Omega)$ as $t \to \infty$.

It is of interest to note that in this case there is an infinity of solutions in the set M and that the use of the Liapunov functional allows a very nice characterization of them; they are the isothermal oscillations of the body, representing pure shear stresses. It should be noted that to obtain the needed compactification it is necessary for the problem to represent a dynamical system in two Banach spaces, here, for example, H_1 and H_0 with the imbedding of H_1 into H_0 completely continuous. The boundedness of the trajectories in H_1 then imply that the trajectory is in a compact set in H_0 and allows the application of the theorem. In this problem, which is linear, the generation of the H_n spaces is quite natural, they are velocity spaces. For nonlinear problems, unfortunately, this is far from easy.

Acknowledgement

This research was supported by the Office of Naval Research under Grant No. NONR N0014-67-A-0191-0009.

REFERENCES

[1] Admvuson, N. R. and L. R. Raymond; AICHE J., 11, 339-362, (1965).

[2] Brayton, R. K.; Quarterly Appl. Math., 24, (1966).

[3] Brayton, R. K. and W. L. Miranker; Arch. Rat. Mech. and Anal, 17, 61-73, (1964).

[4] Chafee, N. and E. F. Infante; Applicable Math., to appear.

[5] Dafermos, C. M.; Arch. Rat. Mech. and Anal., 29, 241-271, (1968).

[6] Eriksen, J. L.; Int. J. Solids and Structures, 2, 573-580, (1966).

[7] Fichera, G.; Lectures on Elliptic Boundary Differential Systems and Eigenvalue Problems, Springer-Verlag, 1965, p. 21.

[8] Hale, J. K.; J. Math. Anal. and Appl., 26, 39-59, (1969).

[9] Hale, J. K. and M. Cruz; J. Diff. Eqns., 7, 334-355, (1970).

[10] Hale, J. K. and E. F. Infante; Proc. Nat. Acad. Sci., 58, 405-409, (1967).

[11] LaSalle, J. P.; Int. Symp. Diff. Eqns. and Dym. Systems, Academic Press, 1967, p. 277.

[12] Moser, J.; Quarterly Appl. Math., 25, 1-9, (1967).

[13] Slemrod, M., J. Math. Anal. and Appl., to appear.

[14] Slemrod, M. and E. F. Infante; Proc. IUTAM Symp. on Inst. Cont. Systems, Springer-Verlag, to appear.

[15] Urabe, M.; Army Math. Res. Center T.S.R. #437, (1963).

GREEN FUNCTIONS OF PSEUDOPERIODIC DIFFERENTIAL OPERATORS

Minoru URABE

Abstract A matrix-valued function $f(t,u)=f(t,u_1,u_2,\dots,u_m)$ is called to be pseudoperiodic in t and $u=(u_1,u_2,\dots,u_m)$ with periods ω_0 and $\omega=(\omega_1,\omega_2,\dots,\omega_m)$ if it is periodic in $u_1, u_2,\dots, u_m$ with periods $\omega_1, \omega_2,\dots, \omega_m$ and in addition it satisfies

$$f(t+\omega_0,u)=f(t,u+\omega_0)=f(t,u_1+\omega_0, u_2+\omega_0, \dots, u_m+\omega_0).$$

Let $A(t,u)$ be a continuous square matrix pseudoperiodic in t and u with periods ω_0 and ω. In the present paper it is shown that the pseudo-periodic differential operator L defined by

$$Ly = \frac{dy}{dt} - A(t,u)y$$

has a Green function $G(t,s,u)$ under the condition related with the exponential dichotomy of solutions to the differential system $Ly=0$ and the additional condition related with period ω_0. Some remarks are made to the above conditions. An application of the above result to quasiperiodic differential operators is also shown.

§1. Introduction

In the present paper a matrix-valued function $f(t,u)=f(t,u_1,u_2,\dots, u_m)$ is called to be pseudoperiodic in t and $u=(u_1,u_2,\dots, u_m)$ with periods ω_0 and $\omega=(\omega_1,\omega_2,\dots,\omega_m)$ if it is periodic in $u_1, u_2,\dots, u_m$ with periods $\omega_1, \omega_2,\dots, \omega_m$ and in addition it satisfies

$$(1.1)\quad f(t+\omega_0,u) = f(t,u+\omega_0) = f(t,u_1+\omega_0, u_2+\omega_0,\dots, u_m+\omega_0).$$

It is easily seen that $f(t,u)$ is pseudoperiodic if and only if $f_0(t,u)$ defined by

$$(1.2)\quad f_0(t,u) = f(t,u-t) = f(t,u_1-t,u_2-t,\dots,u_m-t)$$

is periodic in t, u_1, u_2, ..., u_m with periods $\omega_0, \omega_1, \omega_2,\dots, \omega_m$. From (1.2) it is clear that $f(t,0)$ is quasiperiodic if $f(t,u)$ is pseudoperiodic in t and u.

Throughout the paper it is assumed that

(1.3) $\omega_i > 0$ (i=0, 1, 2, ..., m).

Let A(t,u) be a continuous square matrix pseudoperiodic in t and u with periods ω_0 and ω. In the present paper the pseudoperiodic differential operator L defined by

(1.4) $$Ly = \frac{dy}{dt} - A(t,u)y$$

is called to be <u>regular</u> if there is a continuous square matrix P(u)= $P(u_1, u_2,\ldots, u_m)$ periodic in $u_1, u_2,\ldots, u_m$ with periods $\omega_1, \omega_2, \ldots, \omega_m$ satisfying the conditions as follows:

(1.5) $$\begin{cases} \text{(i)} & P^2(u) = P(u), \\ \text{(ii)} & \|\Phi(t,u)P(u)\| \leq K_0 e^{-\gamma t} \quad \text{for } t \geq 0, \\ & \|\Phi(t,u)[E-P(u)]\| \leq K_0 e^{\gamma t} \quad \text{for } t \leq 0, \\ \text{(iii)} & P(u+\omega_0)\Phi(\omega_0,u) = \Phi(\omega_0,u)P(u), \end{cases}$$

where E is the unit matrix, $\Phi(t,u)$ is the fundamental matrix of the differential system

(1.6) $$Ly = 0$$

satisfying the initial condition

(1.7) $$\Phi(0, u) = E,$$

K_0 and γ are positive numbers, and $\|\cdot\|$ denotes any norm. The conditions (i) and (ii) of (1.5) mean that the exponential dichotomy (see [2]) takes place for solutions to (1.6) for all u. The condition (iii) of (1.5) is concerned with period ω_0 of the pseudoperiodic matrix A(t,u). In what follows, a continuous periodic matrix P(u) associated with the regular pseudoperiodic differential operator L so that (1.5) may be fulfilled will be called briefly <u>continuous periodic matrix associated with the regular pseudoperiodic differential operator</u> L.

In the present paper it will be shown that <u>if the pseudoperiodic differential operator</u> L <u>defined by</u> (1.4) <u>is regular, then</u>

1° L <u>has a Green function</u> G(t,s,u) <u>with the property</u>

(1.8) $\|G(t,s,u)\| \leq Ke^{-\gamma|t-s|}$ <u>for all</u> t,s <u>and</u> u,

2° *for any continuous pseudoperiodic vector-valued function* $f(t,u)$, *the differential system*

$$Lx = f(t,u) \tag{1.9}$$

has a unique pseudoperiodic solution $x=x(t,u)$, *which is given by*

$$x(t,u) = \int_{-\infty}^{\infty} G(t,s,u)f(s,u)ds. \tag{1.10}$$

Some remarks will be made to the regularity of pseudoperiodic differential operators in §5. In §6 the above results concerning regular pseudoperiodic differential operators will be applied to quasiperiodic differential operators and thereby Green functions will be shown to exist also for regular quasiperiodic differential operators (for the definition of regular quasiperiodic differential operators, see §6).

On the basis of the above results one can obtain a series of existence theorems of pseudoperiodic solutions to nonlinear pseudoperiodic differential systems and those of quasiperiodic solutions to nonlinear quasiperiodic differential systems. These results will be published elsewhere in near future.

§2. Lemmas on fundamental matrices

For the fundamental matrix $\Phi(t,u)$ of the pseudoperiodic differential system (1.6) satisfying the initial condition (1.7), one can prove easily the following lemma in a similar way as for fundamental matrices of periodic differential systems.

LEMMA 1. $\Phi(t,u)=\Phi(t,u_1,u_2,\ldots,u_m)$ *is periodic in* $u_1,u_2,\ldots,u_m$ *with periods* $\omega_1,\omega_2,\ldots,\omega_m$ *and in addition it satisfies*

$$\Phi(t+\omega_0,u) = \Phi(t,u+\omega_0)\Phi(\omega_0,u). \tag{2.1}$$

From this lemma follow successively the following lemmas.

LEMMA 2.

$$\Phi(t+p\omega_0,u)=\Phi(t,u+p\omega_0)\Phi(\omega_0,u+(p-1)\omega_0)\times \times\Phi(\omega_0,u+(p-2)\omega_0) \ldots \Phi(\omega_0,u+\omega_0)\Phi(\omega_0,u), \tag{2.2}$$

$$\Phi(p\omega_0,u)=\Phi(\omega_0,u+(p-1)\omega_0) \ldots \Phi(\omega_0,u+\omega_0)\Phi(\omega_0,u), \tag{2.3}$$

$$\Phi(t+p\omega_0,u)=\Phi(t,u+p\omega_0)\Phi(p\omega_0,u) \tag{2.4}$$

(p=0, 1, 2, ...).

LEMMA 3.

$$\Phi^{-1}(\omega_0, u) = \Phi(-\omega_0, u+\omega_0), \tag{2.5}$$

$$\Phi(t-\omega_0, u) = \Phi(t,u-\omega_0)\Phi(-\omega_0, u). \tag{2.6}$$

LEMMA 4.

$$\Phi(t-p\omega_0, u) = \Phi(t,u-p\omega_0)\Phi^{-1}(p\omega_0, u-p\omega_0), \tag{2.7}$$

$$\Phi^{-1}(p\omega_0, u) = \Phi(-p\omega_0, u+p\omega_0), \tag{2.8}$$

$$\Phi(t-p\omega_0, u) = \Phi(t, u-p\omega_0)\Phi(-p\omega_0, u) \tag{2.9}$$

$$(p=0,1,2,\ldots).$$

§3. Lemmas on regular pseudoperiodic operators

Suppose that the pseudoperiodic differential operator L defined by (1.4) is regular, that is, for L there is a continuous square matrix P(u) periodic in $u_1, u_2, \ldots, u_m$ with periods $\omega_1, \omega_2, \ldots, \omega_m$ satisfying (i)-(iii) of (1.5). For convenience' sake, put

$$P(u) = P_+(u), \quad E-P(u) = P_-(u). \tag{3.1}$$

Then from (i)-(iii) of (1.5) readily follow

$$(3.2)\quad \begin{cases} \text{(i)} \quad P_+(u)+P_-(u)=E, \ P_+^2(u)=P_+(u), \ P_-^2(u)=P_-(u), \\ \text{(ii)} \quad \|\Phi(t,u)P_+(u)\| \le K_0 e^{-\gamma t} \qquad \text{for } t \ge 0, \\ \qquad \|\Phi(t,u)P_-(u)\| \le K_0 e^{\gamma t} \qquad \text{for } t \le 0, \\ \text{(iii)} \ P_+(u+\omega_0)\Phi(\omega_0,u)=\Phi(\omega_0,u)P_+(u), \\ \qquad P_-(u+\omega_0)\Phi(\omega_0,u)=\Phi(\omega_0,u)P_-(u). \end{cases}$$

Now since A(t,u) is bounded for all t and u, there is a positive number α such that

$$\|A(t,u)\| \le \alpha \quad \text{for all } t \text{ and } u. \tag{3.3}$$

Hence we have

LEMMA 4.

$$(3.4)\quad \begin{cases} \|P_\pm(u)\Phi^{-1}(t,u)\| \le \|P_\pm(u)\Phi^{-1}(t_0,u)\| \ e^{\alpha\omega_0}, \\ \|\Phi(t,u)P_\pm(u)\| \le \|\Phi(t_0,u)P_\pm(u)\| \ e^{\alpha\omega_0} \end{cases}$$

if

(3.5) $$|t - t_0| \leq \omega_0.$$

Proof. $X(t,u)=P_{\pm}(u)\Phi^{-1}(t,u)$ and $Y(t,u)=\Phi(t,u)P_{\pm}(u)$ satisfy respectively equations

$$\frac{dX}{dt} = -X\cdot A(t,u) \text{ and } \frac{dY}{dt} = A(t,u)Y.$$

Hence by the well-known property of solutions to linear defferential systems we have the conclusions of the lemma. Q.E. D.

As is seen from the proof, the character of $P_{\pm}(u)$ specified in (3.2) is not necessary for the validity of Lemma 4, in other words, Lemma 4 is valid for arbitrary matrices $P_{\pm}(u)$ independent of t.

Making use of Lemma 4, we obtain

LEMMA 5.

(3.6) $$\begin{cases} \text{(i)} \quad \|P_+(u)\Phi^{-1}(t,u)\| \leq K_1 e^{\gamma t} & \text{for } t\leq 0, \\ \text{(ii)} \quad \|P_-(u)\Phi^{-1}(t,u)\| \leq K_1 e^{-\gamma t} & \text{for } t\geq 0, \end{cases}$$

where

(3.7) $$K_1 = K_0 e^{(\alpha+\gamma)\omega_0}.$$

Proof. By (2.8), for any nonnegative integer p,

$$\Phi(-p\omega_0, u) = \Phi^{-1}(p\omega_0, u-p\omega_0).$$

Hence by means of (2.3) and (iii) of (3.2), we successively have

$$\begin{aligned} P_+(u)\Phi^{-1}(-p\omega_0,u) &= P_+(u)\Phi(p\omega_0,u-p\omega_0) \\ &= P_+(u)\Phi(\omega_0,u-\omega_0)\Phi(\omega_0,u-2\omega_0) \ \dots\ \Phi(\omega_0, u-p\omega_0) \\ &= \Phi(\omega_0, u-\omega_0)P_+(u-\omega_0)\Phi(\omega_0, u-2\omega_0) \ \dots\ \Phi(\omega_0, u-p\omega_0) \\ &\dots\dots\dots\dots \\ &= \Phi(\omega_0, u-\omega_0)\Phi(\omega_0, u-2\omega_0) \ \dots\Phi(\omega_0, u-p\omega_0)P_+(u-p\omega_0) \\ &= \Phi(p\omega_0, u-p\omega_0)P_+(u-p\omega_0). \end{aligned}$$

Therefore by (ii) of (3.2) we have

$$(3.8) \qquad \|P_+(u)\Phi^{-1}(-p\omega_0,\ u)\| \le K_0 e^{-\gamma p\omega_0} \qquad (p=0,1,2,\ldots).$$

Now for any $t \le 0$, there is a nonnegative integer p such that

$$-(p+1)\omega_0 < t \le -p\omega_0 \le 0.$$

Then by Lemma 4 we have

$$\|P_+(u)\Phi^{-1}(t,u)\| \le \|P_+(u)\Phi^{-1}(-p\omega_0,\ u)\|\, e^{\alpha\omega_0}.$$

Thus by means of (3.8) we have

$$\begin{aligned}\|P_+(u)\Phi^{-1}(t,u)\| &\le K_0 e^{-\gamma p\omega_0}\cdot e^{\alpha\omega_0}\\ &= K_0 e^{(\alpha+\gamma)\omega_0}\ e^{-\gamma(p+1)\omega_0}\\ &< K_1 e^{\gamma t},\end{aligned}$$

which proves (i) of (3.6). Inequality (ii) of (3.6) can be proved similarly. Q. E. D.

LEMMA 6.

$$(3.9) \qquad \begin{cases} \text{(i)} \quad \|\Phi(t,u)P_+(u)\Phi^{-1}(s,u)\| \le Ke^{-\gamma(t-s)} & \underline{\text{for}}\ t \ge s, \\ \text{(ii)} \quad \|\Phi(t,u)P_-(u)\Phi^{-1}(s,u)\| \le Ke^{-\gamma(s-t)} & \underline{\text{for}}\ t \le s, \end{cases}$$

<u>where</u>

$$(3.10) \qquad K = K_0^2 e^{(2\alpha+\gamma)\omega_0}.$$

Proof. Since (ii) of (3.9) can be proved similarly, only (i) of (3.9) will be proved.

If $t \ge 0 \ge s$, (i) of (3.9) readily follows from (ii) of (3.2) and (i) of (3.6).

If $t \ge s \ge 0$, suppose that

$$0 \le (p-1)\omega_0 \le s < p\omega_0$$

for some positive integer p. Since

$$\begin{aligned}\Phi(s,\ u) &= \Phi(\overline{s-p\omega_0} + p\omega_0,\ u)\\ &= \Phi(s-p\omega_0,\ u+p\omega_0)\Phi(\omega_0,\ u+(p-1)\omega_0)\ldots\Phi(\omega_0,\ u)\end{aligned}$$

by (2.2), by (iii) of (3.2) and (2.3) we successively have

$$P_+(u)\Phi^{-1}(s,u)$$

$$= P_+(u)\Phi^{-1}(\omega_0,u)\ldots\Phi^{-1}(\omega_0,u+(p-1)\omega_0)\Phi^{-1}(s-p\omega_0,u+p\omega_0)$$

$$= \Phi^{-1}(\omega_0,u)P_+(u+\omega_0)\Phi^{-1}(\omega_0,u+\omega_0)\ldots\Phi^{-1}(\omega_0,u+(p-1)\omega_0)\times$$

$$\times\,\Phi^{-1}(s-p\omega_0,u+p\omega_0)$$

....

$$= \Phi^{-1}(\omega_0,u)\ldots\Phi^{-1}(\omega_0,u+(p-1)\omega_0)P_+(u+p\omega_0)\Phi^{-1}(s-p\omega_0,u+p\omega_0)$$

$$= \Phi^{-1}(p\omega_0,u)P_+(u+p\omega_0)\Phi^{-1}(s-p\omega_0,u+p\omega_0).$$

Then by (2.7) we have

$$\Phi(t,u)P_+(u)\Phi^{-1}(s,u) \tag{3.11}$$

$$= \Phi(t-p\omega_0,u+p\omega_0)P_+(u+p\omega_0)\Phi^{-1}(s-p\omega_0,u+p\omega_0).$$

If $t\geq p\omega_0$, then by (ii) of (3.2) and (i) of (3.6) we have

$$\begin{aligned}\|\Phi(t,u)P_+(u)\Phi^{-1}(s,u)\| &\leq K_0e^{-\gamma(t-p\omega_0)}\cdot K_1e^{\gamma(s-p\omega_0)}\\ &= K_0^2e^{(\alpha+\gamma)\omega_0}e^{-\gamma(t-s)}\\ &< Ke^{-\gamma(t-s)}.\end{aligned}$$

If $t<p\omega_0$, then

$$0 \leq (p-1)\omega_0 \leq s \leq t < p\omega_0.$$

Therefore from (3.11), by (ii) of (3.2), (i) of (3.6) and Lemma 4 we have

$$\begin{aligned}\|\Phi(t,u)P_+(u)\Phi^{-1}(s,u)\| &\leq K_0e^{\alpha\omega_0}\cdot K_1e^{\gamma(s-p\omega_0)}\\ &= K_0^2e^{(2\alpha+\gamma)\omega_0}e^{\gamma s}\cdot e^{-\gamma p\omega_0}\\ &< Ke^{-\gamma(t-s)}.\end{aligned}$$

The above results show that (i) of (3.9) is always valid when $t \geq s \geq 0$.

The validity of (i) of (3.9) in the case where $s \leq t \leq 0$ can be proved in a similar way. Hence this completes the proof for (i) of (3.9).

Q. E. D.

§4. Green functions of regular operators

For regular pseudoperiodic differential operators we obtain

THEOREM 1. *If the pseudoperiodic differential operator* L *defined by* (1.4) *is regular, then for any continuous vector-valued function* f(t,u) *bounded for all* t *the differential system*

$$(4.1) \qquad Lx = f(t,u)$$

has a unique solution $x=x(t,u)$ *bounded for all* t, *which is given by*

$$(4.2) \qquad x(t,u) = \int_{-\infty}^{\infty} G(t,s,u)f(s,u)ds.$$

Here

$$(4.3) \qquad G(t,s,u) = \begin{cases} \Phi(t,u)P_+(u)\Phi^{-1}(s,u) & \text{for } t \geq s, \\ -\Phi(t,u)P_-(u)\Phi^{-1}(s,u) & \text{for } t < s, \end{cases}$$

where $\Phi(t,u)$ *is the fundamental matrix of the differential system* (1.6) *satisfying the initial condition* (1.7), *and* $P_+(u)$ *and* $P_-(u)$ *are continuous periodic matrices associated with the regular pseudo-periodic differential operator* L *so that* (3.2) *may be fulfilled.*

Proof. Any solution $x=x(t,u)$ to (4.1) can be written in the form

$$(4.4) \qquad x(t,u) = \Phi(t,u)c(u) + \Phi(t,u)\int_0^t \Phi^{-1}(s,u)f(s,u)ds.$$

Suppose that f(t,u) is bounded for all t. Then since L is regular, by Lemma 5 the integrals

$$\int_{-\infty} P_+(u)\Phi^{-1}(s,u)f(s,u) \quad \text{and} \quad \int^{\infty} P_-(u)\Phi^{-1}(s,u)f(s,u)ds$$

exist. Hence we can write (4.4) as follows:

$$(4.5) \qquad x(t,u) = \Phi(t,u)c(u) + \Phi(t,u)\int_0^t P_+(u)\Phi^{-1}(s,u)f(s,u)ds + \Phi(t,u)\int_0^t P_-(u)\Phi^{-1}(s,u)f(s,u)ds$$

$$= \Phi(t,u)c(u)$$
$$+\Phi(t,u)[\int_{-\infty}^{t} P_+(u)\Phi^{-1}(s,u)f(s,u)ds-\int_{-\infty}^{0} P_+(u)\Phi^{-1}(s,u)f(s,u)ds]$$
$$+\Phi(t,u)[\int_{0}^{\infty} P_-(u)\Phi^{-1}(s,u)f(s,u)ds-\int_{t}^{\infty} P_-(u)\Phi^{-1}(s,u)f(s,u)ds]$$
$$= \Phi(t,u)c_0(u)+\int_{-\infty}^{\infty} G(t,s,u)f(s,u)ds,$$

where

$$(4.6)\quad c_0(u) = c(u)-\int_{-\infty}^{0} P_+(u)\Phi^{-1}(s,u)f(s,u)ds+\int_{0}^{\infty} P_-(u)\Phi^{-1}(s,u)f(s,u)ds.$$

In (4.5), by Lemma 6 we have

$$(4.7)\quad \|\int_{-\infty}^{\infty} G(t,s,u)f(s,u)ds\| \leq \int_{-\infty}^{\infty} \|G(t,s,u)\|ds\cdot \sup_t \|f(t,u)\| \leq M\cdot \sup_t \|f(t,u)\|,$$

where

$$(4.8)\quad M = \int_{-\infty}^{t} Ke^{-\gamma(t-s)}ds+\int_{t}^{\infty} Ke^{-\gamma(s-t)}ds = 2K/\gamma.$$

Now from (4.5) follows

$$c_0(u) = \Phi^{-1}(t,u)x(t,u)-\Phi^{-1}(t,u)\int_{-\infty}^{\infty} G(t,s,u)f(s,u)ds,$$

which implies

$$(4.9)\quad P_{\pm}(u)\ c_0(u) = P_{\pm}(u)\Phi^{-1}(t,u)x(t,u) -P_{\pm}(u)\Phi^{-1}(t,u)\int_{-\infty}^{\infty} G(t,s,u)f(s,u)ds.$$

Suppose that $x(t,u)$ is bounded for all t. Then letting $t\to-\infty$ or $+\infty$ in (4.9), by Lemma 5 we have

$$P_{\pm}(u)c_0(u) = 0,$$

which implies

$$c_0(u) = P_+(u)c_0(u)+P_-(u)c_0(u) = 0.$$

Thus from (4.5) we obtain

$$(4.10)\quad x(t,u) = \int_{-\infty}^{\infty} G(t,s,u)f(s,u)ds.$$

On the other hand, it is easily seen that $x=x(t,u)$ given by (4.10) is a solution to (4.1) bounded for all t. Thus we have the conclusion of the theorem. Q. E. D.

The function $G(t,s,u)$ defined by (4.3) is the Green function of the regular pseudoperiodic differential operator L. By Lemma 6 it is clear that

$$\|G(t,s,u)\| \leq Ke^{-\gamma|t-s|} \quad \text{for all } t, s \text{ and } u. \tag{4.11}$$

From Theorem 1 we can easily get

THEOREM 2. If the pseudoperiodic differential operator L defined by (1.4) with periods ω_0 and ω is regular, then for any continuous pseudoperiodic vector-valued function $f(t,u)$ with periods ω_0 and ω the differential system (4.1) has a unique pseudoperiodic solution $x=x(t,u)$ with periods ω_0 and ω, and it is given by (4.2).

Proof. It is sufficient to prove that the solution given by (4.2) is pseudoperiodic with periods ω_0 and ω when $f(t,u)$ is pseudoperiodic with periods ω_0 and ω.

By (4.3) the solution given by (4.2) is written as follows:

$$\begin{aligned} x(t,u) = &\int_{-\infty}^{t} \Phi(t,u)P_+(u)\Phi^{-1}(s,u)f(s,u)ds \\ &-\int_{t}^{\infty} \Phi(t,u)P_-(u)\Phi^{-1}(s,u)f(s,u)ds. \end{aligned} \tag{4.12}$$

Since $\Phi(t,u)$, $P_{\pm}(u)$ and $f(s,u)$ are all periodic in $u_1, u_2, \ldots, u_m$ with periods $\omega_1, \omega_2, \ldots, \omega_m$, it is clear that $x(t,u)$ is periodic in $u_1, u_2, \ldots, u_m$ with periods $\omega_1, \omega_2, \ldots, \omega_m$. Hence it remains to prove that

$$x(t+\omega_0,u) = x(t,u+\omega_0). \tag{4.13}$$

In order to prove (4.13), consider

$$\Phi(t+\omega_0,u)P_{\pm}(u)\Phi^{-1}(s,u),$$

which, by Lemmas 1 and 3 and (iii) of (3.2), can be written as follows:

$$\begin{aligned} &\Phi(t+\omega_0,u)P_{\pm}(u)\Phi^{-1}(s,u) \\ &= \Phi(t,u+\omega_0)\Phi(\omega_0,u)P_{\pm}(u)\Phi^{-1}(s,u) \\ &= \Phi(t,u+\omega_0)P_{\pm}(u+\omega_0)\Phi(\omega_0,u)\Phi^{-1}(s,u) \\ &= \Phi(t,u+\omega_0)P_{\pm}(u+\omega_0)[\Phi(s,u)\Phi(-\omega_0,u+\omega_0)]^{-1} \end{aligned}$$

$$= \Phi(t,u+\omega_0)P_{\pm}(u+\omega_0)\Phi^{-1}(s-\omega_0,u+\omega_0).$$

Hence from (4.12) we have

$$x(t+\omega_0,u) = \int_{-\infty}^{t+\omega_0} \Phi(t,u+\omega_0)P_+(u+\omega_0)\Phi^{-1}(s-\omega_0,u+\omega_0)f(s,u)ds$$
$$- \int_{t+\omega_0}^{\infty} \Phi(t,u+\omega_0)P_-(u+\omega_0)\Phi^{-1}(s-\omega_0,u+\omega_0)f(s,u)ds$$
$$= \int_{-\infty}^{t} \Phi(t,u+\omega_0)P_+(u+\omega_0)\Phi^{-1}(s,u+\omega_0)f(s+\omega_0,u)ds$$
$$- \int_{t}^{\infty} \Phi(t,u+\omega_0)P_-(u+\omega_0)\Phi^{-1}(s,u+\omega_0)f(s+\omega_0,u)ds.$$

Since $f(s+\omega_0,u)=f(s,u+\omega_0)$ by the pseudoperiodicity of $f(t,u)$, we thus have (4.13). Q. E. D.

§5. Remarks to the regularity of pseudoperiodic operators

1° Pseudoperiodic operators as almost periodic operators. As is stated in §1, a continuous pseudoperiodic matrix $A(t,u)$ can be written as

(5.1) $$A(t,u) = A_0(t,u+t) = A_0(t,u_1+t,u_2+t,\ldots,u_m+t)$$

where $A_0(t,u)=A_0(t,u_1,u_2,\ldots u_m)$ is a continuous matrix periodic in t, $u_1,u_2,\ldots,u_m$ with periods $\omega_0,\omega_1,\omega_2,\ldots,\omega_m$. Hence $A(t,u)$ is almost periodic in t and continuous in u uniformly with respect to t. Let $B(t)$ be a continuous square matrix almost periodic in t and $\mathcal{L}$ be the differential operator defined by

(5.2) $$\mathcal{L}y = \frac{dy}{dt} - B(t)y.$$

In [1], $\mathcal{L}$ is called to be regular if the differential system

(5.3) $$\mathcal{L}y = f(t)$$

has at least one solution bounded for all t for any continuous almost periodic vector-valued function $f(t)$. As is stated in [1], $\mathcal{L}$ is regular, if and only if the exponential dichotomy takes place for solutions to the differential system

(5.4) $$\mathcal{L}y = 0,$$

that is, there is a square matrix P such that

$$(5.5)\qquad \begin{cases} \text{(i)} \quad P^2 = P, \\ \text{(ii)} \quad \|U(t)PU^{-1}(s)\| \leq Ce^{-\sigma(t-s)} \quad \text{for all } t \geq s, \\ \qquad \|U(t)(E-P)U^{-1}(s)\| \leq Ce^{-\sigma(s-t)} \quad \text{for all } t \leq s, \end{cases}$$

where $U(t)$ is the fundamental matrix of (5.4) satisfying the initial condition $U(0)=E$, and C and σ are positive numbers. The function $G(t,s)$ defined by

$$G(t,s) = \begin{cases} U(t)PU^{-1}(s) & \text{for } t \geq s, \\ -U(t)(E-P)U^{-1}(s) & \text{for } t < s \end{cases}$$

is the <u>Green function</u> of the almost periodic differential operator $\mathcal{L}$. Inequalities in (ii) of (5.5) clearly imply

$$(5.6)\qquad \begin{cases} \|U(t)P\| \leq Ce^{-\sigma t} & \text{for all } t \geq 0, \\ \|U(t)(E-P)\| \leq Ce^{\sigma t} & \text{for all } t \leq 0. \end{cases}$$

Hence comparing (i) and (ii) of (1.5) with (i) of (5.5) and (5.6), we see that (i) and (ii) of our regularity conditions (1.5) are fulfilled if the pseudoperiodic differential operator L defined by (1.4) is regular for all u as an almost periodic differential operator containing a parameter u.

2° <u>The invariance of the regularity under pseudoperiodic transformations</u>. Let $T(t,u)$ be a square matrix pseudoperiodic in t and u with periods ω_0 and ω, and suppose that

$$(5.7)\qquad \det T(t,u) \neq 0 \quad \text{for all } t \text{ and } u$$

and $T(t,u)$ is continuously differentiable with respect to t. Then by the transformation

$$(5.8)\qquad y = T(t,u)\tilde{y},$$

the differential system (1.6) is transformed to the differential system

$$(5.9)\qquad \frac{d\tilde{y}}{dt} - \tilde{A}(t,u)\tilde{y} = 0,$$

where

$$(5.10)\qquad \tilde{A}(t,u) = T^{-1}(t,u)A(t,u)T(t,u) - T^{-1}(t,u)\frac{\partial T(t,u)}{\partial t}.$$

Let $\tilde{\Phi}(t,u)$ be the fundamental matrix of (5.9) satisfying the initial condition

(5.11) $$\tilde{\Phi}(0,u) = E,$$

then it is easily seen that

(5.12) $$\tilde{\Phi}(t,u) = T^{-1}(t,u)\Phi(t,u)T(0,u),$$

where $\Phi(t,u)$ is the fundamental matrix of (1.6) satisfying the initial condition (1.7). From (5.12), we can easily see that if the pseudoperiodic differential operator L defined by (1.4) is regular, then the transformed pseudoperiodic differential operator $\tilde{L}$ defined by

(5.13) $$\tilde{L}y = \frac{dy}{dt} - \tilde{A}(t,u)y$$

is also regular and the continuous periodic matrix associated with $\tilde{L}$ is given by

(5.14) $$\tilde{P}(u) = T^{-1}(0,u)P(u)T(0,u)$$

where P(u) is the continuous periodic matrix associated with the original operator L. By this result, in order to verify the regularity of a pseudoperiodic differential operator, it is sufficient to verify it for an operator obtained after a certain pseudoperiodic transformation.

3° Special cases.

CASE I: the case where

(5.15) $$\|\Phi(t,u)\| \leq K_0 e^{-\gamma t} \qquad \text{for} \quad t \geq 0$$

or

(5.16) $$\|\Phi(t,u)\| \leq K_0 e^{\gamma t} \qquad \text{for} \quad t \leq 0.$$

Here $\Phi(t,u)$ is the fundamental matrix of (1.6) satisfying the initial condition (1.7) and K_0 and γ are positive numbers.

In the case (5.15), conditions (1.5) are fulfilled by P(u)=E. In the case (5.16), conditions (1.5) are fulfilled by P(u)=0. Hence in either case the pseudoperiodic differential operator L defined by (1.4) is regular.

In the case (5.15), the trivial solution to (1.6) is exponentially stable for all u, and in the case (5.16), the trivial solution to (1.6) is exponentially stable negatively.

CASE II: the case where

(5.17) $$\Phi(\omega_0,u) = U(u) \oplus V(u)$$

and

(5.18) $$\|U(u)\|, \|V^{-1}(u)\| \leq \theta < 1 \qquad \text{for all } u.$$

Here $\Phi(t,u)$ is the fundamental matrix of (1.6) satisfying the initial condition (1.7) and $\|\cdot\|$ denotes L_p-norm $(p\geq 1)$.

Let

$$E = E_1 \oplus E_2$$

be the decomposition of the unit matrix E corresponding to that of $\Phi(\omega_0,u)$ and put

$$(5.19) \qquad P_+ = E_1 \oplus 0 , \quad P_- = 0 \oplus E_2 .$$

Then it is clear that

$$(5.20) \qquad P_+ + P_- = E , \quad P_+^2 = P_+ , \quad P_-^2 = P_- ,$$

$$(5.21) \qquad \begin{cases} P_+\Phi(\omega_0,u) = \Phi(\omega_0,u)P_+ = U(u) \oplus 0 , \\ P_-\Phi(\omega_0,u) = \Phi(\omega_0,u)P_- = 0 \oplus V(u). \end{cases}$$

Now for any $t\geq 0$, suppose that

$$0 \leq p\omega_0 \leq t < (p+1)\omega_0$$

for some nonnegative integer p. Then by (2.2) and (5.18) we have

$$\begin{aligned} &\|\Phi(t,u)P_+\| \\ &= \|\Phi(\overline{t-p\omega_0}+p\omega_0,u)P_+\| \\ &= \|\Phi(t-p\omega_0,u+p\omega_0)\Phi(\omega_0,u+(p-1)\omega_0)\dots\Phi(\omega_0,u)P_+\| \\ &\leq \|\Phi(t-p\omega_0,u+p\omega_0)P_+\|\cdot\theta^p. \end{aligned}$$

Since $|t-p\omega_0| < \omega_0$, by Lemma 4 we thus have

$$\|\Phi(t,u)P_+\| \leq e^{\alpha\omega_0}\cdot e^{p\log\theta} ,$$

where α is the positive number satisfying (3.3). Put

$$(5.22) \qquad \gamma = -\frac{1}{\omega_0}\log\theta ,$$

then $\gamma>0$ and we have

$$(5.23) \qquad \begin{aligned} \|\Phi(t,u)P_+\| &\leq e^{\alpha\omega_0}\cdot e^{-p\gamma\omega_0} \\ &= e^{(\alpha+\gamma)\omega_0}\cdot e^{-\gamma(p+1)\omega_0} \\ &< K_0 e^{-\gamma t} , \end{aligned}$$

where

(5.24) $$K_0 = e^{(\alpha+\gamma)\omega_0} .$$

Likewise for any $t \leq 0$, we have

(5.25) $$\|\Phi(t,u)P_-\| \leq K_0 e^{\gamma t} .$$

Equalities (5.20) and (5.21) together with (5.23) and (5.25) show that conditions (3.2) are all fulfilled by

$$P_+(u) = P_+ \quad \text{and} \quad P_-(u) = P_- .$$

Hence in this case the pseudoperiodic differential operator L defined by (1.4) is regular.

§6. Application to quasiperiodic operators

In the present paper, by a quasiperiodic matrix-valued function f(t), we mean a matrix-valued function such that

(6.1) $$f(t) = f_0(t,t,\ldots,t)$$

for some continuous matrix-valued function $f_0(t,u)=f_0(t,u_1,u_2,\ldots,u_m)$ periodic in t, u_1, u_2,..., u_m with periods ω_0, ω_1, ω_2,..., ω_m. Put

(6.2) $$f_0(t,u+t) = f_0(t,u_1+t,\ldots,u_m+t) = \overline{f}(t,u),$$

then as is stated in §1, $\overline{f}(t,u)$ is pseudoperiodic in t and u with periods ω_0 and $\omega=(\omega_1,\omega_2,\ldots,\omega_m)$. Thus we see that a matrix-valued function f(t) is quasiperiodic with periods $\omega_0,\omega_1,\ldots,\omega_m$ if and only if

(6.3) $$f(t) = \overline{f}(t,0)$$

for some continuous matrix-valued function $\overline{f}(t,u)$ pseudoperiodic in t and u with periods ω_0 and $\omega=(\omega_1,\omega_2,\ldots,\omega_m)$. In what follows, a continuous pseudoperiodic function $\overline{f}(t,u)$ corresponding to a quasiperiodic function f(t) in the above way will be called briefly a continuous pseuoperiodic function corresponding to a quasiperiodic function f(t).

Let A(t) be a square matrix quasiperiodic in t and $\overline{A}(t,u)$ be a continuous pseudoperiodic matrix corresponding to A(t). In what follows, the quasiperiodic differential operator L defined by

(6.4) $$Ly = \frac{dy}{dt} - A(t)y$$

will be called to be regular if and only if the corresponding pseudoperiodic differential operator $\overline{L}$ defined by

(6.5) $$\bar{L}y = \frac{dy}{dt} - \bar{A}(t,u)y$$

is regular, and the latter will be called briefly the pseudoperiodic differential operator corresponding to the quasiperiodic differential operator L.

From Theorems 1 and 2, we then have

THEOREM 3. *If the quasiperiodic differential operator* L *defined by* (6.4) *with periods* $\omega_0, \omega_1, \ldots, \omega_m$ *is regular, then for any quasiperiodic vector-valued function* f(t) *with periods* $\omega_0, \omega_1, \ldots, \omega_m$ *the differential system*

(6.6) $$Lx = f(t)$$

has a unique quasiperiodic solution x=x(t) *with periods* $\omega_0, \omega_1, \ldots, \omega_m$, *and it is given by*

(6.7) $$x(t) = \int_{-\infty}^{\infty} G(t,s)f(s)ds \ .$$

Here

(6.8) $$G(t,s) = \bar{G}(t,s,0) \ ,$$

where $\bar{G}(t,s,u)$ *is the Green function of the pseudoperiodic differential operator* $\bar{L}$ *corresponding to the given quasiperiodic differential operator* L.

Proof. Let $\bar{f}(t,u)$ be a continuous pseudoperiodic vector-valued function corresponding to the given quasiperiodic vector-valued function f(t). Since the pseudoperiodic differential operator $\bar{L}$ corresponding to the given quasiperiodic differential operator L is regular by the definition of the regularity, by Theorem 2 the differential system

(6.9) $$\bar{L}x = \bar{f}(t,u)$$

has a pseudoperiodic solution $x=\bar{x}(t,u)$ with periods ω_0 and ω, and it is given by

(6.10) $$\bar{x}(t,u) = \int_{-\infty}^{\infty} \bar{G}(t,s,u)\bar{f}(s,u)ds \ .$$

Since $\bar{x}(t,u)$ is pseudoperiodic with periods ω_0 and ω, $x(t)=\bar{x}(t,0)$ is quasiperiodic with periods $\omega_0, \omega_1, \ldots, \omega_m$. Since (6.9) reduces to (6.6) for u=0, x=x(t) is a solution to (6.6). Thus we see that (6.6) has a quasiperiodic solution $x=x(t)=\bar{x}(t,0)$ with periods $\omega_0, \omega_1, \ldots, \omega_m$. By (6.10) it is clear that this solution is given by (6.7).

It remains now to prove the uniqueness of quasiperiodic solutions to (6.6). Let $x=\hat{x}(t)$ be an arbitrary quasiperiodic solution with periods $\omega_0,\omega_1,\ldots,\omega_m$. Then from (6.6) we have

$$\frac{d\hat{x}(t)}{dt} - \bar{A}(t,0)\hat{x}(t) = \bar{f}(t,0) .$$

Since $\hat{x}(t)$ and $\bar{f}(t,0)$ are both quasiperiodic, they are both bounded for all t. Then by Theorem 1, $\hat{x}(t)$ is determined uniquely. This proves the uniqueness of quasiperiodic solutions. Q. E. D.

By (4.3), equality (6.8) implies

$$(6.11) \qquad G(t,s) = \begin{cases} \Phi(t)P_+(0)\Phi^{-1}(s) & \text{for } t\geq s, \\ -\Phi(t)P_-(0)\Phi^{-1}(s) & \text{for } t<s, \end{cases}$$

where $\Phi(t)$ is the fundamental matrix of the differential system

$$(6.12) \qquad Ly = 0$$

satisfying the initial condition

$$(6.13) \qquad \Phi(0) = E .$$

The function G(t,s) defined by (6.11) is the <u>Green function</u> of the regualr quasiperiodic differential operator L. By (4.11) it is clear that

$$(6.14) \qquad \|G(t,s)\| \leq Ke^{-\gamma|t-s|} \qquad \text{for all } t \text{ and } s.$$

<u>REFERENCES</u>

[1] Бурд, В. Ш., Ю. С. Колесов и М. А. Красносельский : Исследование функции Грина дифференциальных операторов с почти периодическими коэффициентами, Изв. Акад. Наук СССР, Сер. Мат., 33(1969), 1089-1119.

[2] Massera, J. L. and J. J. Schäffer: Linear differential equations and function spaces, Academic Press, New York, 1966.

ALMOST PERIODIC SOLUTIONS BY THE METHOD OF AVERAGING

George Seifert *

Consider the scalar equation

$$x'' + \omega^2 x = f(t) \tag{1}$$

where ω is a positive constant, f is real-valued, continuous, and periodic with period $T > 0$, and not identically zero. If the least period $T = 2\pi/\omega$, it is well known that there exist no periodic solutions of (1); in fact, all solutions of (1) are unbounded on $-\infty < t < \infty$. An equivalent condition on f is that its Fourier series contain a nonzero $\sin \omega t$ or $\cos \omega t$ term.

On the other hand, if the Fourier series for f does not contain such a $\sin \omega t$ or $\cos \omega t$ term, all solutions of (1) will be bounded on $-\infty < t < \infty$, and if also $f(t + 2\pi/\omega) = f(t)$ for all t, (1) will have solutions of period $2\pi/\omega$; this follows easily from the so-called Fredholm alternative; cf. [2], p.146.

In case f is almost periodic (a.p. for short) the condition that its Fourier series contain no $\cos \omega t$ or $\sin \omega t$ term is insufficient to guarantee that all solutions are bounded on $-\infty < t < \infty$. If, for example, $f(t) = \sum_{k=1}^{\infty} \frac{1}{k^2} \cos \omega(1 + \frac{1}{k^3})t$, this follows by the variation of constants formula and the fact that any nontrivial linear combination of Φ_1 and Φ_2, where

$$\Phi_1(t) = \sum_{k=1}^{\infty} k \sin \omega t/k^3, \qquad \Phi_2(t) = \sum_{k=1}^{\infty} k \sin^2 \omega t/2k^3,$$

* The author was supported in part by the National Science Foundation under Grant G.P. 24418.

is unbounded on $-\infty < t < \infty$; we omit the details.

If, however, the periods of the nonzero terms in the Fourier series for f are different from $2\pi/\omega$, or do not have $2\pi/\omega$ as a point of accumulation, then all solutions of (1) will be bounded on $-\infty < t < \infty$. This is a consequence of the following result for linear systems; cf. [1]; here x and $b(t)$ are complex n-vectors, A is a complex $n \times n$ matrix, and t is a real variable.

Theorem 1. Let $b(t)$ be almost periodic and have Fourier series

$$\sum c_k e^{i\lambda_k t};$$

let $\{\mu_k\}$, $k = 1, \dots, n$, be the set of eigenvalues of A. If there exists a constant $\eta > 0$ such that $|i\lambda_k - \mu_\ell| \geq \eta$ for $k = 1, 2, \dots$, $\ell = 1, \dots, n$, then the system

$$x' = Ax + b(t)$$

has an a.p. solution.

In fact, under the hypotheses of Theorem 1, there exists a unique a.p. solution whose Fourier exponents are contained in the set $\{\lambda_k\}$, $k = 1, 2, \dots$.

Returning to the case where f is periodic with $2\pi/\omega$, it can be shown that if μ and ϵ are nonzero constants but of sufficiently small absolute value, then the scalar equation

$$x'' + \omega^2 x + \mu x^3 = \epsilon f(t) \tag{2}$$

has solutions of period $2\pi/\omega$. A method due primarily to Cesari and discussed in [2], chapt. 9, can be used to show this. However, if f is a.p. and is such that ω is an accumulation point of its set of Fourier exponents, then the question of the existence of an a.p. solution of (2) even for $|\mu|$ and $|\epsilon|$ sufficiently small, but nonzero, seems open.

It is the purpose of this paper to show that by the use of the so-called method of averaging, cf. [2], pp.186-194, one can obtain some results in that direction.

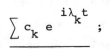

We first recall some definitions and properties of a.p. functions. We denote by R the set of reals and by Z^n the set of complex n-vectors and by $|z|$ some convenient norm for z in Z^n.

Definition 1. A function p on R to Z^n is called a trigonometric polynomial if $p(t) = \sum_{k=1}^{N} c_k e^{i\lambda_k t}$, c_k in Z^n, λ_k in R.

Definition 2. A function f on R to Z^n is called a.p. if it is continuous and if there exists a sequence $\{p_k\}$, $k=1,2,\ldots$, p_k a trigonometric polynomial, such that $|p_k(t) - f(t)| \to 0$ as $k \to \infty$, uniformly for t in R.

With each a.p. function f we can associate a mean value

$$m(f) = \lim_{T\to\infty} \int_0^T f(t)\,dt$$

and a Fourier series

$$f \sim \sum c_k e^{i\lambda_k t}$$

where λ_k is in R, c_k is in Z^n. If $f \not\equiv 0$, then $c_k \neq 0$, and the series may be finite. The set $\{\lambda_k\}$ is called the set of Fourier exponents of f. If $f \equiv 0$, its Fourier series is taken to be the zero vector and its set of Fourier exponents to be empty.

The method of averaging applies to systems of the form

$$x' = \epsilon f(t,x,\epsilon) \; ; \tag{3}$$

here x and $f(t,x,\epsilon)$ are real n-vectors, the set of which we denote by R^n, $\epsilon \geq 0$, and f and its first partial derivatives with respect to the components of x are continuous in (t,x,ϵ). We also assume that f is a.p. in t for fixed (x,ϵ) and for fixed $\epsilon \geq 0$ is continuous in x uniformly for (t,x) in each set $R \times K$, K a compact subset of R^n. Finally, we require that $f(t,y,\epsilon) \to f(t,x,0)$ as $(y,\epsilon) \to (x,0)$ uniformly for (t,x) in $R \times K$, K as above.

Our aim is to find an $\epsilon_o > 0$ and a function $u(t,y,\epsilon)$ on $R \times R^n \times (0,\epsilon_o)$ to R^n such that the equation

$$(4) \qquad x = y + \epsilon\, u(t,y,\epsilon)$$

defines a suitable change of variables taking (3) into a system which is essentially an a.p. perturbation of $y' = \epsilon\, f_o(y)$, where $f_o(y) = m(f(t,y,0))$. To this end, u and its first partial derivatives with respect to t and the components y_i of y should exist and be continuous in (t,y,ϵ), and in fact for fixed $\epsilon > 0$, continuous in y uniformly for (t,y) in sets of the form $R \times K$, K a compact subset of R^n. We also want y and its first partials to be a.p. in t for fixed y and $\epsilon > 0$, and that $\epsilon \frac{\partial u_i}{\partial y_j} \to 0$ as $\epsilon \to 0$ uniformly for (t,y) in any set $R \times K$, K compact.

Formally, from (3) and (4) we then obtain

$$(5) \qquad (I + \epsilon \frac{\partial u}{\partial y})y' = \epsilon[f(t,y + \epsilon\, u(t,y,\epsilon), \epsilon) - \frac{\partial u}{\partial t}] ;$$

here I is the identity $n \times n$ matrix, and $\frac{\partial u}{\partial y}$ is the matrix with elements $\frac{\partial u_i}{\partial y_j}$. Since

$$f(t,y+\epsilon u, \epsilon) = f(t,y+\epsilon u, \epsilon) - f(t,y,0) + f(t,y,0) - f_o(y) + f_o(y)$$

where $f_o(y)$ is as defined above, we see that if we can find a function u with the properties specified above and such that

$$(5.1) \qquad \frac{\partial u}{\partial t} = f(t,y,0) - f_o(y) ,$$

then for ϵ_o sufficiently small, (5) can be put into the form

$$(6) \qquad y' = \epsilon[I + \epsilon \frac{\partial u}{\partial y}]^{-1} [f_o(y) + g(t,y,\epsilon)]$$

$$= \epsilon[f_o(y) + h(t,y,\epsilon)] , \quad 0 \le \epsilon < \epsilon_o .$$

It can be shown that h will be a.p. in t for fixed (y,ϵ) and have the same continuity properties that f has, and that $h(t,y,\epsilon) \to h(t,y,0) \equiv 0$ as $\epsilon \to 0$ uniformly for (t,y) in $R \times K$, K any compact subset of R^n; we omit the details.

We note the right side of eq. (5.1) has mean value zero. However, this does not guarantee the existence of an a.p. function u satisfying (5.1). Therefore the following result is of great importance in the use of this method; cf. [2], Lemma 3.1, p.187.

Lemma. Let the function $F: R \times R^n \to R^n$ be continuous for (t,y) in $R \times R^n$. Let F be a.p. in t for each fixed y in R^n, continuous in y uniformly for (t,y) in $R \times K$, K any compact subset of R^n, and $m(F) = 0$. Finally, let $\frac{\partial F}{\partial y}$ exist and be continuous for (t,y) in $R \times R^n$. Then there exists an $\epsilon_o > 0$ and a function $u(t,y,\epsilon): R \times R^n \times (0,\epsilon_o) \to R^n$, such that

(i) u, $\frac{\partial u}{\partial t}$, and $\frac{\partial u}{\partial y}$ are continuous in (t,y,ϵ), and a.p. in t for fixed (y,ϵ) in $R^n \times (0,\epsilon_o)$;

(ii) for fixed ϵ, $0 < \epsilon < \epsilon_o$, u, $\frac{\partial u}{\partial t}$, and $\frac{\partial u}{\partial y}$ are continuous in y uniformly for (t,y) in $R \times K$, K as above; and

(iii) with r defined by

$$r(t,y,\epsilon) = \frac{\partial u}{\partial t} - F(t,y) , \tag{7}$$

r, $\frac{\partial r}{\partial y}$, ϵu, and $\epsilon \frac{\partial u}{\partial y}$ all approach 0 as $\epsilon \to 0$ uniformly for (t,y) in any set $R \times K$, K as above.

If we take $F(t,y) = f(t,y,0) - f_o(y)$, and use the function u given by this lemma in (4), this equation can again be used to transform (3) into an equation of the form of (6). Clearly, the existence of an a.p. solution of (6) will imply the existence of an a.p. solution of (3).

A well-known condition sufficient for the existence of an a.p. solution of (6) is that there exist a $\bar{y}$ in R^n such that

$f_o(\bar{y}) = 0$, and that all the eigenvalues of $\frac{\partial f_o}{\partial y}(\bar{y})$ have nonzero real parts.

Using the preceding results, the following theorem has been obtained for a system of the form

$$(8) \qquad x' = (A + \epsilon C(t)x + \epsilon g(x,\epsilon) + \epsilon b(t)$$

where x, $g(x,\epsilon)$ and $b(t)$ are in R^n, $\epsilon > 0$, A is a real $n \times n$ matrix similar to a diagonal matrix with pure imaginary entries, the elements of the $n \times n$ matrix $C(t)$ and $b(t)$ are a.p., and g and its first partials with respect to x are continuous on $R^n \times [0,\infty)$. The proof of this result, as well as its application to an undamped Duffing equation with a.p. forcing, will appear in a forthcoming paper [3].

Theorem 2. Let A, $C(t)$, $g(x,\epsilon)$ and $b(t)$ be as in (8), and let

(i) $C_o = m(e^{-tA} C(t) e^{tA})$,

(ii) $g_o(y) = m(e^{-tA} g(e^{tA} y, 0))$, and

(iii) $b_o = m(e^{-tA} b(t))$.

Suppose the equation

$$(9) \qquad C_o y + g_o(y) + b_o = 0$$

has a solution $\bar{y}$ such that the eigenvalues of $C_o + \frac{\partial g_o}{\partial y}(\bar{y})$ all have real parts different from zero. Then there exists a $\epsilon_o > 0$ such that for each ϵ, $0 \leq \epsilon < \epsilon_o$, eq. (8) has an a.p. solution $x(t,\epsilon)$ such that $|x(t,\epsilon) - e^{tA}\bar{y}| \to 0$ as $\epsilon \to 0$ uniformly for t in R.

As an application of Theorem 2, we consider the 2-dimensional system

$$(10)\qquad \begin{cases} x_1' = x_2 \\ x_2' = -x_1 + \epsilon\nu x_1 + \epsilon^m g(x_1) + p(t) \end{cases}$$

where $g(x) = \sum_{j=1}^{m} a_j x^j$, $\nu a_m < 0$, $m = 2k+1$, k a positive integer, and p is a.p. with Fourier series

$$\sum_{j=1}^{\infty} A_j \cos\lambda_j t + B_j \sin\lambda_j t, \qquad \lambda_1 = 1, \quad A_1^2 + B_1^2 > 0.$$

Theorem 3. For (10) as above, there exists a $\nu_o = \nu_o(A_1, B_1, a_m) > 0$ such that for each ν with $|\nu| > \nu_o$, there exists a $\epsilon_o = \epsilon_o(\nu) > 0$ such that if $0 < \epsilon < \epsilon_o$, then (10) has an a.p. solution $x(t,\epsilon) = (x_1(t,\epsilon), x_2(t,\epsilon))$ such that $|\epsilon\, x(t,\epsilon) - \bar{x}(t)| \to 0$ as $\epsilon \to 0$ uniformly for t in R; here $\bar{x}(t) = (\alpha \sin(t+\delta), \alpha\cos(t+\delta)$ for some constants α and δ.

The special case of (10) where $g(x) = -x^3$ is proved in the paper [3] mentioned above. The proof of Theorem 3 is quite similar to the proof of this special case, and thus we give only a sketch.

To put (10) into the form of (8) we fix $\epsilon > 0$, multiply each side of (10) by ϵ, and put $y = \epsilon\, x$. We obtain

$$(11)\qquad \begin{cases} y_1' = y_2 \\ y_2' = -y_1 + \epsilon\nu y_1 + \epsilon h(y_1,\epsilon) + \epsilon\, p(t) \end{cases}$$

where $y = (y_1, y_2)$ and $h(y_1,\epsilon) = \sum_{j=1}^{m} a_j\, \epsilon^{m-j}\, y_1^j$. We note that (11) is well-defined for $\epsilon = 0$.

Clearly (11) is in the form of (8) with

$$A = \begin{pmatrix} 0 & 1 \\ -1 & 0 \end{pmatrix}, \quad C(t) = \begin{pmatrix} 0 & 0 \\ \nu & 0 \end{pmatrix}, \quad b(t) = \begin{pmatrix} 0 \\ p(t) \end{pmatrix},$$

and $g(y,\epsilon) = \begin{pmatrix} 0 \\ h(y_1,\epsilon) \end{pmatrix}$.

We assume without loss of generality that $A_1 = 0$ and $B_1 > 0$; if not, we replace t by $t + \omega$ with a proper choice of ω.

By straightforward calculations, we obtain

$$c_o = \frac{\nu}{2}\begin{pmatrix} 0 & -1 \\ 1 & 0 \end{pmatrix}, \quad b_o = \frac{1}{2}\begin{pmatrix} B_1 \\ 0 \end{pmatrix} \quad \text{and}$$

$$g_o(y) = a_{2k+1}\begin{pmatrix} b_1 y_1^{2k} y_2 + b_3 y_1^{2k-2} y_2^{3} + \ldots + b_{2k+1} y_2^{2k+1} \\ b_o y_1^{2k+1} + b_2 y_1^{2k-1} y_2^{3} + \ldots + b_{2k} y_2^{2k} y_1 \end{pmatrix},$$

and (9) becomes in this case

$$(12) \quad \begin{cases} -\nu y_2/2 + a(b_1 y_1^{2k} y_2 + b_3 y_1^{2k-2} y_2^{3} + \ldots + b_{2k+1} y_2^{2k+1}) - b/2 = 0 \\ \nu y_1/2 + a(b_o y_1^{2k+1} + b_2 y_1^{2k-1} y_2^{2} + \ldots + b_{2k} y_2^{2k} y_1) = 0 \end{cases}$$

where $a_{2k+1} = a$, and $B_1 = b$. Clearly (12) has a solution $(y_1, y_2) = (0, \bar{y})$, provided $\bar{y}$ satisfies

$$(13) \qquad 2a\, b_{2k+1}\, \bar{y}^{\,2k+1} - \nu \bar{y} = b.$$

We find that

$$b_{2k+1} = -\frac{1}{2\pi}\int_0^{2\pi} (\sin t)^{2k+2}\, dt$$

$$= -\frac{1}{2} \cdot \frac{3}{4} \cdot \frac{5}{6} \cdots \frac{2k+1}{2k+2} \quad .$$

By hypothesis $\nu a < 0$. Assume first $a < 0$, $\nu > 0$; the other case follows analogously. Thus $c = 2ab_{2k+1} > 0$. Consider the graph of $(y, f(y))$, where $f(y) = c\, y^{2k+1} - \nu y$.

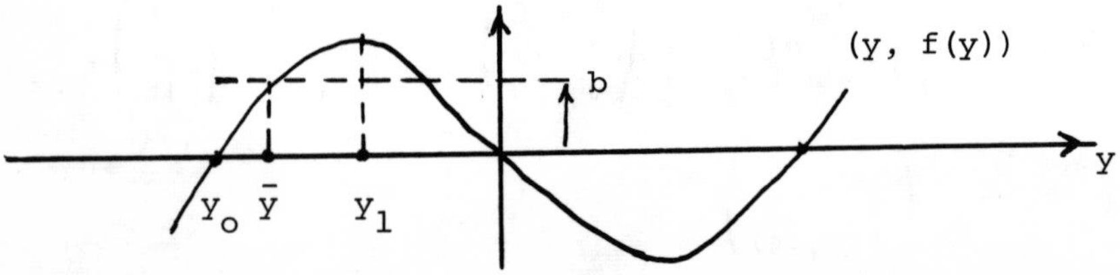

Let y_1 be the negative point at which $f(y)$ has a maximum value and y_0 be the negative zero of f. By elementary calculations

$$y_0 = -\left(\frac{\nu}{c}\right)^{1/2k}, \quad y_1 = -\left(\frac{\nu}{(2k+1)c}\right)^{1/2k}, \quad \text{and}$$

$$f(y_1) = \frac{2k\,\nu}{2k+1}\left(\frac{\nu}{(2k+1)c}\right)^{1/2k}.$$

If we take ν_0 to be the positive root of

$$\frac{2k\,\nu}{2k+1}\left(\frac{\nu}{(2k+1)c}\right)^{1/2k} = b\,, \quad \text{i.e.,} \quad f(y_1) = b\,,$$

it follows that for $\nu > \nu_0$, 3 real solutions of $f(y) = b$ exist. We choose $\bar{y}$ to be the solution such that $y_0 < \bar{y} < y_1$. By direct calculations, we obtain $b_{2k} = -b_{2k+1}$. Hence

$$\frac{\partial g_0}{\partial y}(0,\bar{y}) = \frac{1}{2}\begin{pmatrix} 0 & (2k+1)\,c\,\bar{y}^{2k} \\ -c\,\bar{y}^{2k} & 0 \end{pmatrix}, \text{ and thus}$$

$$(14) \qquad C_0 + \frac{\partial g_0}{\partial y}(0,\bar{y}) = \frac{1}{2}\begin{pmatrix} 0 & -\nu + (2k+1)\,c\,\bar{y}^{2k} \\ \nu - c\,\bar{y}^{2k} & 0 \end{pmatrix}$$

But $y_0^{2k} > \bar{y}^{2k} > y_1^{2k}$; i.e., $\nu/c > \bar{y}^{2k} > \nu/(2k+1)c$. Thus the nonzero elements in the matrix (14) are both positive and hence its eigenvalues are real, non-zero, and, in fact, of opposite sign.

Clearly

$$\left[\exp\begin{pmatrix} 0 & 1 \\ -1 & 0 \end{pmatrix}\right]\begin{pmatrix} 0 \\ \bar{y} \end{pmatrix} = \begin{pmatrix} \cos t & \sin t \\ -\sin t & \cos t \end{pmatrix}\begin{pmatrix} 0 \\ \bar{y} \end{pmatrix} = \bar{y}\begin{pmatrix} \sin t \\ \cos t \end{pmatrix},$$

and the last conclusion of Theorem 4 follows with $\alpha = \bar{y}$ and $\delta = 0$.

As mentioned above, the case $a > 0$, $\nu < 0$ is treated similarly.

The fact that the eigenvalues of (14) are real and of opposite sign also tells us that the a.p. solution $x(t,\epsilon)$ has a saddlepoint type of stability; i.e., in a sufficiently small neighborhood of this a.p. solution there exist 2 one-parameter families of solutions Γ_+

and Γ_- such that if x is in Γ_+, $|x(t) - x(t,\epsilon)| \to 0$ as $t \to +\infty$, while if x is in Γ_-, $|x(t) - x(t,\epsilon)| \to 0$ as $t \to -\infty$. For futher details, cf. [2].

Clearly, if we put $x_1 = x$ and $x_2 = x'$ (10) is equivalent to the scalar equation

$$(15) \qquad x'' + (1 - \epsilon\nu)x - \epsilon^m g(x) = p(t).$$

Suppose p has a sequence of Fourier exponents $\{\lambda_j\}$ such that $\lambda_j^2 < 1$, $j = 1,2,\ldots$, and $\lambda_j \to 1$ as $j \to \infty$. Let $\nu > \nu_o$ and $\epsilon_o(\nu)$ be fixed, and as in Theorem 3. Then for j fixed and sufficiently large, $1 - \epsilon_o(\nu)\nu < \lambda_j^2$, and if $\epsilon = \epsilon_j = (1 - \lambda_j^2)/\nu$, it follows that $\epsilon_j < \epsilon_o(\nu)$, and an a.p. solution of (15) exists. But $1 - \epsilon_j\nu = \lambda_j^2$, and we see that the Fourier exponent λ_j of p is in resonance with the linear part of (15). Apparently, the presence of the nonlinear term in (15) to some extent destroys the resonance effect.

References

[1] Coppel, W. A., Almost periodic properties of ordinary differential equations, Ann. Mat. Pura Appl. 76(1967), 27-50.

[2] Hale, J. K., Ordinary Differential Equations, Wiley-Interscience, (1969).

[3] Seifert, G., On almost periodic solutions for undamped systems with almost periodic forcing, (to appear in Proc. American Math. Soc.).

ANALYTIC EXPRESSIONS OF SOLUTIONS WITH THE ORDER OF POSITIVE POWERS OF x NEAR AN IRREGULAR TYPE SINGULARITY

Masahiro Iwano

Dedicated to Mr. Suketada Aoki on his Seventy-Second Birthday

0. Introduction. We consider a system of two nonlinear ordinary differential equations of the form

$$\text{(A)} \qquad x^{\sigma+1} y' = f(x, y, z), \quad x z' = g(x, y, z),$$

where σ is a positive integer; x is a complex independent variable; y and z are both scalar; $f(x, y, z)$ and $g(x, y, z)$ are holomorphic functions of (x, y, z) at the origin of the product space of three complex x-, y-, z-planes and they vanish at the origin:

$$f(0, 0, 0) = 0, \qquad g(0, 0, 0) = 0.$$

1. Two problems on boundary layer equation. To explain our purpose of this paper, we shall explain an introductory story about the reason why the author is interested in the study of this system.

The starting point of the present work was two problems concerning the boundary layer equation:

$$\text{(B)} \qquad f''' + f f'' + \lambda(k^2 - f'^2) = 0, \quad \lambda < 0, \quad k > 0,$$

$$\text{(1.1)} \qquad f(0) = 0, \quad f'(0) = 0, \quad f'(\infty) = k,$$

$$\text{(1.2)} \qquad 0 < f'(t) < k \qquad \text{for} \quad 0 < t < \infty .$$

The following two problems were proposed by physicists at a symposium on fluid dynamics which was held at the University of Tokyo about eleven years ago:

i) Does this boundary value problem have solutions ?

ii) When the boundary value problem has solutions $f(t)$, can we construct asymptotic formulas of the functions $f'(t)-k$ for large t ?

Since the equation (B) does not involve the independent variable t explicitly, if we take f as a new independent variable, the order of the differential equation is reduced by unity. Observe that

$$f'' = \frac{1}{2} \frac{d\,(f'^2)}{df}, \qquad f''' = \frac{1}{2} f' \frac{d^2\,(f'^2)}{df^2}.$$

Then, by taking $f'(t)^2 = g(f)$ as a new dependent variable, the equation (B) becomes

(Ba) $$\sqrt{g}\,\ddot{g} + f\,\dot{g} + 2\lambda(k^2 - g) = 0,$$

(1.1a) $$g(0) = 0, \qquad g(\infty) = k^2,$$

(1.2a) $$0 < g(f) < k^2 \qquad \text{for} \qquad 0 < f < \infty .$$

Instead of solving Problem i) and Problem ii) directly, Professor Masuo Hukuhara proposed the following two problems:

ia) <u>Does the boundary value problem</u> (Ba)-(1.1a)-(1.2a) <u>have solutions</u> ?

iia) <u>Can we construct an analytic expression of a local general solution</u> $g(f)$ <u>such that</u> $g(f)$ <u>tends to</u> k^2 <u>as</u> f <u>approaches infinity</u> ?

It is easy to verify that, if Problem ia) has a solution, then Problem i) also has a solution. However, these two problems are still open problems. The following result is the latest one obtained by R. Iglisch and F. Kemnitz (see P. Hartman [1]):

<u>Let</u> $k > 0$, $0 \leqq \beta < k$ <u>and</u> $\lambda < 0$ <u>be arbitrarily fixed constants</u>. <u>Then there exists a constant</u> $A(\lambda, \beta, k)$ <u>and an increasing continuous function</u> $\gamma(\alpha)$ <u>of</u> α <u>defined over a half line</u> $\alpha \geqq A(\lambda, \beta, k)$ <u>and vanishing at</u> $\alpha = A(\lambda, \beta, k)$ <u>with the property that the solution</u> $f(t)$ <u>of the initial value problem</u>

(B) $$f''' + f f'' + \lambda(k^2 - f'^2) = 0$$

<u>with initial conditions</u>

$$f(0) = \alpha, \qquad f'(0) = \beta, \qquad f''(0) = \gamma$$

<u>is extended to infinity satisfying inequalities</u>

$$\beta < f'(t) < k\,, \qquad 0 < f''(t) \qquad \underline{\text{for}} \quad 0 < t < \infty$$

and tends to the quantity k if and only if the initial values α, β and γ satisfy the inequalities

$$\alpha \geqq A(\lambda, \beta, k), \qquad 0 \leqq \gamma \leqq \gamma(\alpha).$$

For a solution $f(t)$ of this boundary value problem, P.Hartman [1] solved completly the Problem ii) as follows:

There exist constants $c_0 > 0$, c_1 such that $f(t)$ behaves as

$$k - f'(t) \simeq c_0\, t^{-1-2\lambda} \exp\left(-\frac{1}{2}t^2 - c_1 t\right),$$

$$f''(t) \simeq t(k - f'(t))$$

if and only if $f''(0) = \gamma(\alpha)$; for other solutions $f(t)$ with $\alpha > A(\lambda, \beta, k)$ and $0 \leqq f''(0) < \gamma(\alpha)$, asymptotic relations

$$k - f'(t) \simeq c_0\, t^{2\lambda}, \qquad f''(t) \simeq -2\lambda\, c_0\, t^{2\lambda - 1}$$

hold with a suitable constant $c_0 > 0$.

Immediately after these problems were proposed, the author tried to solve Problem iia) by applying a general theory(M. Iwano [6,7]) about how to construct analytic expressions of bounded solutions of nonlinear ordinary differential equations. Unfortunately my method did not work successfully. However, Hartman's beautiful result encouraged the author to find an improvement of the method. After many years of study, he succeeded in solving the Problem iia). But, it is still difficult to construct analytic expressions of local solutions $f(t)$ of equation (B) such that $f'(t)$ tend to k as t approaches infinity.

To study Problem iia), we put

$$g(f) = k^2 + G(f).$$

Then equation (Ba) is reduced to the equation

$$\sqrt{k^2 + G}\,\ddot{G} + f\dot{G} - 2\lambda G = 0, \qquad G(\infty) = 0.$$

We consider the second order linear differential equation

$$k\ddot{H} + f\dot{H} - 2\lambda H = 0, \quad H(\infty) = 0$$

which has an irregular singularity at infinity. By applying Hukuhara-Turrittin theory(M. Hukuhara [2,3]; H.L. Turrittin [12]), we can construct an analytic expression of a local general solution at infinity. After a standard calculation, we have an expression of the form

$$H(f) = C_1 f^{-2\lambda+1} \exp\left(-\frac{f^2}{2k}\right)\cdot(1 + O(f^{-1})) + C_2 f^{2\lambda}(1 + O(f^{-1})),$$

where C_1 and C_2 are arbitrary constants, the symbol $O(f^{-1})$ represents a holomorphic function of f which admits an asymptotic expansion in powers of f^{-1} as f tends to infinity through one of the sectors

$$\left|\arg f \pm \frac{\pi}{4}\right| < \frac{\pi}{2}, \qquad |f| > R.$$

It is naturel to expect that the expression $H(f)$ represents an approximate solution of a solution $G(f)$. Actually this is true. To show this fact, we put

$$g(f) = k^2 + h(x), \qquad x = f^{-1}.$$

We first introduce another variable by xh', so that the given single differential equation becomes a system of two unknowns $\eta = h$ and $\zeta = xh'$. Following the Hukuhara-Turrittin theory, in order to reduce the linear parts of the nonlinear differential equations to a normal form, we next make a transformation of the form

$$\begin{pmatrix} h \\ xh' \end{pmatrix} = \begin{pmatrix} 1 & 0 \\ -2\lambda & 1 \end{pmatrix} \begin{pmatrix} 1 & kx^2 \\ 2\lambda k(2\lambda-1)x^2 & 1 \end{pmatrix} \begin{pmatrix} z \\ y \end{pmatrix}.$$

Then the equations satisfied by y and z become a system of the form (A), where

$$\sigma = 2, \qquad f_y(x, 0, 0) = \frac{1}{k} + (2\lambda - 1)x^2 + O(x^4),$$

$$g_z(x, 0, 0) = -2\lambda + O(x^2) .$$

The system is written as

$$x^3 y' = \left(\frac{1}{k} + (2\lambda - 1)x^2\right)y + A(x)y + B(x)z + \sum_{j+k \geqq 2} a_{jk}(x)y^j z^k ,$$

$$x z' = -2\lambda z + C(x)y + D(x)z + \sum_{j+k \geqq 2} b_{jk}(x)y^j z^k ,$$

where the coefficients are all holomorphic functions in a full neighborhood of $x = 0$ and satisfy conditions

$$A(x),\ B(x),\ C(x) = O(x^4);\quad D(x) = O(x^2);\quad a_{jk}(x),\ b_{jk}(x) = O(1),$$

and the power series in the right-hand sides are uniformly convergent whenever values of x, y, z are sufficiently small in absolute value.

By applying Hukuhara theory (M. Hukuhara [4]) about how to construct formal solutions depending on several arbitrary constants, we have a formal solution of the form

$$y \sim U(x) + \sum_{j,k} P_{jk}(x)U(x)^j V(x)^k, \quad z \sim V(x) + \sum_{j,k} Q_{jk}(x)U(x)^j V(x)^k,$$

where the coefficients are expressed by formal power series of x and satisfy formally conditions

$$P_{10}(x),\ P_{01}(x),\ Q_{10}(x) \sim O(x^4) ; \qquad Q_{01}(x) \sim O(x^2),$$

and

$$U(x) = C_1\, x^{2\lambda - 1} \exp\left(-\frac{1}{2kx^2}\right) , \qquad V(x) = C_2\, x^{-2\lambda} .$$

The author (M. Iwano [8, 9]) has proved that <u>this formal solution is uniformly convergent whenever values of</u> $x, U(x)$ <u>and</u> $V(x)$ <u>satisfy inequalities of the form</u>

$$0 < |x| < a'', \quad \left|\arg x \pm \frac{\pi}{4}\right| < \frac{\pi}{2}, \quad |U(x)| < b'', \quad |V(x)| < b''.$$

The coefficients are holomorphic functions of x admitting asymptotic expansions in powers of x as x tends to the origin through one of the sectors

$$\left|\arg x \pm \frac{\pi}{4}\right| < \frac{\pi}{2}, \qquad 0 < |x| < a'' .$$

When he applied the old theory, he could construct an analytic expression like this. But the opening angle of the sector in which x is restricted was too small to contain the positive real axis. Since the variable x is originally real, if the sector of x in which the analytic expression is valid does not contain the positive real axis near the origin, the expression will be useless.

2. Assumptions. We will return to the original equations. We assume that the Jacobian $f_y(0, 0, 0) \equiv \nu$ is different from zero. Then we can assume without loss of generality that

$$\text{(2.1)} \qquad f_x(0, 0, 0) = 0, \quad f_z(0, 0, 0) = 0, \quad g_y(0, 0, 0) = 0.$$

Indeed, it is sufficient to apply, if it is necessary, a transformation of the form

$$y = Y - \frac{1}{\nu} f_z(0) Z - \frac{1}{\nu} f_x(0) x, \qquad z = \frac{1}{\nu} g_y(0) x^{\sigma} Y + Z.$$

We assume further that the real part of the Jacobian $g_z(0, 0, 0) \equiv \mu$ is positive.

Then we can find a formal solution of the form

$$\text{(F)} \qquad y \sim \sum_{r,k} P_{rk} x^r V(x)^k, \qquad z \sim \sum_{r,k} Q_{rk} x^r V(x)^k,$$

where $V(x)$ is a general solution of the equation either

$$\text{(I)} \qquad x v' = \mu v, \qquad V(x) = C x^{\mu}$$

or

$$\text{(II)} \qquad x v' = \mu v + c x^{\mu}, \quad V(x) = x^{\mu}(C + c \log x) \quad (c \neq 0).$$

When $c \neq 0$, the quantity μ is necessarily a positive integer. This case was first studied by H. Poincaré [11] when the first equation of system (A) did not appear.

3. Main result. We shall prove the following theorem:

Theorem. There exists a solution of the form $Y(x, V(x)), Z(x, V(x))$ whenever values of x and $V(x)$ belong to a domain of the form

$$(3.1) \qquad 0 < |x| < a'', \quad \left|\arg(-\nu) - \sigma \arg x\right| < \frac{3\pi}{2} - \varepsilon', \quad |v| < b''.$$

Here $Y(x, v)$ and $Z(x, v)$ are holomorphic functions of (x, v) in domain (3.1) and have uniformly convergent expansions of the form

$$(3.2) \qquad Y(x, v) = \sum_{k=0}^{\infty} Y_k(x)\, v^k, \qquad Z(x, v) = \sum_{k=0}^{\infty} Z_k(x)\, v^k,$$

where the coefficients have asymptotic expansions of the form

$$(3.3.k) \qquad Y_k(x) \simeq \sum_{r=0}^{\infty} P_{rk}\, x^r, \qquad Z_k(x) \simeq \sum_{r=0}^{\infty} Q_{rk}\, x^r$$

as x tends to the origin through

$$(3.4) \qquad \left|\arg(-\nu) - \sigma \arg x\right| < \frac{3\pi}{2} - \varepsilon', \qquad 0 < |x| < a'',$$

ε' being a sufficiently small positive number.

For the proof of this theorem, Case I and Case II must be treated separately. This theorem is already proved by M. Iwano [8, 9, 10] in which the following more general cases have been studied:

$$(A') \qquad x^{\sigma+1}\, y' = f(x, y, z), \qquad x\, z' = g(x, y, z),$$

where y and z are respectively m- and n-column vectors (M. Iwano [8, 9]);

$$(A'') \qquad x\, 1_n(x^{\sigma})\, y' = f(x, y, z), \qquad x\, z' = g(x, y, z),$$

where y and z are m- and n-column vectors and

$$1_n(x^{\sigma}) = \operatorname{diag}(x^{\sigma_1}, \ldots, x^{\sigma_n})$$

with positive integers σ_j (M. Iwano [10]).

4. Brief sketch of proof of the theorem for case I. If we formally arrange the double power series (F) into the form of a single power series of $V(x)$, we have a formal solution of the form

$$(F_1) \qquad y \sim \sum_{k=0}^{\infty} P_k(x)\, V(x)^k , \quad z \sim \sum_{k=0}^{\infty} Q_k(x)\, V(x)^k ,$$

where the coefficients are formal power series of x. When we try to give an analytic meaning to this formal solution, it is first necessary that the coefficients are already given some analytic interpretation (convergent or asymptotic expansions) . We insert the formal solution into both sides of system (A) and arrange formally resulting equations in powers of $V(x)$. If we equate the coefficients of the terms $V(x)^k$ $(k = 0,1,\dots)$, then we get differential equations which are expected to determine the coefficients. A simple calculation shows that these equations are given by

$$(4.1) \qquad x^{\sigma+1} P_0' = f(x, P_0, Q_0), \quad x Q_0' = g(x, P_0, Q_0),$$

$$(4.2.k) \qquad \begin{cases} x^{\sigma+1} P_k' = K(x) P_k + L(x) Q_k + S_k(x), \\ x Q_k' = M(x) P_k + N(x) Q_k + T_k(x), \end{cases}$$

where

$$K(x)= f_y(x, P_0(x), Q_0(x)), \quad L(x)= f_z(x, P_0(x), Q_0(x)),$$

$$M(x)= g_y(x, P_0(x), Q_0(x)), \quad N(x)= g_z(x, P_0(x), Q_0(x)),$$

and $S_k(x)$, $T_k(x)$ are polynomials of $P_1(x), \dots, P_{k-1}(x), Q_1(x), \dots, Q_{k-1}(x)$ and of their derivatives and, consequently, they are regarded as known functions of x. Since $x = 0$ is an irregular singularity for each system, we can obtain

The coefficients of (F_1) are determined uniquely and successively as solutions of nonlinear (4.1) or linear (4.2.k) differential equations in such a way that they are holomorphic functions of x and admit the asymptotic expansions

$$P_k(x) \simeq \sum_{r=0}^{\infty} P_{rk}\, x^r , \qquad Q_k(x) \simeq \sum_{r=0}^{\infty} Q_{rk}\, x^r$$

as x tends to the origin through

$$\left|\arg(-\nu) - \sigma \arg x\right| < \frac{3\pi}{2} - \varepsilon' , \qquad 0 < |x| < a_1 .$$

In order to study an analytic meaning of the formal solution (F_1), the author applies fixed point technique which was deviced by Prof. M. Hukuhara[5] based on a well known Tihonof fixed point theorem (see also M. Iwano [8]). Then we can conclude that the formal solution (F_1) is uniformly convergent and its sum $Y(x, V(x))$, $Z(x, V(x))$ is a solution of equations (A) whenever values of x and $V(x)$ belong to a domain of the form. However the proof of the uniform convergence needs a lengthy reasoning. I would like to omit the proof and to leave the reader to refer my paper (M. Iwano [8]). If we use majorant technique for the proof (I have seldom used it), I would say that the key of the proof is to prove the following fact:

There exist two functions $\omega(\arg x)$ and $\chi(\arg x)$, which are defined on a closed interval $|\arg(-\nu) - \sigma \arg x| \leq 3\pi/2 - \varepsilon'$ and are strictly positive valued and continuous functions of $\arg x$, with the property that inequalities of the form

$$|P_k(x)| , \quad |Q_k(x)| \quad \leq K \left|\frac{1}{b' \chi(\arg x)}\right|^k$$

hold for

$$0 < |x| < a' \omega(\arg x), \quad \left|\arg(-\nu) - \sigma \arg x\right| < \frac{3\pi}{2} - \varepsilon'$$

for suitably chosen positive constants a', b', ε' and K.

Then the positive constants a'' and b'' appearing in (3.1) are defined by the minimum of the functions $a' \omega(\arg x)$ and $b'\chi(\arg x)$ on the closed interval respectively. It is not so easy to find such functions when we use majorant technique. The author has shown how to construct those functions (M. Iwano [8, 9, 10]). I should be noticed that if we replace those functions by constants then the sector of x is forced to be replaced by much smaller one. This is the reason why my method did not work successfully when I applied the old theory to the boundary layer equation.

<u>5. Brief sketch of proof of the theorem for case II.</u> If we apply the same reasoning to the case II, a difficulty comes out when we try to determine the coefficients $P_k(x)$ and $Q_k(x)$. Indeed, a simple calculation shows that the equations

$$x^{\sigma+1}\, P_0' = f(x,\, P_0,\, Q_0) - c\, x^{\sigma+\mu}\, P_1,$$

$$x\, Q_0' = g(x,\, P_0,\, Q_0) - c\, x^{\mu}\, Q_1$$

must be substituted for equations (4.1). The above equations which are expected to determine unknowns $P_0(x)$, $Q_0(x)$ do contain another pair of unknowns P_1 , Q_1 . Therefore <u>the coefficients can not be determined uniquely or successively in this way</u>. Recently, the author has found a way to give an analytic meaning to the formal solution (F).

Instead of arranging the formal solution (F) in powers of $V(x)$, we arrange formally it in powers of x, so that we have a formal solution of the form

$$(F_2) \qquad y \sim \sum_{r=0}^{\infty} p_r(V(x))\, x^r, \qquad z \sim \sum_{r=0}^{\infty} q_r(V(x))\, x^r,$$

where $p_r(v)$ and $q_r(v)$ are expressed by formal power series of v. By substituting (F_2) for y and z into equations (A) and equating the coefficients of the terms of x^r $(r = 0,1,\dots)$, we find differential equations of the form

$$(5.1) \qquad f(0,\, p_0,\, q_0) = 0, \qquad \mu\, v\, \frac{d\, q_0}{dv} = g(0,\, p_0,\, q_0),$$

$$(5.2.r) \qquad E(v)p_r + F(v)q_r = I_r(v), \qquad \mu\, v\, \frac{dq_r}{dv} = G(v)p_r + H(v)q_r + J_r(v),$$

where

$$E(v) = f_y(0,\, p_0(v),\, q_0(v)), \qquad F(v) = f_z(0,\, p_0(v),\, q_0(v)),$$

$$G(v) = g_y(0,\, p_0(v),\, q_0(v)), \qquad H(v) = g_z(0,\, p_0(v),\, q_0(v))$$

and $I_r(v)$, $J_r(v)$ are polynomials of $p_1(v)$, ..., $p_{r-1}(v)$, $q_1(v)$,

..., $q_{r-1}(v)$ and of their derivatives, so that these polynomials are considered as known functions of v. It is noted that these differential equations possess formal power series solutions of v.

By our assumption, the Jacobian $f_y(0, 0, 0)$ differs from zero. Hence we can solve the first equation of (5.1) with respect to p_0, say $p_0 = A(q_0)$, where $A(q)$ is a holomorphic function of q at $q = 0$. If we replace p_0 by $A(q_0)$, the second equation of (5.1) becomes an equation of Briot-Bouquet type with a regular singularity at $v = 0$. Since this equation admits a formal power series solution of v, a well known theorem in Briot-Bouquet theory implies that this power series has a positive raius of convergence and, consequently, its sum $q_0(v)$ defines a holomorphic function at $v = 0$. Hence, $p_0(v) = A(q_0(v))$ is a holomorphic function of v at $v = 0$ and, by virtue of conditions (2.1), vanishes at $v = 0$. Thus equations (5.1) possess a holomorphic solution $\{p_0(v), q_0(v)\}$. Thus we can arrive at the proposition:

The coefficients $p_r(v)$ and $q_r(v)$ are determined uniquely and successively as solutions of nonlinear (5.1) or linear (5.2.r) differential equations in such a way that they are holomorphic functions in a full neighborhood of the origin in the complex v-plane.

Concerning an analytic meaning of formal solution (F_2), by using fixed point technique, we can prove that

There exist two functions $Y(x, v)$ and $Z(x, v)$ defined on

$$0 < |x| < a'\omega(\arg x), \qquad |v| < b'\chi(\arg x),$$

$$\left|\arg(-\nu) - \sigma \arg x\right| < \frac{3\pi}{2} - \varepsilon' \tag{5.3}$$

with properties that:

i) they are holomorphic functions of (x, v) in domain (5.3);

ii) they admit uniformly asymptotic expansions in powers of x:

$$\left| Y(x, v) - \sum_{r=0}^{N-1} p_r(v)\, x^r \right| \leq K_N |x|^N ,$$

where N is any positive integer and K_N is a constant;

iii) $Y(x, V(x))$, $Z(x, V(x))$ is a solution of equations (A) whenever values of x and $V(x)$ belong to domain (5.3).

Here $\omega(\arg x)$ and $\chi(\arg x)$ are the same functions as those in the case I.

Let a" and b" be defined as before. By means of Cauchy theorem, we have uniformly convergent expansions

$$Y(x, V(x)) = \sum_{k=0}^{\infty} \hat{P}_k(x)\, V(x)^k , \quad Z(x, V(x)) = \sum_{k=0}^{\infty} \hat{Q}_k(x)\, V(x)^k$$

whenever values of x and $V(x)$ belong to the domain

$$0 < |x| < a'', \quad \left| \arg(-\nu) - \sigma \arg x \right| < \frac{3\pi}{2} - \varepsilon' , \quad |v| < b'' .$$

We can assume without loss of generality that $Y(x, v)$ and $Z(x, v)$ are continuous functions in the closure of the above domain.

We shall prove that the functions $\hat{P}_k(x)$ and $\hat{Q}_k(x)$ have the same asymptotic expansions as those of the coefficients $P_k(x)$ and $Q_k(x)$: namely,

$$(5.4) \qquad \hat{P}_k(x) \simeq \sum_{r=0}^{\infty} P_{rk}\, x^r , \qquad \hat{Q}_k(x) \simeq \sum_{r=0}^{\infty} Q_{rk}\, x^r .$$

In fact, owing to Cauchy theorem, we have

$$\frac{1}{2\pi i} \int_{|v|=b''} \frac{Y(x, v)}{v^{k+1}}\, dv = \hat{P}_k(x).$$

On the other hand, from the convergent expansion

$$p_r(v) = \sum_{k=0}^{\infty} P_{rk}\, v^k$$

it follows that

$$\frac{1}{2\pi i} \int_{|v|=b''} \frac{p_r(v)}{v^{k+1}}\, dv = P_{rk} .$$

Hence we have

$$\hat{P}_k(x) - \sum_{r=0}^{N-1} P_{rk}\, x^r = \frac{1}{2\pi i} \int_{|v|=b''} \frac{Y(x, v) - \sum_{r=0}^{N-1} p_r(v)\, x^r}{v^{k+1}}\, dv.$$

By virtue of the uniform asymptoticity of the function $Y(x, v)$, the numerator of the integrand is of order $O(x^N)$ uniformly for $|v| \leq b''$. This proves that the asymptotic expansions (5.4) hold.

By taking $Y_k(x) = \hat{P}_k(x)$, $Z_k(x) = \hat{Q}_k(x)$, the Theorem has been proved.

The detailed study of existence theorems of this type will be published elsewhere.

The idea of domains like (3.1) or (5.3) is already obtained implicitly in the study of existence theorems and of stability conditions by notable mathematicians A.M. Liapunof, P. Hartman, M. Nagumo, M. Hukuhara etc.

6. Remark. The author has studied for many years singularities of Briot-Bouquet type. A problem of constructing formal solutions which depend on several arbitrary parameters is not so hard. However, a problem of giving analytic meanings to the formal solutions thus obtained is not so easy and must be treated case by case although we have a very useful technique named Hukuhara's fixed point method. The author hopes that in future a very powerful technique for the study of analytic meanings of formal solutions will invented.

REFERENCES

1. P. Hartman, Ordinary Differential Equations, John Wiley & Sons, New York, 1964.

2. M. Hukuhara, Sur les points singuliers des équations différentielles linéaires, II. Jour. Fac. Sci. Hokkaido Imp. Univ. 5(1937), 123-166.

3. M. Hukuhara, Sur les points singuliers des équations différentielles linéaires, III, Mem. Fac. Sci. Kyusyu Imp. Univ., Ser. A, 2(1941), 125-137.

4. M. Hukuhara, Intégration formelle d'un système d'équations différentielles non linéaires dans le voisinage d'un point singulier, Ann. Mat. Pura Appl. (4) 19(1940), 35-44.

5. M. Hukuhara, Renzokuna kansû no zoku to syazô (Family and mapping of continuous functions), Mem. Fac. Sci. Kyusyu Univ. 5(1950), 61-63.

6. M. Iwano, Intégration analytique d'un système d'équations différentielles non linéaires dans le voisinage d'un point singulier,I, Ann. Mat. Pura Appl. (4) 44(1957), 261-292.

7. M. Iwano, Intégration analytique d'un système d'équations différentielles non linéaires dans le voisinage d'un point singulier, II, Ann. Mat. Pura Appl. (4) 47(1959), 91-150.

8. M. Iwano, Analytic expressions for bounded solutions of nonlinear ordinary differential equations with an irregular type singular point, Ann. Mat. Pura Appl. (4) 82(1969), 189-256.

9. M. Iwano, Bounded solutions and stable domains of nonlinear ordinary differential equations, Analytic Theory of Differential Equations, Lecture notes in mathematics, Springer-Verlag, 183 (1971), 59-127.

10. M. Iwano, Analytic integration of a system of nonlinear ordinary differential equations with an irregular type singularity (to appear), Ann. Mat. Pura Appl.

11. H. Poincaré, Note sur les propriétés des fonctions définies par les équations différentielles, Jour. École Poly., cahier 45(1878), 13-26.

12. H.L. Turrittin, Convergent solutions of ordinary linear homogenous differential equations in the neighborhood of an irregular singular point, Acta Math. 93(1955), 27-66.

PERTURBATION AT AN IRREGULAR SINGULAR POINT

Yasutaka Sibuya

Although some new results will be presented, this report is expository for the most part, being based on the papers which are listed in Bibliography. Topics which are selected in this report are

(a) representation of a fundamental matrix in terms of a monodromy matrix;

(b) analytic reduction to a system with polynomial coefficients;

(c) asymptotic solutions.

These three subjects (a), (b) and (c) will be discussed in Sections 1, 2 and 3 respectively.

1. A generalization of a result due to T. Saito.

We shall consider a system

$$(1.1) \qquad dy/dx = A(x)y$$

where x is a complex independent variable, y is an n-dimensional vector whose components are unknown quantities, and $A(x)$ is an n-by-n matrix whose components are single-valued and holomorphic in x in a domain

$$(1.2) \qquad 0 < |x| < \delta_0 .$$

It is known that a fundamental matrix $\Phi(x)$ of system (1.1) admits a representation of the form

$$(1.3) \qquad \Phi(x) = P(x)x^R$$

where R is an n-by-n constant matrix and $P(x)$ is an n-by-n matrix whose components are single-valued and holomorphic in x in domain (1.2). The representation (1.3) is based on the fact that

$$\Phi(x)^{-1}\Phi(xe^{2\pi i}) \equiv e^{2\pi iR} \quad .$$

In fact, R is determined by

$$(1.4) \qquad R = \frac{1}{2\pi i} \log[\Phi(x)^{-1}\Phi(xe^{2\pi i})] \quad .$$

In case when components of $A(x)$ depend smoothly on parameters, components of $P(x)$ and R also depend smoothly on parameters. Hence $P(x)$ and R may be computed by a perturbation method under suitable assumptions. Various perturbation methods which are based on Floquet's theorem are available for a periodic system, and those methods may be effectively utilized for system (1.1) as well. (For such a method for a periodic system, see M. Hukuhara [7].) In this section, we shall not discuss such methods. Instead, being motivated by some properties of the matrix

$$(1.5) \qquad U_o(x) = \Phi(xe^{2\pi i})\Phi(x)^{-1} \quad ,$$

we shall present a method for constructing $P(x)$ and R . The motivation was derived from T. Saito's paper [13], and his result will be reproduced at the end of this section.

By virtue of (1.3), we have

$$(1.6) \qquad U_o(x) = P(x)e^{2\pi iR}P(x)^{-1} \quad .$$

Therefore components of $U_o(x)$ are single-valued and holomorphic in x in domain (1.2). On the other hand, $U = U_o(x)$ satisfies a differential equation

$$(1.7) \qquad dU/dx = A(x)U - UA(x) \quad .$$

This means that equation (1.7) has a non-trivial solution which is single-valued and holomorphic in (1.2). Let us find a general form of such a solution. To do this, observe that the general solution of (1.7) is given by

$$(1.8) \qquad U(x) = \Phi(x)C\Phi(x)^{-1}$$

where C is an arbitrary n-by-n constant matrix. By using (1.3) the right-hand side of (1.8) is written as $P(x)x^{R}Cx^{-R}P(x)^{-1}$. Therefore (1.8) gives a solution which is single-valued and holomorphic in (1.2) if and only if the constant matrix C satisfies a condition

$$(1.9)\qquad C = e^{2\pi iR}\, C\, e^{-2\pi iR} .$$

<u>Assume that the monodromy matrix</u> $e^{2\pi iR}$ <u>has</u> n <u>mutually distinct eigenvalues</u> and that $e^{2\pi iR}$ is a diagonal matrix:

$$(1.10)\qquad e^{2\pi iR} = \begin{bmatrix} e^{2\pi i\lambda_1} & & 0 \\ & \ddots & \\ 0 & & e^{2\pi i\lambda_n} \end{bmatrix} .$$

Then (1.9) is satisfied if and only if C is diagonal. Thus the general form of a single-valued and holomorphic solution of (1.7) in domain (1.2) is given by

$$(1.11)\qquad U(x) = P(x)CP(x)^{-1}$$

where C is an arbitrary n-by-n constant diagonal matrix. From (1.11) we conclude that diagonal components of C are eigenvalues of U(x) and that column vectors of P(x) are eigenvectors of U(x) . <u>Assume that diagonal components of</u> C <u>are mutually distinct</u>. Let $\varphi_1(x), \ldots, \varphi_n(x)$ be eigenvectors of U(x) which are single-valued and holomorphic in domain (1.2). Denote also by $p_1(x), \ldots, p_n(x)$ column vectors of P(x) . Then there must be n scalar functions $\alpha_1(x), \ldots, \alpha_n(x)$ such that

$$p_j(x) = \alpha_j(x)\varphi_j(x) \qquad (j = 1, \ldots, n) .$$

Therefore, if we put

$$\beta_j(x) = x^{\lambda_j}\alpha_j(x) \qquad (j = 1, \ldots, n) ,$$

n linearly independent solutions of system (1.1) are given by

$$(1.12)\qquad y = \beta_j(x)\,\varphi_j(x) \qquad (j = 1, \ldots, n) .$$

Suppose that we have found a solution U(x) of (1.7) which is single-valued and holomorphic in (1.2). Assume also that eigenvalues of U(x) are mutually distinct. Let us construct corresponding eigenvectors $\varphi_1(x),\ldots,\varphi_n(x)$ of $U(x)$. Then by substituting (1.12) into system (1.1) we get

$$(1.13) \qquad \varphi_j(x)\ d\beta_j(x)/dx + \beta_j(x)d\varphi_j(x)/dx = [A(x)\varphi_j(x)]\beta_j(x) \quad .$$

This is a vector relation, and by using any component of this vector relation we can find $\beta_j(x)$. This provides us desired n linearly independent solutions of system (1.1). In case when components of $A(x)$ depend on parameters, we can convert this process into a perturbation method.

We shall illustrate the method by utilizing it to

$$(1.14) \qquad d^2\eta/dx^2 + q(x)\eta = 0$$

where η is a scalar and $q(x)$ is single-valued and holomorphic in (1.2). Equation (1.14) was studied in the paper of T. Saito [13]. Put

$$(1.15) \qquad y = \begin{bmatrix} \eta \\ d\eta/dx \end{bmatrix} , \quad A(x) = \begin{bmatrix} 0 & 1 \\ -q(x) & 0 \end{bmatrix} \quad .$$

Then (1.14) is equivalent to

$$(1.16) \qquad dy/dx = A(x)y \quad .$$

If we put

$$U = \begin{bmatrix} u & v \\ w & z \end{bmatrix} ,$$

equation (1.7) is given by

$$(1.17) \qquad \begin{cases} du/dx = w + q(x)v \quad , & dv/dx = z - u \quad , \\ dw/dx = q(x)\ (z-u) & dz/dx = -(w + q(x)v) \quad . \end{cases}$$

From (1.17) we derive

$$(1.18) \qquad d^3v/dx^3 + 4q(x)dv/dx + 2(dq(x)/dx)\ v = 0 \ .$$

Then as it was done by T. Saito, multiplying v on both sides and integrating, eqution (1.18) becomes

$$v\, d^2v/dx^2 - \tfrac{1}{2}(dv/dx)^2 + 2q(x)v^2 = \gamma_1 \tag{1.19}$$

where γ_1 is an arbitrary constant. From (1.17) we further derive

$$\begin{cases} u = -\tfrac{1}{2}\, dv/dx + \gamma_2 \quad , \quad z = \tfrac{1}{2}\, dv/dx + \gamma_2 \ , \\ w = -\tfrac{1}{2} d^2v/dx^2 - q(x)v \end{cases} \tag{1.20}$$

where γ_2 is also an arbitrary constant. By using (1.20) we can write (1.19) as

$$uz - vw - \gamma_2^2 = \tfrac{1}{2}\gamma_1 \quad . \tag{1.21}$$

This implies that eigenvalues of U are

$$\gamma_2 \pm i\sqrt{\frac{\gamma_1}{2}} \quad , \tag{1.22}$$

and corresponding eigenvectors are

$$\varphi_\pm(x) = \begin{bmatrix} 1 \\ \dfrac{-u + \gamma_2 \pm i\sqrt{\dfrac{\gamma_1}{2}}}{v} \end{bmatrix}$$

which become

$$\varphi_\pm(x) = \begin{bmatrix} 1 \\ v^{-1}\{\tfrac{1}{2}\, dv/dx \pm i\sqrt{\tfrac{\gamma_1}{2}}\} \end{bmatrix} \tag{1.23}$$

by virtue of (1.20). Now we can find two linearly independent solutions of system (1.16) in the form

$$y = \beta_\pm(x)\varphi_\pm(x) \tag{1.24}$$

where $\beta_\pm(x)$ are scalar functions which must satisfy

$$\varphi_\pm\, d\beta_\pm/dx + \beta_\pm\, d\varphi_\pm/dx = [A(x)\varphi_\pm]\beta_\pm$$

or

$$(1.25)\qquad d\beta_{\pm}/dx = \left[\frac{\frac{1}{2}\,dv/dx \pm i\sqrt{\frac{\gamma_1}{2}}}{v}\right]\beta_{\pm} \quad ,$$

Hence

$$\beta_{\pm}(x) = v^{\frac{1}{2}}\exp\left\{\pm i\sqrt{\frac{\gamma_1}{2}}\int^{x} v^{-1}d\tau\right\} \quad .$$

The quantities $\beta_{\pm}(x)$ being the first components of vectors (1.24), they are independent solutions of (1.14). Thus we obtained two linearly independent solutions

$$(1.26)\qquad \eta = v^{\frac{1}{2}}\exp\left\{\pm i\sqrt{\frac{\gamma_1}{2}}\int^{x} v^{-1}d\tau\right\}$$

where v is a single-valued and holomorphic function to be determined by equation (1.19). This is the result which has been presented in T. Saito's paper [13]. He utilized this method in various examples and he showed how to use this process as a perturbation method.

2. Analytic reduction to a system with polynomial coefficients. A result which is intimately related with monodromy matrices and which was originally proved by G. D. Birkhoff [1,2] is the following theorem:

Let be given a system

$$(2.1)\qquad dy/dx = x^{k}A(x)y$$

where k *is a non-negative integer and* $A(x)$ *is an n-by-n matrix whose components are convergent power series in* x^{-1}. *Then there exists an n-by-n matrix* $P(x)$ *such that*

(i) *components of* $P(x)$ *are convergent power series in* x^{-1} ;

(ii) $\lim_{x\to\infty} P(x)$ *is non-singular;*

(iii) *the transformation*

$$(2.2)\qquad y = P(x)z$$

reduces system (2.1) *to*

$$(2.3)\qquad dz/dx = x^{k}B(x)z$$

where components of $B(x)$ are polynomials in x^{-1} ;

(iv) $x = 0$ is at worst a regular singular point of system (2.3) .

If requirement (iv) is replaced by the requirement that $x = 0$ be at worst a singular point of the first kind of system (2.3), then this result is in general not true. (See F. R. Gantmacher [3] and P. Masani [11].)

Requirement (iv) indicates that the construction of $P(x)$ is equivalent to the construction of monodromy matrices of system (2.1) at ∞ . Therefore the construction of $P(x)$ is extremely difficult. If we omit requirement (iv) , the construction of $P(x)$ becomes considerably simpler. In this section we shall discuss such a problem in a (non-commutative) Banach algebra over the field of complex numbers which admits a unit element, in order to treat the case when $A(x)$ depends on parameters as well. Having examined the proof of Birkhoff's theorem, we found it rather impossible to prove a result which is similar to Birkhoff's result for differential equations in a Banach algebra if requirement (iv) ought to be retained. (For Birkhoff's result, see H. L. Turrittin [17]; for differential equations in a Banach algebra, see E. Hille [4].)

In this section we shall prove the following theorem.

THEOREM 1: Let be given an equation

$$(2.4) \qquad dy/dx = x^k a(x) y$$

where x is a complex independent variable, k is a non-negative integer, and y is an unknown element in a Banach algebra $\mathfrak{A}$ with a unit element I over the field of complex numbers. We assume that

$$(2.5) \qquad a(x) = \sum_{m=0}^{\infty} a_m x^{-m}$$

where $a_m \in \mathfrak{A}$ and the series converges in norm for large $|x|$. Then, if a positive integer N is sufficiently large, there exist a convergent (in norm) power series

$$(2.6) \qquad p(x) = \sum_{m=0}^{\infty} p_m x^{-m} \qquad (p_m \in \mathfrak{A})$$

<u>and</u> $k+1$ <u>elements</u> $b_0, b_1, \ldots, b_k$ <u>of</u> $\mathfrak{A}$ <u>such that the transformation</u>

$$(2.7)\qquad y = [I + x^{-N-1} p(x)]u$$

<u>reduces</u> (2.4) <u>to</u>

$$(2.8)\qquad du/dx = x^k \left[\sum_{m=0}^{N} a_m x^{-m} + x^{-N-1} \sum_{m=0}^{k} b_m x^{-m} \right] u \quad .$$

<u>Proof</u>: Assume that $N \geqq k$. Put

$$(2.9)\qquad \begin{cases} \tilde{a}(x) = \sum_{m=0}^{N} a_m x^{-m} , \quad \hat{a}(x) = \sum_{m=0}^{\infty} a_{m+N+1} x^{-m} , \\ b(x) = \sum_{m=0}^{k} b_m x^{-m} . \end{cases}$$

Then from (2.4), (2.7) and (2.8) we derive

$$(2.10)\qquad (N+1)x^{-k-1}p(x) - x^{-k} dp(x)/dx + \tilde{a}(x)p(x) - p(x)\tilde{a}(x) + \hat{a}(x)$$
$$- b(x) + x^{-N-1}\{\hat{a}(x)p(x) - p(x)b(x)\} = 0 \quad .$$

Equation (2.10) is equivalent to

$$(2.11)\qquad b_m = a_{m+N+1} + \sum_{\ell=0}^{m} (a_\ell p_{m-\ell} - p_{m-\ell} a_\ell) , \quad m = 0,1,\ldots,k ,$$

$$(2.12)\qquad (N+1+m)p_m + \sum_{\ell=0}^{k+1+m} (a_\ell p_{k+1+m-\ell} - p_{k+1+m-\ell} a_\ell) + a_{k+1+m+N+1} = 0 , \quad m = 0,1,\ldots,N-k-1 ,$$

and

$$(2.13)\qquad (N+1+m)p_m + \sum_{\ell=0}^{k+1+m} a_\ell p_{k+1+m-\ell} - \sum_{\ell=0}^{N} p_{k+1+m-\ell} a_\ell$$
$$- \sum_{\ell=0}^{m^*} p_{m-(N-k)-\ell} b_\ell + a_{k+1+m+N+1} = 0 , \quad m \geqq N-k ,$$

where

$$(2.14)\qquad m^* = \min(k, m-(N-k)) \quad .$$

Inserting (2.11) into (2.13), we obtain equations (2.12) and (2.13) for $p_m (m \geqq 0)$. If we can solve these equations so that the power series (2.6) be convergent in norm for large $|x|$, then we define b_m by (2.11) to complete the proof of Theorem 1.

In order to solve (2.12) and (2.13), we consider the following Banach space $\mathfrak{B}$. The space $\mathfrak{B}$ consists of all infinite sequences

$$(2.15) \qquad (c_0, c_1, c_2, \dots) \quad , \quad c_m \in \mathfrak{A}$$

such that

$$\sum_{m=0}^{\infty} \rho^m |c_m| < \infty ,$$

where ρ is a fixed positive number, and $|\ |$ denotes the norm in $\mathfrak{A}$. We denote by c the sequence (2.15) and put

$$(2.16) \qquad \|c\| = \sum_{m=0}^{\infty} \rho^m \ |c_m| \quad .$$

With this norm, $\mathfrak{B}$ is a Banach space. We choose $\rho > 0$ so that the sequence

$$(2.17) \qquad a = (a_0, a_1, a_2, \dots)$$

belongs to $\mathfrak{B}$. For a given element (2.15) of $\mathfrak{B}$, define another element $\gamma(c)$ by

$$(2.18) \qquad \gamma_m(c) = \begin{cases} \sum_{\ell=0}^{k+1+m} (a_\ell \, c_{k+1+m-\ell} - c_{k+1+m-\ell} \, a_\ell) + a_{k+1+m+N+1}, & m=0,1,\dots,N-k-1 , \\ \sum_{\ell=0}^{k+1+m} a_\ell \, c_{k+1+m-\ell} - \sum_{\ell=0}^{N} c_{k+1+m-\ell} \, a_\ell & \\ \quad - \sum_{\ell=0}^{m^*} c_{m-(N-k)-\ell} \, \beta_\ell(c) + a_{k+1+m+N+1} , & m \geqq N-k , \end{cases}$$

where m^* is given by (2.14) and

$$\beta_m(c) = a_{m+N+1} + \sum_{\ell=0}^{m} (a_\ell \, c_{m-\ell} - c_{m-\ell} \, a_\ell) \ , \ m=0,\dots,k .$$

Then we can write (2.12) and (2.13) as

$$(2.19) \qquad p_m = - \frac{1}{N+1+m} \gamma_m(p) \quad , \quad m \geqq 0 \quad ,$$

where

$$p = (p_0, p_1, p_2, \dots) .$$

If we choose N so large that

(2.20) $$\frac{\rho^{-(k+1)}}{N+1} [3 \|a\| + 4 \|a\|^2 + 1] < 1 \quad ,$$

we can find a solution p of (2.19) such that

$$\|p\| \leq \sum_{m=0}^{\infty} \rho^m |a_{N+1+m}|$$

by Banach fixed point theorem. This completes the proof.

Theorem 1 is not so sharp as Birkhoff's theorem, because $x=0$ may still be an irregular singular point of equation (2.8). However, the first $N+1$ terms of (2.4) are unchanged. Therefore, Theorem 1 may be conveniently used in the study of asymptotic solutions of (2.4) at ∞. If $k \leqq -1$, we can prove a similar result with $b=0$.

Let $\mathfrak{D}$ be a domain of r parameters $\mu_1,\ldots,\mu_r$, and let $\mathfrak{A}$ be a Banach algebra consisting of all n-by-n matrices whose components are holomorphic and bounded in $\mathfrak{D}$. Then Theorem 1 yields a result concerning a system of n linear differential equations containing parameters $\mu_1,\ldots,\mu_r$.

Now we shall state without proof a slightly more general result which we shall use in the next section.

THEOREM 2: *Let be given an equation*

(2.21) $$dy/dx = x^k a(x,\epsilon) y$$

where x *is a complex independent variable,* ϵ *is a complex parameter,* k *is a non-negative integer, and* y *is an unknown element in a Banach algebra* $\mathfrak{A}$ *with a unit element* I *over the field of complex numbers. We assume that*

(2.22) $$a(x,\epsilon) = \sum_{m=0}^{\infty} a_m(\epsilon) x^{-m}$$

where $a_m(\epsilon) \in \mathfrak{A}$ *and these quantities are holomorphic and bounded with respect to* ϵ *in a sector* $\mathfrak{S}$ *defined by*

(2.23) $$0 < |\epsilon| \leqq \delta_0 \quad , \quad |\arg \epsilon| \leqq \delta_1 \quad ,$$

and the series (2.22) is convergent in norm for

$$(2.24) \qquad |x| \geqq R_0$$

uniformly for ϵ in $\mathcal{S}$. We further assume that $a(x,\epsilon)$ admits an asymptotic expansion in powers of ϵ uniformly for x in (2.24) as ϵ tends to zero in $\mathcal{S}$. Then, if a positive integer N is sufficiently large, there exist elements $p(x,\epsilon)$, $b_o(\epsilon),\ldots,b_M(\epsilon)$ of $\mathfrak{A}$ such that

(i) $p(x,\epsilon)$ is holomorphic and bounded for (2.23) and large $|x|$ and admits an asymptotic expansion in powers of ϵ uniformly for large $|x|$ as ϵ tends to zero in $\mathcal{S}$;

(ii) M is a positive integer and $b_m(\epsilon)$ are holomorphic in $\mathcal{S}$ and admit asymptotic expansions in powers of ϵ as ϵ tends to zero in $\mathcal{S}$;

(iii) the transformation

$$(2.25) \qquad y = [I + x^{-N-1}p(x,\epsilon)]u$$

reduces (2.21) to

$$(2.26) \qquad du/dx = x^k \left[\sum_{m=o}^{N} a_m(\epsilon)x^{-m} + x^{-N-1} \sum_{m=o}^{M} b_m(\epsilon)x^{-m}\right] u .$$

We could not determine M explicitly, although we can prove that $b_m(\epsilon)$ are asymptotically equal to zero as ϵ tends to zero in $\mathcal{S}$ for $m \geqq k+1$.

3. Perturbation of asymptotic solutions. Let us consider a system of form (2.1). Assume that the matrix $\lim_{x\to\infty} A(x)$ has n mutually distinct eigenvalues. Then there exists an n-by-n matrix $P(x)$ such that $P(x)$ and $P(x)^{-1}$ are convergent power series in x^{-1} and such that

$$\Lambda(x) = P(x)^{-1}A(x)P(x)$$

is diagonal. Diagonal components of $\Lambda(x)$ are eigenvalues of $A(x)$ and column vectors of $P(x)$ are eigenvectors of $A(x)$. Let us change the unknown quantity y in (2.1) by $y = P(x)z$ to obtain

$$dz/dx = [x^k\Lambda(x) - P(x)^{-1}dP(x)/dx]z \quad .$$

Assume that a direction: $\arg x = \theta$ be given in the x-plane, and put $x = te^{i\theta}$. Then $P(x)^{-1}dP(x)/dx$ is integrable as t tends to $+\infty$. Therefore, as long as the asymptotic behavior of solutions of system for z as t tends to $+\infty$ is concerned, the system

$$du/dx = x^k \Lambda(x)u$$

will give a reasonable approximation. This result can be generalized in the case when $A(x)$ depends on parameters. We shall state such a result.

Consider a system of a form

$$(3.1) \qquad dy/dx = x^k A(x,\mu,\epsilon)y$$

where k is a non-negative integer, x is a complex independent variable, μ is an r-dimensional vector whose components are parameters, ϵ is also a complex parameter, y is an n-dimensional vector, and A is an n-by-n matrix. Let $\mathcal{S}$ be a closed sector in the x-plane, and let R_o, δ_o, δ_1 and δ_2 be positive numbers. We assume that, for every non-negative integer L, the matrix A can be written in a form

$$(3.2) \qquad A(x,\mu,\epsilon) = \sum_{\ell=0}^{L} A_\ell(x,\mu)\epsilon^\ell + \epsilon^{L+1}B_L(x,\mu,\epsilon)$$

where

(I) components of matrices $A_\ell(x,\mu)$ are holomorphic for $x \in \mathcal{S}$, $|x| \geqq R_o$, $|\mu| \leqq \delta_o$, and

$$A_\ell(x,\mu) \cong \sum_{m=0}^{\infty} A_{\ell m}(\mu)x^{-m} \qquad (x \to \infty \text{ in } \mathcal{S})$$

uniformly for $|\mu| \leqq \delta_o$;

(II) components of the matrix $B_L(x,\mu,\epsilon)$ are holomorphic for $x \in \mathcal{S}$, $|x| \geqq R_o$, $|\mu| \leqq \delta_o$, $0 < |\epsilon| \leqq \delta_1$, $|\arg \epsilon| \leqq \delta_2$, and

$$B_L(x,\mu,\epsilon) \cong \sum_{m=0}^{\infty} B_{Lm}(\mu,\epsilon)x^{-m} \qquad (x \to \infty \text{ in } \mathcal{S})$$

uniformly for $|\mu| \leqq \delta_o$, $0 < |\epsilon| \leqq \delta_1$, $|\arg \epsilon| \leqq \delta_2$;

(III) components of matrices $B_{Lm}(\mu,\epsilon)$ are holomorphic for $|\mu| \leqq \delta_o$,

$0 < |\epsilon| \leqq \delta_1$, $|\arg \epsilon| \leqq \delta_2$, and

$$B_{Lm}(\mu,\epsilon) \cong \sum_{\ell=0}^{\infty} B_{Lm\ell}(\mu)\epsilon^{\ell}$$

uniformly for $|\mu| \leqq \delta_0$ as ϵ tends to zero in the sector: $0 < |\epsilon| \leqq \delta_1$, $|\arg \epsilon| \leqq \delta_2$;

(IV) components of $A_{\ell m}(\mu)$ and $B_{Lm\ell}(\mu)$ are holomorphic for $|\mu| \leqq \delta_0$.

The notation $\cong$ means that the right-hand side is the asymptotic expansion of the left-hand side in the usual sense.

THEOREM 3: Assume that the matrix $A_{oo}(0)$ has n mutually distinct eigenvalues $\lambda_1,\dots,\lambda_n$ and that, in the sector $\mathfrak{S}$, we have

$$(3.3) \qquad |\arg(\lambda_j - \lambda_h) + (k+1)\arg x| \leqq \frac{3\pi}{2}$$

for every ordered pair (j,h) such that $j \neq h$, if $\arg(\lambda_j - \lambda_h)$ are chosen in a suitable manner. Let $\mathfrak{S}_0$ be a closed sector which is contained in the interior of $\mathfrak{S}$. Then if positive constant $\tilde{R}_0$ is sufficiently large and if positive constants $\tilde{\delta}_0$ and $\tilde{\delta}_1$ are sufficiently small, there exists an n-by-n matrix $T(x,\mu,\epsilon)$ such that, for every non-negative integer L, $T(x,\mu,\epsilon)$ can be written in a form

$$(3.4) \qquad T(x,\mu,\epsilon) = \sum_{\ell=0}^{L} T_{\ell}(x,\mu)\epsilon^{\ell} + \epsilon^{L+1}V_L(x,\mu,\epsilon)$$

where

(i) $T_{\ell}(x,\mu)$ and $V_L(x,\mu,\epsilon)$ satisfy conditions (I), (II), (III) and (IV) if we replace there A_{ℓ}, $A_{\ell,m}$, B_L, B_{Lm}, $B_{Lm\ell}$, $\mathfrak{S}$, R_0, δ_0, δ_1 by T_{ℓ}, $T_{\ell m}$, V_L, V_{Lm}, $V_{Lm\ell}$, $\mathfrak{S}_0$, $\tilde{R}_0$, $\tilde{\delta}_0$, $\tilde{\delta}_1$ respectively;

(ii) the matrix $T_{oo}(0)$ is non-singular,

and such that the transformation

$$(3.5) \qquad y = T(x,\mu,\epsilon)z$$

reduces system (3.1) to

$$(3.6) \qquad dz/dx = x^k \Lambda(x,\mu,\epsilon)z$$

<u>where</u> $\Lambda(x,\mu,\epsilon)$ <u>is an n-by-n diagonal matrix whose diagonal components are eigenvalues of</u> $A(x,\mu,\epsilon)$.

We shall explain credibility of Theorem 3 without entering into details.

(1) If we assume that the matrix A does not depend on μ and ϵ , the general theory of asymptotic solutions at an irregular singular point provides Theorem 3. For such a theory, see M. Hukuhara [6], J. Malmquist [10], H. L. Turrittin [16] and W. Wasow [18].

(2) If we assume that A does not depend on ϵ , the same method as in case (1) can be used to prove Theorem 3. In this case, T is naturally independent of ϵ . The method in case (1) is based on manipulation with formal power series and successive approximations. Because of the assumption on $A_{oo}(0)$ and because of the fact that analytic property with respect to μ is not destroyed through uniform convergence of successive approximations, the same method as in case (1) can be utilized in case (2) as well. (See P. F. Hsieh and Y. Sibuya [5].)

(3) In general case, it is not difficult to find $T(x,\mu,\epsilon)$ which admits an asymptotic expansion in powers of x^{-1} such that transformation (3.5) reduces (3.1) to (3.6). It is also not difficult to find such a T which admits an asymptotic expansion in powers of ϵ . The main problem is, however, to identify the one of those T with the other so that we can find $T(x,\mu,\epsilon)$ which admits an asymptotic expansion in x^{-1} and ϵ . This means that the main problem involved here is not the existence of $T(x,\mu,\epsilon)$, but its uniqueness. Such a uniqueness problem for asymptotic solutions is in general not simple, and we must treat it carefully. We can avoid such a uniqueness problem by using Theorem 2 of the previous section, if components of $A(x,\mu,\epsilon)$ are convergent power series in x^{-1} . In such a case, by virtue of Theorem 2 of Section 2, we can assume without loss of generality that components of $A(x,\mu,\epsilon)$ are polynomials in x^{-1} . Therefore, let

$$A(x,\mu,\epsilon) = \sum_{m=o}^{N} \{A_{om}(0) + C_m(\mu,\epsilon)\}x^{-m}$$

where $C_m(\mu,\epsilon)$ are n-by-n matrices which tend to 0 as μ and ϵ tend to 0 .

Consider a system

$$(3.7) \qquad dv/dx = x^k \left[\sum_{m=o}^{N} \{A_{om}(0) + X_m\} x^{-m} \right] v$$

where X_m are n-by-n matrices whose components we regard as small parameters. We can apply Theorem 3 to system (3.7), since this is one of case (2). Thereafter by replacing X_m by $C_m(\mu, \epsilon)$, we get Theorem 3 for systems (3.1). If components of $A(x,\mu,\epsilon)$ are not convergent power series in x^{-1}, this method is not applicable. In such a general case, we must examine the uniqueness property of $T(x,\mu,\epsilon)$ in detail. For such a treatment, see Y. Sibuya [14, 15].

There is no general result which is applicable to the case where $A_{oo}(0)$ admits multiple eigenvalues and the matrix A does depend on parameters. In order to illustrate some aspects in such a case, let us consider a single second order equation of a form

$$(3.8) \qquad d^2\eta/dx^2 - [P(x) + \epsilon\, Q(x)]\eta = 0$$

where $P(x)$ and $Q(x)$ are polynomials in x. Let us denote by p and q the degrees of $P(x)$ and $Q(x)$ in x respectively, and put

$$\lim_{x\to\infty} x^{-p} P(x) = P_o , \quad \lim_{x\to\infty} x^{-q} Q(x) = Q_o ,$$

and

$$k = \max\{p,q\} .$$

If we put

$$y = \begin{bmatrix} \eta \\ d\eta/dx \end{bmatrix}, \quad F(x,\epsilon) = \begin{bmatrix} 0 & 1 \\ P(x) + \epsilon\, Q(x) & 0 \end{bmatrix},$$

equation (3.8) is equivalent to

$$(3.9) \qquad dy/dx = F(x,\epsilon) y .$$

Let us change x and y by

$$x = \xi^2 \quad , \quad y = \begin{bmatrix} 1 & 0 \\ 0 & \xi^k \end{bmatrix} u \ .$$

Then system (3.9) is reduced to

$$(3.10) \qquad du/d\xi = \xi^{k+1} A(\xi,\varepsilon)u \quad ,$$

where

$$(3.11) \qquad A(\xi,\varepsilon) = 2 \begin{bmatrix} 0 & 1 \\ (P(x) + \varepsilon\, Q(x))x^{-k} & -\frac{k}{2}\,\xi^{-k-2} \end{bmatrix} \ .$$

Notice that we have

$$(3.12) \qquad A(\xi,\varepsilon) = 2 \begin{bmatrix} 0 & 1 \\ \delta(\varepsilon) & 0 \end{bmatrix} + O(x^{-1})$$

where

$$(3.13) \qquad \delta(\varepsilon) = \begin{cases} P_0 + O(\varepsilon) & \text{if } p \geqq q \ , \\ \varepsilon\, Q_0 & \text{if } p < q \ . \end{cases}$$

This shows that the leading coefficient matrix

$$(3.14) \qquad 2 \begin{bmatrix} 0 & 1 \\ \delta(0) & 0 \end{bmatrix}$$

has two distinct eigenvalues $\pm\, 2\sqrt{P_0}$ if $p \geqq q$, but it has a multiple eigenvalue 0 if $p < q$.

Assume that $p < q$. If $\varepsilon \neq 0$, the matrix

$$(3.15) \qquad \begin{bmatrix} 0 & 1 \\ \delta(\varepsilon) & 0 \end{bmatrix}$$

has two distinct eigenvalues $\pm\, \sqrt{\delta(\varepsilon)}$ which reduce to a multiple eigenvalue 0 at $\varepsilon = 0$. Two eigenvalues of the matrix $A(\xi,\varepsilon)$ are

$$(3.16) \qquad \lambda_{\pm}(\xi,\epsilon) = -\frac{k}{2}\xi^{-k-2} \pm \sqrt{\frac{k^2}{4}x^{-k-2} + 4x^{-k}(P(x)+\epsilon Q(x))} \ .$$

According to the general theory of asymptotic solutions at an irregular singular point, there exists a transformation

$$(3.17) \qquad u = T(\xi,\epsilon)v$$

which takes system (3.10) to

$$(3.18) \qquad dv/d\xi = \xi^{k+1}\Lambda(\xi,\epsilon)v$$

where $\Lambda(\xi,\epsilon)$ is a 2-by-2 diagonal matrix whose diagonal components are $\lambda_{\pm}(\xi,\epsilon)$. Components of $T(\xi,\epsilon)$ admit asymptotic expansions in powers of ξ^{-1} . Put

$$(3.19) \qquad T(\xi,\epsilon) \cong \sum_{m=0}^{\infty} T_m(\epsilon)\xi^{-m} \ .$$

If we expand $\lambda_{\pm}(\xi,\epsilon)$ in powers of ξ^{-1} , then coefficients of such expansions are polynomials in $\epsilon^{-\frac{1}{2}}$. Since components of the matrix $A(\xi,\epsilon)$ do not have such singularities at all, the quantities $T_m(\epsilon)$ ought to have singularities in ϵ . This means that the asymptotic expansion of $T(\xi,\epsilon)$ in powers of ξ^{-1} is not smooth at $\epsilon = 0$. Therefore the general theory of asymptotic solutions at an irregular singular point is not satisfactory if we want to apply it to the present problem. We can also show that, if we try to expand $T(\xi,\epsilon)$ in powers of ϵ , coefficients will have singularities at $\xi = \infty$.

It has been known for long time that equation (3.8) has two linearly independent solutions of the form

$$(3.20) \qquad \eta = [P(x)+\epsilon Q(x)]^{-\frac{1}{4}} \exp\left\{\pm\int^x [P(t)+\epsilon Q(t)]^{\frac{1}{2}} dt\right\}[1+o(1)] \ .$$

In order to assure the validity of formula (3.20), certain requirements must be satisfied. For example, we must avoid zeros of $P(x)+\epsilon Q(x)$. In various cases, a careful investigation of formula (3.20) leads to results more satisfactory than that which we derive from a straightforward applicaton of the general theory of asymptotic solutions at an irregular singular point. Such an investigation tends to become a study of equation (3.8) in the large.

In seeking asymptotic representations of the largest zeros of the Hermite polynomials, F. Zernike [19] was led to the differential equation

$$(3.21) \qquad d^2\eta/dx^2 - (x + \epsilon x^2)\eta = 0$$

for which he sought solutions satisfying

$$(3.22) \qquad \eta(x,\epsilon) \to 0 \quad \text{as} \quad x \to \infty \ .$$

He studied this problem for x and ϵ real and non-negative. Notice that equation (3.21) is a special case of (3.8) with $p=1$ and $q=2$, and hence a case $p<q$. Zernike's scheme was to reduce (3.21) to

$$(3.23) \qquad d^2w/dx^2 - xw = 0$$

by a formal transformation

$$(3.24) \qquad \eta = p(x,\epsilon)w + \epsilon\, q(x,\epsilon)dw/dx \ ,$$

where $p(x,\epsilon)$ and $q(x,\epsilon)$ are formal power series in ϵ whose coefficients are polynomials in x. Polynomials in x have singularities at $x=\infty$. Hence transformation (3.24) will not give any satisfactory result in an immediate neighborhood of $x=\infty$. However, Zernike's method can be used in a domain where formula (3.20) is not valid. F. E. Mullin [12] generalized Zernike's method and he investigated a differential equation of a form

$$(3.25) \qquad x^2d^2\eta/dx^2 - [P(x) + \epsilon\, Q(x,\epsilon)]\eta = 0 \ ,$$

where $P(x)$ and $Q(x,\epsilon)$ are polynomials in x such that deg $P <$ deg Q. Mullin reduced equation (3.25) to

$$(3.26) \qquad x^2d^2w/dx^2 - [P(x) + \epsilon\, S(x,\epsilon)]w = 0$$

by a transformation of form (3.24), where $S(x,\epsilon)$ is a polynomial in x such that deg $P \geq$ deg S. In Mullin's result, the quantities $p(x,\epsilon)$ and $q(x,\epsilon)$ are, as in Zernike's result, formal power series in ϵ whose coefficients are polynomials in x. For equation (3.26) $x=0$ is at worst a regular singular point. According to Mullin's remark, if $x=0$ is an irregular singular point, Zernike's method can not be generalized. In particular, in such a case, the quantities $p(x,\epsilon)$ and $q(x,\epsilon)$ ought to contain negative powers of x. So far, we have discussed only formal parts of Zernike's and Mullin's results. Their results, however, also contain analytic justification of formal results which we shall not discuss in this report.

From formula (3.20), we can infer that asymptotic behavior of solutions of equation (3.8) as x tends to infinity changes radically as ϵ tends to zero if $p<q$, whereas this is not true if $p \geq q$. This is a sharp difference between these two cases. D. A. Lutz [8,9] has investigated the general situation under which a pertubation does not change asymptotic behavior of solutions at an irregular singular point. Problems which we discussed in this section were mainly concerned with smoothness of asymptotic representations of solutions at an irregular singular points. Frankly speaking, we do not know how our problem is related with Lutz's problem.

Acknowledgment: The research related to this report was supported in part by National Science Foundation under Grant GP-7041X. This report was prepared for the most part at the University of Edinburgh, Scotland. The author's research in Edinburgh was partly supported by the Science Research Council of Great Britain.

BIBLIOGRAPHY

1. G. D. Birkhoff, Singular points of ordinary linear differential equations, Trans. Amer. Math. Soc., 10(1909) 436-470;
2. G. D. Birkhoff, Equivalent singular points of ordinary linear differential equations, Math. Ann., 74 (1913) 134-139;
3. F. R. Gantmacher, The theory of matrices, II, Chelsea Publ. Co., 1959;
4. E. Hiller, Lectures on ordinary differential equations, Addison-Wesley, 1969;

5. P. F. Hsieh and Y. Sibuya, Note on regular perturbations of linear ordinary differential equations at irregular singular points, Funk. Ekva., 8 (1966) 99-108;

6. M. Hukuhara, Sur les points singuliers des équations différentielles linéaries, II and III, J. Fac. Sci., Hokkaido Univ., 5 (1937) 157-166 and Mem. Fac. Sci., Kyushu Univ., 2 (1942) 125-137;

7. M. Hukuhara, Sur les équations différentielles linéaires à coefficients périodiques et contenant un paramètre, J. Fac. Sci., Univ. Tokyo, Sec. I, 7 (1954) 69-85;

8. D. A. Lutz, Linear perturbations of irregular singular systems, Proc. U.S.-Japan Seminar on Diff. and Func. Equations, Benjamin, 1967, 555-558.

9. D. A. Lutz, Perturbations of matrix differential equations in the neighborhood of a singular point, Funk. Ekva., 13 (1970) 97-107;

10. J. Malmquist, Sur l'étude analytique des solutions d'un système d'équation différentielles dans le voisinage d'un point singulier d'indétermination, I, II and III, Acta Math., 73 (1940) 87-129, 74 (1941) 1-64 and 74 (1941) 109-128;

11. P. Masani, On a result of G. D. Birkhoff on linear differential systems, Proc. Amer. Math. Soc., 10 (1959) 696-698;

12. F. E. Mullin, On the regular perturbation of the subdominant solution to second order linear ordinary differential equations with polynomial coefficients, Funk. Ekva., 11 (1968) 1-38;

13. T. Saito, On a singular point of a second order linear differential equation containing a parameter, Funk. Ekva., 5 (1963) 1-29;

14. Y. Sibuya, Simplification of a system of linear ordinary differential equations about a singular point, Funk Ekva., 4 (1962) 29-56;

15. Y. Sibuya, Perturbation of linear ordinary differential equations at irregular singular points, Funk. Ekva., II (1968) 235-246;

16. H. L. Turrittin, Convergent solutions of ordinary linear homogeneous differential equations in the neighborhood of an irregular singular point, Acta Math. 93 (1955) 27-66;

17. H. L. Turrittin, Reduction of ordinary differential equations to the Birkhoff canonical form, Trans. Amer. Math. Soc., 107 (1963) 485-507;

18. W. Wasow, Asymptotic expansions for ordinary differential equations, Interscience, 1965;

19. F. Zernike, Eine asymptotische Entwicklung für die grösste Nullstelle der Hermiteschen Polynome, Koninklijke Akademie van Wetenschappen te Amsterdam, Proc. of the Section of Sci., 34 (1931) 673-680.

ON CONTINGENT EQUATIONS

Norio KIKUCHI

The purpose of this lecture is to exhibit some results about contingent equations and to show some of their applications to optimal control problems. We shall begin with explaining notations and definitions concerning contingent equations. Let I be a compact interval $[t_0,T]$. By X we denote the collection of all nonempty subsets of n-euclidean space R^n. A map $F(t,x)$ of $I \times R^n$ into X will be called an **orientor field.** A relation of the form

$$\dot{x}(t) \in F(t,x(t))$$

will be called a **contingent equation.** A function $x(t)$ will be called a **solution** of an orientor field $F(t,x)$ if $x(t)$ is absolutely continuous on I and satisfies the contingent equation stated above almost everywhere on I. While ordinary differential equations correspond to vector fields, contingent equations correspond to set-valued fields (that is, orientor fields) approximately known up to a given accuracy. Hence in this case we have to deal with the more general theory than that of differential inequalities. The theory of a contingent equation was originated by M. Hukuhara [9] and developed by A. Marchaud [19], [20] and S. K. Zaremba [33]. On the other hand the notion of contingent equations arises naturally in control theory, which is noticed by T. Ważewski [25]. That is : assume that there is given a differential equation containing control parameters

$$\dot{x}(t) = f(t,x(t),u(t)),$$

$$u(t) \in U,$$

where $f(t,x,u)$ is a mapping of $I \times R^n \times R^r$ into R^n and $u(t)$ is a measurable function defined on I with values in R^r and U is a set in R^r. A function $x(t)$ which satisfies the relation stated above also satisfies the following contingent equation

$$\dot{x}(t) \in F(t,x(t)),$$

where $F(t,x)$ is a set-valued function defined by

$$F(t,x) \equiv f(t,x,U).$$

By using a suitable implicit function theorem we can consider the converse problem. That is : for a function $x(t)$ which satisfies the contingent equation

$$\dot{x}(t) \in f(t,x(t),U),$$

we can select a measurable function $u(t)$ such that

$$\dot{x}(t) = f(t,x(t),u(t)),$$
$$u(t) \in U.$$

In the sense stated above the theory of contingent equations seems to be identical with that of differential equations with control parameters. However, it may happen that

$$f(t,x,U) = g(t,x,V)$$

holds, even if the pairs $\{f,U\}$ and $\{g,V\}$ are not necessarily identical. Hence we can say that the theory of contingent equations treats systematically differential equations with control parameters. Furthermore, contingent equations include differential equations containing more than two control parameters which have different characteristics each other as in the differential equations used in differential games.

In this lecture we first treat the existence problem of solutions for an orientor field and examine the properties of the family of those solutions.

Next, we generalize a notion of solutions for the orientor field for which we can not necessarily find a solution and we investigate the relation among those solutions.

Finally, we apply the theory of contingent equations to optimal control problems and derive Pontryagin's Maximum Principle and the bang - bang property.

§1 In this chapter we shall give some fundamental properties of set-valued functions [1], [10], [11], [16], [17].

Let $x \cdot y$ denote the usual scalar product and let $\text{dist}(x,y)$ denote the euclidean distance between two points $x,y \in R^m$. For a set $A \subset R^m$ and a point $x \in R^m$ we put

$$x \cdot A = \sup\{x \cdot a \ ; \ a \in A\}.$$

For two sets $A,B \subset R^m$ we define the Hausdorff distance between them as follows

$$\text{Dist}(A,B) = \inf\{\varepsilon>0; \ V(A,\varepsilon) \supset B, \ V(B,\varepsilon) \supset A\},$$

where $V(A,\varepsilon)$ means a closed ε-neighborhood of A defined by

$$V(A,\varepsilon) = \{x \in R^m; \ \text{dist}(x,A) \leqq \varepsilon\}.$$

We put for a set $A \subset R^m$

$$|A| = \text{Dist}(A,0),$$

where 0 is the origin of R^m. $\text{co}A$ denotes the convex closure of A, i.e., the smallest closed and convex set which contains A. $\text{bdry}A$ denotes the boundary of A. The collection of all nonempty compact subsets of R^m with topology induced by the Hausdorff distance is a complete metric space, which we shall denote by

$Comp(R^m)$. $Conv(R^m)$ denotes the subset of $Comp(R^m)$ whose elements are convex sets.

Let E be a measurable set in R^ℓ. a.e. E means 'almost everywhere on E'. In this chapter we assume that $F(t)$ is a function defined on E with values in $Comp(R^m)$.

Definition 1. $F(t)$ is said to be **upper(resp. lower) semi-continuous at** $t_0 \in E$ if for every positive ε we can find some neighborhood V of t_0 such that

$$V(F(t_0),\varepsilon) \supset F(t)$$

$$(\text{resp. } V(F(t),\varepsilon) \supset F(t_0))$$

holds for every $t \in V$. When $F(t)$ is upper (resp. lower) semi-continuous at every point of E, then $F(t)$ is said to be **upper(resp. lower) semi-continuous on E**. $F(t)$ is said to be **continuous at** t_0 **(resp. on E)** when $F(t)$ is upper and lower semi-continuous at t_0 (resp. on T).

We put

$$\liminf_{t\to a} F(t) = \{x\in R^m ; \lim_{t\to a} \operatorname{dist}(x,F(t)) = 0\},$$

$$\limsup_{t\to a} F(t) = \{x\in R^m ; \varliminf_{t\to a} \operatorname{dist}(x,F(t)) = 0\}.$$

When

$$\liminf_{t\to a} F(t) = \limsup_{t\to a} F(t),$$

then we write it as follows

$$\lim_{t\to a} F(t).$$

Definition 2. $F(t)$ is said to be **measurable** on E if the set

$$\{t\in E ; F(t) \cap C \neq \phi\}$$

is measurable for every $C \in Comp(R^m)$.

Lemma 1. Let $F(t)$ be measurable on E. Then there exists a function $f(t)$ measurable on E such that

$$f(t) \in F(t) \quad \text{for} \quad t \in E.$$

Let $F(t)$ be measurable on E. If there exists a function $m(t)$ integrable on E such that

$$|F(t)| \leqq m(t) \qquad t \in E,$$

we say that $F(t)$ is **integrable** on E.

Let $F(t)$ be an integrable function with values in $Conv(R^m)$. Then the integral

$$\int_E F(t)dt$$

of the Lebesgue type has been investigated by Hukuhara [11].

On the other hand the integral of set-valued functions which are not necessarily convex-valued has been investigated by Aumann [1]. The definition of integral is as follows. Let M be the set

$$M = \{f(t) \varepsilon R^m \ ; \ f(t) \text{ integrable on } E, \ f(t) \varepsilon F(t) \text{ a.e. } E\}.$$

Define an integral of $F(t)$ by

$$\int_E F(t)dt = \{\int_E f(t)dt; \ f(t) \varepsilon M\},$$

i.e., the set of all Lebesgue integrals of members of M.

For an integrable function $F(t) \varepsilon \text{Conv}(R^m)$, this integral can be verified to be identical with the one of Lebesgue type mentioned earlier.

Let $\{A_k\}$ be a sequence of subsets of R^m. We put

$$\limsup_{k\to\infty} A_k = \{x \varepsilon R^m; \ \underline{\lim}_{k\to\infty} \text{dist}(x, A_k) = 0\}.$$

<u>Lemma 2.</u> Let $\{F_k(t)\} \subset \text{Comp}(R^m)$ $(k = 1,2,\ldots)$ be a sequence of functions integrable on E and suppose that there exists an integrable function $F_0(t) \varepsilon \text{Comp}(R^m)$ such that

$$F_k(t) \subset F_0(t) \quad \text{a.e. } E$$

for every k, then $\limsup_{k\to\infty} F_k(t)$ is integrable on E and there holds the relation

$$\limsup_{k\to\infty} \int_E F_k(t)dt \subset \int_E \limsup_{k\to\infty} F_k(t)dt.$$

<u>Remark.</u> In case when $\{F_k(t)\}$ is a sequence of functions with values in $\text{Conv}(R^m)$, we give in [13] the similar result as Lemma 2, but the proof is incomplete. Lemma 2 is due to Aumann [1]. Since both integrals stated above are identical for integrable functions with values in $\text{Conv}(R^m)$ [17], Lemma 2 also holds for the integrals of the Lebesgue type investigated by Hukuhara [11].

<u>**Lemma 3.**</u> Let $F(t)$ be integrable on E. Then there holds the relation

$$\int_E F(t)dt = \int_E \mathrm{co}F(t)dt.$$

The following definitions and property of the derivative of set-valued functions have been given by Hukuhara [11].

Let $A,B,C \in \mathrm{Conv}(R^m)$ such that

$$A = B + C,$$

where

$$B + C = \{b+c;\ b\in B,\ c\in C\}.$$

Since such C is uniquely determined, we define $A - B$ by

$$A - B = C.$$

<u>**Definition 3.**</u> Let $F(t)$ be a function defined on an interval I with values in $\mathrm{Conv}(R^m)$. If there exists a $K \in \mathrm{Conv}(R^m)$ such that

$$\lim_{h\to 0} (F(t_0+h) - F(t_0))/h = K,$$

we say that $F(t)$ is **differentiable at** $t_0 \in I$ and we denote K a **derivative** of $F(t)$ at t_0 and write it as follows

$$\frac{d}{dt}F(t_0).$$

<u>**Lemma 4.**</u> Let $F(t) \in \mathrm{Conv}(R^m)$ be a function integrable on I. Then there holds

$$\frac{d}{dt}\int_{t_0}^{t} F(t)dt = F(t) \qquad \text{a.e. } I.$$

Let $F(t,x) \in \mathrm{Comp}(R^n)$ be an orientor field defined on $I\times R^n$ and suppose that there exists a function $m(t)$ integrable on I such that

$$|F(t,x)| \leqq m(t) \qquad (t,x) \in I\times R^n.$$

We shall assume the following assumptions on $F(t,x)$ as needed.

Hypothesis H(F). $F(t,x)$ is measurable in t for each fixed $x \in R^n$ and is upper semi-continuous in x for each fixed $t \in I$.

Hypothesis H_1(F). $F(t,x)$ is measurable in t for each fixed $x \in R^n$ and is continuous in x for each fixed $t \in I$.

§2. In this chapter we shall assume that an orientor field is convex-valued and give some fundamental properties of solutions of the contingent equation [2], [4], [13], [14], [15], [22], [32].

Theorem 1. Let $F(t,x)$ satisfy the hypothesis $H(F)$. Then there exists a function $x(t)$ absolutely continuous on I which satisfies the contingent equation

$$\frac{dx(t)}{dt} \in F(t, x(t)) \quad \text{a.e. } I.$$

Proof. Let D be a subdivision of I

$$t_0 < t_1 < \dots < t_p = T.$$

We put

$$\delta(D) = \max\{t_{i+1} - t_i \ ; \ 0 \leqq i \leqq p-1\}.$$

Let x_0 be a point of R^m. Since $F(t,x_0)$ is measurable in t, we can select a measurable function $f_0(t)$ such that

$$f_0(t) \in F(t,x_0) \quad t \in I.$$

For $t \in [t_0, t_1]$ we define

$$x(t;D) = x_0 + \int_{t_0}^{t} f_0(t)dt$$

and put

$$x_1 = x(t_1;D).$$

We define inductively $\{x_k\}$ and $\{f_k(t)\}$ $(k = 0,1,2,\dots,p)$ as follows. Suppose that x_k is defined and then for $t \in [t_k,t_{k+1}]$ we define

$$x(t;D) = x_k + \int_{t_k}^{t} f_k(t)dt$$

and put

$$x_{k+1} = x(t_{k+1};D),$$

where $f_k(t)$ is a measurable function such that

$$f_k(t) \varepsilon F(t,x_k) \quad t \varepsilon I.$$

A Cauchy polygon (more precisely, Cauchy curve) $x(t;D)$ has thus been defined on I. By defining $y(t;D)$ as follows

$$y(t;D) = \begin{cases} x_k & t \varepsilon [t_k,t_{k+1}) \ (0 \leqq k \leqq n-2), \\ x_{n-1} & t \varepsilon [t_{n-1},t_n], \end{cases}$$

we have the relation

$$x(t;D) \varepsilon x(\tau;D) + \int_\tau^t F(t,y(t;D))dt \quad t,\tau \varepsilon I.$$

Let $\{D_k\}$ $(k = 1,2,...)$ be a sequence of subdivisions of I such that

$$\lim_{k\to\infty} \delta(D_k) = 0.$$

Since there holds for every k

$$\left|\frac{dx(t;D_k)}{dt}\right| \leqq m(t) \quad \text{a.e. } I,$$

$\{x(t;D_k)\}$ is a normal family on I. Hence we can assume without loss of generality that $\{x(t;D_k)\}$ and $\{y(t;D_k)\}$ converges uniformly to a function $x(t)$ continuous on I. By using Lemma 3 we have

$$x(t) \varepsilon x(\tau) + \int_\tau^t F(t,x(t))dt \quad t,\tau \varepsilon I.$$

Since $x(t)$ can be verified to be absolutely continuous on I, we have by Lemma 4 that

$$\frac{dx(t)}{dt} \varepsilon F(t,x(t)) \quad \text{a.e. } I.$$

Hence we can construct a solution of the contingent equation.

Let $C(I)$ be the Banach space of all functions $x(t) \varepsilon R^n$ continuous on I with the sup-norm.
Let K be a compact set in R^n. We denote by subspace $T(K,F)$ of $C(I)$ the collection of all functions $x(t)$ absolutely continuous on I such that

$$\frac{dx(t)}{dt} \varepsilon F(t,x(t)) \quad \text{a.e. } I,$$

$x(t_0) \in K$.

<u>Theorem 2.</u> Let $F(t,x)$ satisfy the hypothesis $H(F)$. Then $T(K,F)$ is a compact set in $C(I)$.

We denote the funnel $Z(K,F) \subset R^{n+1}$ the union of the graphs of functions belonging to $T(K,F)$, i.e.,

$$Z(K,F) = \{(t,x(t));x(t) \in T(K,F),t \in I\}.$$

<u>Theorem 3.</u> Let $F(t,x)$ satisfy the hypothesis $H_1(F)$. Then every point $(t_1,x_1) \in \text{bdry}Z(K,F)$ can be peripherally attainable from bdryK, i.e., there exists a function $x(t) \in T(K,F)$ such that

$$(t,x(t)) \in \text{bdry}Z(K,F) \quad t \in [t_0,t_1],$$

$$x(t_1) = x_1$$

and such $x(t)$ satisfies the following relation

$$\frac{dx(t)}{dt} \in \text{bdry}F(t,x(t)) \quad \text{a.e. } [t_0,t_1].$$

§3. In the previous chapter we considered the contingent equation which corresponds to a convex-valued orientor field. In control problems an orientor field $F(t,x)$ defined by

$$F(t,x) \equiv f(t,x,U)$$

cannot be necessarily convex-valued, even if U is convex. Hence in this chapter we shall treat the case when orientor field $F(t,x)$ is not necessarily convex-valued [5], [17], [29], [30], [31], [32]. The contingent equation corresponding to a non-convex orientor field has been treated by T. Ważewski [32], who considers the problem by extending the notion of solutions. A. F. Filippov [5] considers the contingent equation which corresponds to a Lipschitz-continuous orientor field.

In this chapter we shall assume that orientor field $F(t,x)$ satisfies the hypothesis $H(F)$ stated in Chapter 1.

<u>Definition 3 (Ważewski).</u> A function $x(t)$ will be called a **quasi-trajectory** of $F(t,x)$ if there exists a sequence $\{x_k(t)\}$ of functions absolutely continuous on I such that

$$|\dot{x}_k(t)| \leq m(t) \quad \text{a.e. } I,$$

$$\lim_{k\to\infty} x_k(t) = x(t) \quad t \in I,$$

$$\lim_{k\to\infty} \text{dist}\,(\dot{x}_k(t),F(t,x_k(t))) = 0 \quad \text{a.e. } I.$$

We denote by $WG(F)$ the collection of all quasi-trajectories of $F(t,x)$.

<u>**Theorem 4.**</u> There exists a function $x(t)$ absolutely continuous on I which satisfies the relation

$$x(t) \in x(t_0) + \int_{t_0}^{t} F(t,x(t))dt \quad t \in I.$$

<u>**Definition 4.**</u> A function which satisfies the relation stated in Theorem 4 will be called a **generalized solution** of $F(t,x)$. We denote by $G(W)$ the set of all generalized solutions of $F(t,x)$.

<u>**Theorem 5.**</u> Let $F(t,x)$ satisfy the hypothesis $H_1(F)$. Then

$$T(coF) = G(F) = WG(F),$$

holds, where $T(coF)$ denotes the set of all solutions of $coF(t,x)$.

§4 Having in the previous chapters developed the general theory of contingent equations, we investigate a time optimal process in this chapter [13], [17].

Let $F(t,x,u)$ be a mapping of $I\times R^n\times R^r$ into $Comp(R^n)$ and U be a compact set in R^r. By a control system $S(F,U)$ we mean orientor fields $F(t,x,u)$ and a set U.

In this chapter we assume that the control system $S(F,U)$ satisfies the following hypothesis.

<u>**Hypothesis H(F,U).**</u> $F(t,x,u)$ is measurable in t for each fixed $(x,u) \in R^n\times R^r$ and is continuous in (x,u) for each fixed $t \in I$. There exists a function $m(t)$ integrable on I such that

$$|F(t,x,u)| \leqq m(t) \quad (t,x,u) \in I\times R^n\times U.$$

$F(t,x,U)$ is a function defined on $I\times R^n$ with values in $Comp(R^n)$. $F(t,x,U)$ is measurable in t for each fixed x and continuous in x for each fixed t. Hence there exists a function $x(t)$ absolutely continuous on I which satisfies the relation

$$x(t) \in x(t_0) + \int_{t_0}^{t} F(t,x(t),U)dt \quad t \in I.$$

We say that such $x(t)$ is a **generalized solution of** $S(F,U)$.

Let $F_U(t,x)$ be an orientor field defined by

$$F_U(t,x) \equiv F(t,x,U),$$

which we will call an orientor field associated with the system $S(F,U)$.

We notice that a generalized solution of $S(F,U)$ is equivalent to that of $F_U(t,x)$.

Let $K \in \mathrm{Comp}(R^n)$ and $K(t) \in \mathrm{Comp}(R^n)$ be a function upper semi-continuous on I. A generalized solution $x(t)$ of $S(F,U)$ such that for some $\bar{t} \in I$

$$x(t_0) \in K, \quad x(\bar{t}) \in K(\bar{t})$$

will be called a generalized solution **attainable from K to K(t).**

We shall consider the problem that we seek a generalized solution $x(t)$ attainable from K to $K(t)$ and a time T such that

$$x(T) \in K(T)$$

and

$$y(t) \notin K(t)$$

for any $t \in [t_0,T)$ and any generalized solution $y(t)$ of $S(F,U)$. Such $x(t)$ will be called an **optimal generalized solution** and T will be called an **optimal time.**

We denote by $Z(K,F,U)$ the funnel of generalized solutions $x(t)$ of $S(F,U)$ such that

$$x(t_0) \in K.$$

Theorem 6. Suppose that there exists at least one generalized solution attainable from K to $K(t)$. Then there exists at least an optimal generalized solution $x(t)$ with optimal time t_1, which satisfies the following relation

$$(1) \quad (t,x(t)) \in \mathrm{bdry}Z(K,F,U) \quad t \in [t_0,t_1],$$

and there exists a vector $\eta(t) \in R^n$ for almost every $t \in [t_0,t_1]$ such that

$$(2) \quad \eta(t)\cdot\frac{dx(t)}{dt} = \sup\{\eta(t)\cdot F(t,x(t),u); u\in U\} \quad \text{a.e. } [t_0,t_1].$$

If we assume that $F(t,x,U)$ is convex-valued, there exists a function $u(t)$ measurable on I such that

$$(3) \quad \eta(t)\cdot\frac{dx(t)}{dt} = \eta(t)\cdot F(t,x(t),u(t))$$
$$= \sup\{\eta(t)\cdot F(t,x(t),u); u\in U\} \quad \text{a.e. } [t_0,t_1],$$

$$u(t) \in U.$$

Furthermore, if we assume that $F(t,x,u)$ is an open mapping of u for each fixed $(t,x) \in I\times R^n$, the function $u(t)$ stated above is found to have the property

$$(4) \quad u(t) \in \mathrm{bdry}U \quad \text{a.e. } [t_0,t_1].$$

<u>Proof.</u> Since the relation

$$T(K,coF_U) = G(K,F_U)$$

holds, the existence of an optimal generalized solution follows from the fact that $T(K,coF_U)$ is compact in $C(I)$ and $K(t)$ is upper semi-continuous on I. For an optimal generalized solution $x(t)$ with optimal time t_1 the relation

$$(t_1,x(t_1)) \in \text{bdry}Z(K,coF_U)$$

can be verified. Since the funnel $Z(K,F,U)$ of generalized solutions of $S(F,U)$ is identical with $Z(K,coF_U)$, the result (1) follows from Theorem 3, and hence we have the relation

$$\frac{dx(t)}{dt} \in \text{bdry}coF(t,x(t),U) \quad \text{a.e. } [t_0,t_1].$$

Hence we can select a vector $\eta(t)$ for almost every $t \in [t_0,t_1]$ such that the result (2) holds. If $F(t,x,U)$ is convex-valued, the optimal generalized solution satisfies by Theorem 3

$$\dot{x}(t) \in \text{bdry}F(t,x(t),U) \quad \text{a.e. } [t_0,t_1]$$

and hence the set

$$U(t) = \{u \in U;\ \dot{x}(t) \in \text{bdry}F(t,x(t),u)\}$$

is not empty for almost every $t \in [t_0,t_1]$. Since U is compact and $F(t,x,u)$ is continuous in u, the set $U(t)$ can be verified to be in $\text{Comp}(R^r)$. By using Scorza-Dragoni type property for set-valued functions [15], [24], $U(t)$ can be verified to be measurable on $[t_0,t_1]$. Hence we can select a function $u(t)$ measurable on $[t_0,t_1]$ such that

$$u(t) \in U(t) \quad \text{a.e. } [t_0,t_1]$$

and hence the result (3) is obtained. The result (4) follows from (3) and the openness of $F(t,x,u)$.

The result (3) is a generalization to contingent equations of **Pontryagin's Maximum Principle** for differential equations and the result (4) means the **bang-bang** property which has been known for some linear differential equations. The result (1) is the well known property as **Hukuhara's phenomenon** and the result (2) is a generalized form of Pontryagin's Maximum Principle

In concluding this lecture I would like to make the following corrections to my paper [15].

Page 365, line 17 replace "/2n" by "$/2^n$".

Page 367, line 20 delete "Let L be an integral of $k(t)$ on I.".

Page 368, line 21 replace "the Lipschitz constant L" by "$\dot{x}(t)$ such that $|\dot{x}(t)| \leq k(t)$".

Page 369, line 11 replace "the same Lipschitz constant L" by "derivatives whose absolute values do not exceed k(t)".
Page 369, line 19 replace "the Lipschitz constant L" by "$\dot{x}^*(t)$, $|\dot{x}^*(t)| \leqq k(t)$".

References

[1] Aumann, R. J., Integrals of set values functions, J. Math. Anal. and Appl. 12 (1965), 1-12.

[2] Castaing, Ch., Sur les équations différentielles multivoques, C. R. Acad. Sc. Paris, t.263 (1966), 63-66.

[3] Filippov, A. F., On certain questions in the theory of optimal control, J. SIAM Control, 1 (1962), 76-84.

[4] ______________, Differential equations with many-valued discontinuous right-hand side, Soviet Math. Dokl., 4 (1963), 941-945.

[5] ______________, Classical solutions of differential equations with multivalued right-hand side, SIAM J. Control, 5 (1967), 609-621.

[6] Hermes, H., A note on the range of a vector measure; applications to the theory of optimal control, J. Math. Anal. and Appl., 8 (1964), 78-83.

[7] _________, Calculus of set valued functions and control, J. Math. and Mech., 18 (1968), 47-59.

[8] _________, The generalized differential equation $\dot{x} \varepsilon R(t,x)$, Advances in Mathematics, 4 (1970), 149-169.

[9] Hukuhara, M., Sur les systèmes d'équations différentielles ordinaires, Jap. Journ. of Math. 6 (1930), 269-299.

[10] ___________, Sur l'application semi-continue dont la valeur est un compact convexe, RIMS-11, Res. Inst. Math. Sci., Kyoto Univ., (1966).

[11] ___________, Intégration des applications mesurables dont la valeur est un compact convexe, RIMS-15, Res. Inst. Math. Sci., Kyoto Univ., (1966).

[12] Kikuchi, N., Existence of optimal controls, RIMS-20, Res. Inst. Math. Sci., Kyoto Univ., (1967).

[13] __________, Control problems of contingent equation, Publ. RIMS, Kyoto Univ. Ser. A, 3 (1967), 85-99.

[14] __________, On some fundamental theorems of contingent equations in connection with the control problems, Publ. RIMS, Kyoto Univ. Ser. A, 3 (1967), 177-201.

[15] __________, On contingent equations satisfying the Carathéodory type conditions, Publ. RIMS, Kyoto Univ. Ser. A, 3 (1967), 361-371.

[16] __________, On contingent equations and control problems, Funkcial. Ekvac. (Japanese) I, II, 21 (1969), 3-43; 22(1969), 1-24.

[17] __________, On control problems for functional-differential equations, Funkcial. Ekvac. 14 (1971), 1-23.

[18] LaSalle, J. P., The time optimal control problem, Contr. Theory Nonlinear Oscillations (Princeton), 5 (1961), 1-24.

[19] Marchaud, A., Sur les champs des demi-droites et les équations différentielles du premier ordre, Bull. Soc. Math. France 63 (1934), 1-38.

[20] ___________, Sur les champs continus de demi-cônes convexes et leurs intégrales, Compositio Math. 3 (1936), 89-127.

[21] Pliś, A., Remarks on measurable set-valued functions, Bull. Acad. Polon. Sci., 9 (1961), 857-859.

[22] ________, Measurable orientor fields, Bull. Acad. Polon. Sci., 13 (1965), 565-569.

[23] Pontryagin, L. S., et al, The mathematical theory of optimal processes, Interscience, New York, 1962.

[24] Scorza-Oragoni, G., Un teorem sulle funzioni continue rispetto ad une e misurabili rispetto ad un altra variable, Rend. Sem. Mat. Univ. Padova, 17 (1948), 102-106.

[25] Ważewski, T., Systèmes de commande et équations au contingent, Bull. Acad. Polon. Sci., 9 (1961), 151-155.

[26] ____________, Sur une condition d'existence des fonctions implicites measurables, Bull. Acad. Polon. Sci., 9 (1961), 861-863.

[27] ____________, Sur une condition équivalent à l'équation au contingent, Bull. Acad. Polon. Sci., 9 (1961), 865-867.

[28] ____________, Sur la sémicontinuité inférieure du "tendeur" d'un ensemble compact, variant d'un façon continue, Bull. Acad. Polon. Sci., 9 (1961), 869-872.

[29] ____________, Sur une généralisation de la notion des solution d'une équation au contingent, Bull. Acad. Polon. Sci., 10 (1962), 11-15.

[30] ____________, Sur les systèmes de commande non linéaires dont le contredomaine de commande n'est pas forcément convexe, Bull. Acad. Polon. Sci., 10 (1962), 17-21.

[31] ____________, Sur quelques définitions équivalentes des quasitrajectoires des systèmes de commande, Bull. Acad. Polon. Sci., 10 (1962), 469-474.

[32] ____________, On an optimal control problem, Differential Equations and Their Applications : Proceeding of the Conference held in Prague, September 1962 (1963), 229-242.

[33] Zaremba, S. C., Sur les équations au paratingent, Bull. Sci. Math. (12) 60(1936), 139-160.

A NONLINEAR STURM-LIOUVILLE THEOREM

Paul H. Rabinowitz

A well known theorem of Sturm-Liouville in the theory of ordinary differential equations concerns the eigenvalue problem:

(1) $$\mathcal{L}u \equiv -(p(x)u')' + q(x)u = \lambda a(x)u \ , \quad 0 < x < \pi$$

together with the separated boundary conditions (henceforth denoted by B.C.)

(2) $$a_0 u(0) + b_0 u'(0) = 0 \ , \quad a_1 u(\pi) + b_1 u'(\pi) = 0$$

where $(a_0^2 + b_0^2)(a_1^2 + b_1^2) \neq 0$ and p, q, a are respectively continuously differentiable and positive, continuous, and continuous and positive on the interval $[0, \pi]$. Under the above assumptions, the eigenvalues μ_n of (1) - (2) are simple and form an increasing sequence with $\mu_n \to \infty$ as $n \to \infty$. In addition, any eigenvector v_n corresponding to μ_n has exactly $(n-1)$ simple zeroes in $(0,\pi)$. We are interested in obtaining a nonlinear version of this result together with some applications.

First the result will be stated in a somewhat different fashion. As a technical convenience for what follows suppose that 0 is not an eigenvalue for $\mathcal{L}$ under the B.C. (2). Then (1) - (2) can be converted to an equivalent integral equation:

(3) $$u(x) = \lambda \int_0^{\pi} g(x,y)\, a(y)\, u(y)\, dy$$

where g is the Greens function for $\mathcal{L}$ together with the B. C. Let $E = C^1[0,\pi] \cap$ B. C. under the usual maximum norm:

$$\|u\|_1 = \max_{x \in [0,\pi]} |u(x)| + \max_{x \in [0,\pi]} |u'(x)| \ .$$

By a solution of (3) we mean a pair $(\lambda, u) \in \mathbb{R} \times E$. Let S_k^+ denote the set of $\varphi \in E$ such that φ has exactly $k-1$ simple zeroes in $(0,\pi)$, all zeroes of φ in $[0,\pi]$ are simple, and φ is positive in a deleted neighborhood of $x = 0$. Set $S_k^- = -S_k^+$ and $S_k = S_k^+ \cup S_k^-$. Then S_k^+, S_k^-, and S_k are open subsets of E and any eigenfunction v_k corresponding to μ_k belongs to S_k. We make v_k unique by requiring that $\|v_k\|_1 = 1$ and $v_k \in S_k^+$.

Using the terminology just introduced, the Sturm-Liouville theorem can be reformulated as follows: (3) possesses a line of trivial solutions $\{(\lambda, 0) | \lambda \in \mathbb{R}\}$ and in addition for each integer $k > 0$, a line of nontrivial solutions given by $\{(\mu_k, \alpha v_k) | \alpha \in \mathbb{R}\}$. We will obtain a nonlinear analogue of this result. Consider

$$\mathcal{L}u = \lambda F(x, u, u') , \quad 0 < x < \pi \tag{4}$$

together with the B.C. (2). It is assumed that F is a continuous function of its arguments in $[0, \pi] \times \mathbb{R}^2$ and $F(x, \xi, \eta) = a(x)\xi + 0((\xi^2 + \eta^2)^{1/2})$ near $(\xi, \eta) = (0, 0)$. As above, (4), (2) can be converted to the equivalent integral equation:

$$u(x) = \lambda \int_0^{\pi} g(x, y) F(y, u(y), u'(y)) dy . \tag{5}$$

Because of the form of F, (5) also possesses the line of trivial solutions $\{(\lambda, 0) | \lambda \in \mathbb{R}\}$. Let $\mathcal{S}$ denote the closure in $\mathbb{R} \times E$ of the set of nontrivial solutions of (5). By a theorem of Krasnoselski ([4]), the only possible trivial solutions belonging to $\mathcal{S}$ are the points $(\mu_k, 0)$, $k \in \mathbb{N}$, i.e. the possible bifurcation points.

Concerning the structure of $\mathcal{S}$, we have:

<u>Theorem 6:</u> For each integer $k > 0$, $\mathcal{S}$ contains a component, $\mathcal{C}_k$, which meets $(\mu_k, 0)$ and is unbounded in $\mathbb{R} \times S_k$.

(By a component of $\mathcal{S}$ we mean a maximal closed connected subset). Thus the statement of Theorem 6 contains in particular the linear case (3) where $\mathcal{C}_k$ is a line. To prove Theorem 6, a general theorem from nonlinear functional analysis which will be stated below and two lemmas are employed.

Let $\hat{E}$ be a real Banach space and $G : \mathbb{R} \times \hat{E} \to \hat{E}$ be compact, i.e. be continuous and map bounded sets into relatively compact sets. Suppose further that $G(\lambda, u) = \lambda Lu + H(\lambda, u)$ where L is a compact linear map and $H(\lambda, u) = 0(\|u\|)$ near $u = 0$ uniformly on bounded λ intervals. Consider the equation

$$u = G(\lambda, u) . \tag{7}$$

A solution of (7) is a pair $(\lambda, u) \in \mathbb{R} \times \hat{E}$. Then (7) possesses the line of trivial solutions. Let $\hat{\mathcal{S}}$ denote the closure of the set of nontrivial solutions of (7) and let μ be a real characteristic value of L. Then we have:

<u>Theorem 8</u>: If μ is a real characteristic value of L of odd multiplicity, $\hat{\mathcal{S}}$ contains a component, $\mathcal{C}$, containing $(\mu, 0)$ and which is either unbounded or meets $(\hat{\mu}, 0)$

where $\mu \neq \hat{\mu}$ is a real characteristic value of L.

For a proof of Theorem 8, see Rabinowitz [7] or [8].

Note that (5) has the form (7) and μ_k being a simple eigenvalue of $\mathcal{L}$ is a simple characteristic value of the corresponding linear integral operator. Hence the hypotheses of Theorem 8 are satisfied here and C_k, the component of $\mathcal{S}$ containing $(\mu_k, 0)$ is either unbounded in $\mathbb{R} \times E$ or meets $(\mu_j, 0)$, $j \neq k$. Actually C_k is unbounded in $\mathbb{R} \times S_k$ as we shall see via the following two lemmas.

Lemma 9: There exists a neighborhood $\mathcal{N}_j$ of $(\mu_j, 0)$ such that $(\lambda, u) \in \mathcal{N}_j \cap \mathcal{S}$ implies $u \in S_j$ or $u \equiv 0$.

Proof: If not, there exists a sequence $(\lambda_n, u_n) \in \mathcal{S}$ such that $(\lambda_n, u_n) \to (\mu_j, 0)$ as $n \to \infty$ and $u_n \notin S_j$. From (5) or equivalently (7),

$$\frac{u_n}{\|u_n\|_1} = \lambda_n L \frac{u_n}{\|u_n\|_1} + \frac{H(\lambda_n, u_n)}{\|u_n\|_1} . \tag{10}$$

The $0(\|u\|_1)$ condition on H implies $\|u_n\|_1^{-1} H(\lambda_n, u_n) \to 0$ as $n \to \infty$. Moreover the compactness of L and boundedness of $\{u_n / \|u_n\|_1\}$ implies that $\{Lu_n / \|u_n\|_1\}$ possesses a convergent subsequence $\{Lu_{n_p} / \|u_{n_p}\|_1\}$. From (10) this subsequence converges in E to v with $\|v\|_1 = 1$ and satisfying

$$v = \mu_j L v . \tag{11}$$

Hence $v = v_j$ or $v = -v_j$. In either event $v \in S_j$ and since S_j is open, $u_{n_p} / \|u_{n_p}\|_1 \in S_j$ for all j large. But this implies $u_{n_p} \in S_j$, a contradiction. Thus the lemma is established.

Lemma 12: Suppose (λ, u) is a solution of (4) and u has a double zero, i.e. there exists $\tau \in [0, \pi]$ such that $u(\tau) = 0 = u'(\tau)$. Then $u \equiv 0$.

Proof: If $F(x, \xi, \eta)$ is locally Lipschitz continuous with respect to (ξ, η), the result follows immediately from the uniqueness theorem for the initial value problem for (4). The more general case amounts to reproving the uniqueness result for the special initial data (0, 0) using the "0-condition" on $F - a\xi$. See [8] for a proof.

Proof of Theorem 6: Fix $k > 0$. Suppose $C_k \subset (\mathbb{R} \times S_k) \cup \{(\mu_k, 0)\}$. Then by Theorem 8, C_k must be unbounded in this set and we are through. Thus suppose

$\mathcal{C}_k \not\subset (\mathbb{R}\times S_k) \cup \{(\mu_k, 0)\}$. By Lemma 9, $\mathcal{C}_k \cap \mathcal{N}_k \subset (\mathbb{R}\times S_k) \cup \{(\mu_k, 0)\}$. Since $\mathcal{C}_k$ is connected, there exists $(\bar{\lambda}, \bar{u}) \in \mathcal{C}_k \cap (\mathbb{R}\times \partial S_k)$, $(\bar{\lambda}, \bar{u}) \neq (\mu_k, 0)$, and $(\bar{\lambda}, \bar{u}) = \lim_{n\to\infty} (\lambda_n, u_n)$ with $(\lambda_n, u_n) \in \mathcal{C}_k \cap (\mathbb{R}\times S_k)$. This implies $\bar{u}$ has a double zero and by Lemma 12, $\bar{u} \equiv 0$. Hence $(\bar{\lambda}, \bar{u}) = (\mu_j, 0)$ for some $j \neq k$. (Recall that the only trivial solutions belonging to $\mathcal{S}$ correspond to bifurcation points). But this implies $(\lambda_n, u_n) \in \mathcal{N}_j$ and therefore $u_n \in S_j$ for n large. Since this is impossible, the theorem is proved.

Remarks: If 0 is an eigenvalue of $\mathcal{L}$, Theorem 6 still obtains with the aid of an approximation argument. See Rabinowitz [6] or [8]. A sharper version of Theorem 8 for μ a simple characteristic value (see [7]) shows that $\mathcal{C}_k = \mathcal{C}_k^+ \cup \mathcal{C}_k^-$ where $\mathcal{C}_k^+ \cap \mathcal{C}_k^- = \{(\mu_k, 0)\}$ and $\mathcal{C}_k^+$, $\mathcal{C}_k^-$ are unbounded in $\mathbb{R}\times S_k^+$, $\mathbb{R}\times S_k^-$ respectively. Thus we get an even nicer correspondence with the linear case (3). If F is smooth near $(\xi, \eta) = (0, 0)$, then a general theorem on bifurcation from simple eigenvalues (see e.g. [2] or [3]) implies that $\mathcal{C}_k$ near $(\mu_k, 0)$ is a smooth curve of the form $(\lambda, u) = (\mu_k + 0(1), \alpha v_k + 0(|\alpha|))$ for α near 0. Lastly we note that Theorem 6 can readily be generalized to permit $\mathcal{L}$ to be nonlinear and a more general dependence of $\mathcal{L}$ and F on λ (see [6]).

To give some idea of what the sets $\mathcal{C}_k$ may look like, consider the following example:

$$(13) \qquad \begin{cases} -u'' = \lambda(1 + f(u^2 + (u')^2))u & 0 < x < \pi \\ u(0) = 0 = u(\pi) \end{cases}$$

where f is continuous and $f(0) = 0$. The linearization of (13) about $u = 0$ is:

$$(14) \qquad -v'' = \mu v, \quad 0 < x < \pi, \quad v(0) = 0 = v(\pi)$$

which possesses eigenvalues $\mu_n = n^2$ and eigenfunctions $v_n = \alpha_n \sin nx$. We will study $\mathcal{C}_1$ for (14) and in particular try for solutions of the form $(\lambda, u) = (\lambda, c \sin x)$. This leads to the equation:

$$(15) \qquad 1 = \lambda(1 + f(c^2))$$

relating λ and c. The freedom we have in choosing f leads to a wide range of possible behavior for $\mathcal{C}_1$.

Next some qualitative consequences of Theorem 6 will be studied.

Corollary 16: Suppose in (4) $F(x, \xi, \eta) = a(x)\xi + f(x, \xi, \eta)\xi$ where f is continuous in its arguments and $f(x, 0, 0) \equiv 0$. If $\mu_k > 0$ and $f \geq 0$, then $\mathcal{C}_k$ lies in $[0, \mu_k] \times S_k$ while if $\mu_k > 0$ and $f \leq 0$, $\mathcal{C}_k$ lies in $[\mu_k, \infty) \times S_k$.

Proof: The result is an immediate consequence of Theorem 6 together with a comparison argument which shows if e.g. $f \geq 0$ and $(\lambda, u) \in (\mathbb{R} \times S_k) \cap \mathcal{S}$, then $\lambda \in [0, \mu_k]$. See [6] for details.

The effect of a priori bounds on the sets $\mathcal{C}_k$ will be studied next.

Corollary 17: Suppose there exists a continuous real valued function $M(\lambda)$ for $\lambda \in \mathbb{R}^+$ such that (λ, u) a solution of (5) with $\lambda \geq 0$ implies $\|u\|_1 \leq M(\lambda)$. If $\mu_k > 0$, then for all $\lambda \in (\mu_k, \infty)$ there exists $u \in S_k$ such that $(\lambda, u) \in \mathcal{C}_k$.

Proof: By Theorem 6, $\mathcal{C}_k$ is unbounded in $\mathbb{R} \times S_k$. Note that $(0, u)$ cannot be a solution of (5). Hence $\mathcal{C}_k$ lies in $\mathbb{R}^+ \times S_k$. The existence of $M(\lambda)$ implies the projection of $\mathcal{C}_k$ on $\mathbb{R}^+$ cannot be bounded. Hence the result follows from the connectedness of $\mathcal{C}_k$.

Conditions under which such a priori bounds may be obtained can be found in Crandall and Rabinowitz [2] and Wolkowiskey [8]. As another application of Theorem 6 involving a different kind of a priori bound, we will prove a generalized version of a theorem of Nehari [5]. Consider

$$(18) \qquad -u'' = f(x, u)u, \quad 0 < x < \pi, \quad u(0) = 0 = u(\pi)$$

where f is continuous on $[0, \pi] \times \mathbb{R}$, $f(x, 0) = 0$, $f(x, u) > 0$ if $u \neq 0$, and there exists a continuous function $\rho : \mathbb{R} \to \mathbb{R}^+$ with $\rho(s) \to \infty$ as $|s| \to \infty$ and such that $|u| > s$ implies $f(x, u) > \rho(s)$.

Note that (18) differs from the equations treated earlier in that its right hand side has no linear part at $u = 0$.

Theorem 19: Under the above hypotheses on f, for each integer $k > 0$ there exists $u_k \in S_k$ such that u_k satisfies (18).

To prove Theorem 19, we require the following lemma which will be proved in the Appendix.

Lemma 21: Consider the equation

(22) $\quad -u'' = \lambda(b(x) + f(x,u))u \ , \quad 0 < x < \pi \ , \quad u(0) = 0 = u(\pi)$

where f is as above and $b \geq 0$ is continuous in $[0,\pi]$. Then for each integer $k > 0$ there exists a continuous function $M_k(\lambda) : (0,\infty) \to \mathbb{R}^+$ such that if (λ, u) is a solution of (22) with $u \in S_k$, then $\|u\|_1 \leq M_k(\lambda)$.

Proof of Theorem 19: An approximation argument is used. Let $\theta \in (0,1)$. Consider

$(23)_\theta \quad -u'' = \lambda(\theta + f(x,u))u \ , \quad 0 < x < \pi \ ; \quad u(0) = 0 = u(\pi) \ .$

The eigenvalues $\omega_k(\theta)$ of

(24) $\quad -w'' = \omega\theta w \ , \quad 0 < x < \pi \ ; \quad w(0) = 0 = w(\pi)$

are $\omega_k(\theta) = k^2/\theta > 1$ for all $k \geq 1$. By Corollary 16 $(23)_\theta$ possesses a component $\mathcal{C}_k(\theta)$ of solutions which is unbounded in $[0, \omega_k(\theta)] \times S_k$. By Lemma 21, the projection of $\mathcal{C}_k(\theta)$ on $\mathbb{R}$ contains $(0, \omega_k(\theta))$. In particular there exists $u_k(\theta) \in S_k$ such that $(1, u_k(\theta)) \in \mathcal{C}_k(\theta)$ and $\|u_k(\theta)\|_1 \leq M_k(1,\theta)$. The proof of Lemma 21 shows that M_k can be chosen independent of θ. From $(23)_\theta$, $\max_{x\in[0,\pi]} |u_k''(\theta)|$ can be bounded independently of θ. These bounds, the Arzela Ascoli Theorem, and $(23)_\theta$ imply there is a sequence $\theta_n \to 0$ as $n \to \infty$ such that $u_k(\theta_n)$ converges in $C^2[0,\pi]$ to a solution u_k of (18) with $u_k \in \overline{S_k}$. It only remains to show that $u_k \not\equiv 0$. But this follows by the argument of Lemma 9. The theorem is proved.

Remark 25: By using the ideas contained in the above proof, a version of Theorem 19 can be obtained for (22) with $\lambda = 1$.

Many people have studied nonlinear eigenvalue problems such as (4), (2) by examining a corresponding initial value problem and using shooting techniques. It seems unlikely that such methods can be used to obtain Theorem 6. On the other hand, Theorem 6 can be employed to shed some light on the corresponding initial value problem. For convenience we replace (2) by the B.C.

(26) $\quad u(0) = 0 = u(\pi) \ .$

Moreover we assume F in (4) is Lipschitz continuous in ξ and η and therefore the initial value problem for (4) possesses a unique solution.

Consider the map $\Psi : \mathcal{S} \to \mathbb{R}^2$, $\Psi(\lambda, u) = (\lambda, u'(0))$. The map Ψ is 1-1 via uniqueness of solutions to the initial value problem and is continuous. Therefore $\Psi(\mathcal{C}_k) \equiv \mathcal{J}_k$ is a connected subset of $\mathbb{R}^2$ and $\mathcal{J}_k \cap \mathcal{J}_j = \emptyset$ if $k \neq j$. Note that even though $\mathcal{C}_k$ is unbounded in $\mathbb{R}\times E$, $\mathcal{S}_k$ may be bounded in $\mathbb{R}^2$. By the remarks following the proof of Theorem 6, $\mathcal{S}_k = \mathcal{S}_k^+ \cup \mathcal{S}_k^- \equiv \Psi(\mathcal{C}_k^+) \cup \Psi(\mathcal{C}_k^-)$. As an interesting consequence of Corollary 16 we get

Corollary 27: Suppose in addition to the hypotheses of Corollary 17, F is Lipschitz continuous in ξ and η and (26) obtains. Let μ_r be the smallest positive eigenvalue of (3), (2). If $\lambda > \mu_k > \mu_r$, then there exist constants $C_r^+ > \ldots > C_k^+ > 0 > C_k^- > \ldots > C_r^-$ such that $(\lambda, C_j^{\pm}) \in \mathcal{S}_j^{\pm}$, $r \leq j \leq k$.

Proof: The result follows immediately from Corollary 16 and the properties of the sets $\mathcal{S}_j^{\pm}$.

Since $C_r^+, -C_r^- \leq M(\lambda)$, one could use a shooting technique in the interval of initial derivatives $[-M(\lambda), M(\lambda)]$ to find the solutions whose existence is given by Corollary 27.

We conclude with some remarks on periodic B.C. Suppose all functions involved in (1) and (4) are π periodic in x and (2) is replaced by

(28) $$u(0) = u(\pi), \quad u'(0) = u'(\pi) .$$

This case is interesting because some of the important structure obtained for the separated B.C. case is lost. The linear theory here again gives an increasing sequence of eigenvalues ζ_n with $\zeta_n \to \infty$ as $n \to \infty$. However the eigenvalues need not be simple although they are of multiplicity at most 2 and then have two corresponding linearly independent eigenvectors. More precisely (see Coddington-Levinson [1]) $\zeta_1 < \zeta_2 \leq \zeta_3 < \zeta_4 \leq \zeta_5 < \ldots$ etc. Any eigenfunction corresponding to ζ_1 has no zeroes in $[0,\pi]$; any eigenfunctions corresponding to ζ_{2k}, ζ_{2k+1}, $k \geq 1$ have exactly 2k simple zeroes in $[0,\pi]$. Thus in particular, ζ_1 is a simple eigenvalue. We again set up a family of open sets to take advantage of the nodal properties. Let $\hat{E}$ denote the subset of $C^1[0,\pi]$ of π periodic functions. Let $T_0^+ = \{\varphi \in \hat{E} \mid \varphi > 0\}$, $T_0^- = -T_0^+$, $T_0 = T_0^+ \cup T_0^-$, and $T_k = \{\varphi \in \hat{E} \mid \varphi$ has exactly 2k simple zeroes in $[0,\pi)\}$.

Consider now (4), (28). Again as a convenience we assume 0 is not an eigenvalue of $\mathcal{L}$. Hence (4), (28) can be converted to an operator equation of the form (7). Let $\hat{\mathcal{S}}$ denote the closure of the set of nontrivial solutions of this equation. With a small modification, the proof of Lemma 9 gives us:

Lemma 29: There exists a neighborhood $\mathfrak{M}_j$ of $(\zeta_{2j}, 0)$ (and $(\zeta_{2j+1}, 0)$ if $j \neq 0$)

such that $(\lambda, u) \in \mathfrak{m}_j \cap \hat{\mathcal{S}}$ implies $u \in T_j$ or $u \equiv 0$.

Since ζ_1 is a simple eigenvalue of (1), (28), a combination of Lemmas 29 and 12, and Theorem 8 yields:

Theorem 30: $\hat{\mathcal{S}}$ contains a component $\hat{\mathcal{C}}_1$ which meets $(\zeta_1, 0)$ and is unbounded in $\mathbb{R} \times T_0$.

Whenever ζ_{2k}, $k \geq 1$ is simple, the above lemmas and Theorem 8 implies that $\hat{\mathcal{S}}$ contains a component $\hat{\mathcal{C}}_k$ in $(\mathbb{R} \times T_k) \cup \{(\zeta_{2k}, 0)\}$ meeting $(\zeta_{2k}, 0)$. However $\hat{\mathcal{C}}_k$ need not be unbounded in $\mathbb{R} \times T_k$ but may also meet $(\zeta_{2k+1}, 0)$. Moreover if $\zeta_{2k} = \zeta_{2k+1}$, i.e. we have an eigenvalue of multiplicity 2, bifurcation need not occur at all. A simple such example is given by:

(31) $$-u'' + u = \lambda(u + (u')^3) \qquad 0 < x < \pi$$

with u satisfying (28). Multiplying (31) by u' and integrating over a period yields:

$$\lambda \int_0^{2\pi} (u')^4 \, dx = 0 \quad .$$

Since the equation possesses no solutions when $\lambda = 0$, (31), (28) possesses only the trivial solutions and the line of solutions $\{(1, \alpha) \mid \alpha \in \mathbb{R}\}$ in $\mathbb{R} \times T_0$.

Thus in general other than for $(\zeta_1, 0)$ results analogous to Theorem 6 do not obtain for (4), (19) and even to obtain bifurcation more hypotheses will have to be made. One way to guarantee bifurcation and even some sort of global result is to impose variational structure on the problem. More precisely suppose that F in (4) is independent of u', $q \geq 0$, and $F(x, u) = \frac{\partial}{\partial u} \hat{F}(x, u)$ with $\hat{F}(x, 0) \equiv 0$. Then (4), (19) is the Euler equation of the variational problem:

$$\text{Extremize} \int_0^{\pi} \hat{F}(x, \varphi) \, dx$$

over the class of $\varphi \in \hat{E}$ with

$$\int_0^{\pi} (p(\varphi')^2 + q\varphi^2) \, dx = R \,, \qquad R \text{ a constant} \quad .$$

By a theorem of Krasnoselski [4], each point $(\zeta_k, 0)$ will be a bifurcation point for (4), (28). Moreover if $\hat{F}$ is odd in u and appropriate technical conditions are satisfied, it follows from a theorem of Ljusternik [4] that for each $R > 0$, there exist infinitely many distinct solutions $(\lambda_n(R), u_n(R))$ of (4), (28) with

$$\int_0^{\pi} (p|u_n'|^2 + qu_n^2) \, dx = R \ .$$

An interesting open question is whether for all integers $k \geq 0$, there exists a solution $(\lambda_k(R), u_k(R))$ with $u_k(R) \in T_k$.

APPENDIX

Proof of Lemma 21: The proof consists of two steps. First we show there exists $M_k(\lambda, c)$ such that if (λ, u) is as in the statement of the lemma with $|u'(0)| \leq c$, then $\|u\|_1 \leq M_k(\lambda, c)$. Then we show $M_k(\lambda, c)$ can be chosen independently of c.

Suppose $u'(0) > 0$. (The argument for $u'(0) < 0$ is the same.) Let y_1 be the first zero of u'. Then u is a monotone increasing function and from (18), u'' is a monotone decreasing function in $[0, y_1]$. Hence u' is monotone decreasing in $[0, y_1]$ and $u'(0) = \max_{[0,y_1]} u'(x) \leq c$, $\max_{[0,y_1]} u \leq c\, y_1$. Let z_1 be the first zero of u in $(0, \pi)$. Then from (18), u and u' are monotone decreasing in $[y_1, z_1]$ so $\max_{[y_1, z_1]} u \leq c\, y_1$. Moreover integrating (18):

$$-u'(z_1) = \lambda \int_{y_1}^{z_1} (b + f)u \, dx \quad . \tag{32}$$

The bounds obtained for u in $[y_1, z_1]$ and (32) give a bound for $|u'(x)|$ in $[y_1, z_1]$. Continuing in this fashion leads to an estimate $\|u\|_1 \leq M_k(\lambda, c)$ where M_k is continuous in λ and c. Note also that if u' is known at any zero z_j of u, an estimate of the same form for $\|u\|_1$ obtains with c replaced by $|u'(z_j)|$.

It remains to show that $M_k(\lambda, c)$ can be chosen independently of c. If not, there exists a sequence (λ, u_n) satisfying (18) with $u_n \in S_k$ and $|u_n'(0)| \to \infty$. Let $\sigma_{j,n}$ denote the jth zero of u_n in $[0, \pi]$, $0 \leq j \leq k$. By our above remarks, $|u_n'(\sigma_{j,n})| \to \infty$ as $n \to \infty$. Let $I_{j,n} = [\sigma_{j,n}, \sigma_{j+1,n}]$. Since u_n' has a zero in $I_{j,n}$, the Mean Value Theorem implies that $\max_{I_{j,n}} |u_n''(x)| \to \infty$ as $n \to \infty$. Hence from (18), $\max_{I_{j,n}} |u_n(x)| \to \infty$ as $n \to \infty$. Therefore for any $s > 0$, if $n = n(s)$ is sufficiently large, there exists $x_{j,n} \in I_{j,n}$ such that $u_n(x_{j,n}) > s$ and $f(x_{j,n}, u_n(x_{j,n})) > \rho(s)$.

Consider the subinterval of $I_{j,n}$ in which $s = s(\lambda, k)$ is so large that $\rho(s) > \frac{4k^2}{\lambda}$ and therefore $\lambda(b + f) > 4k^2$. By the Sturm Comparison Theorem the length of this subinterval is less than $\frac{\pi}{2k}$. At least one of the intervals $I_{j,n}$, say $I_{p,n}$ has length $\geq \frac{\pi}{k}$. Consequently we can find a subinterval $[\gamma_n, \delta_n]$ of length $\geq \frac{\pi}{4k}$ such that $|f| \leq \frac{k^2}{\lambda} = \rho(\bar{s})$, $|u| \leq \bar{s}$ with $|u| = \bar{s}$ at one end of the subinterval and $u = 0$ at the other end. For convenience suppose $u(\gamma_n) = 0$, $u(\delta_n) = \bar{s}$. From (18),

$$(33) \qquad |u_n'(\gamma_n) - u_n'(\delta_n)| = \left|\int_{\gamma_n}^{\delta_n} \lambda(b + f)u \, dx\right| \leq (\lambda \|b\| + k^2)\bar{s} .$$

This implies:

$$(34) \qquad u(\delta_n) = \bar{s} = \int_{\gamma_n}^{\delta_n} u'(x)dx \geq \frac{\pi}{4k} [|u_n'(\gamma_n)| - (\lambda \|b\| + k^2)\bar{s}] .$$

But the right hand side of (34) is unbounded as $n \to \infty$ while the left hand side is bounded. Hence we have a contradiction. The uniformity of the argument in λ on bounded λ intervals gives the continuity of M_k in λ and the lemma is proved.

Remark: Note that if b is replaced by θb, $\theta \in [0, y]$, $M_k(\lambda)$ can be chosen independently of θ. Note also that as $\lambda \to 0$, $M_k(\lambda) \to \infty$.

BIBLIOGRAPHY

[1] Coddington, E. A. and Levinson, N., Theory of Ordinary Differential Equations, Mcgraw-Hill, New York, 1955.

[2] Crandall, M. G. and Rabinowitz, P. H., Nonlinear Sturm-Liouville eigenvalue problems and topological degree, J. Math. Mech., 19, 1083-1102 (1970).

[3] Crandall, M. G. and Rabinowitz, P. H., Bifurcation from simple eigenvalues, to appear, J. Funct. Anal.

[4] Krasnoselski, M. A., Topological Methods in the Theory of Nonlinear Integral Equations, Macmillan, New York, 1965.

[5] Nehari, Zeev, Characteristic values associated with a class of nonlinear second order differential equations, Acta. Math., 105, 141-175 (1961).

[6] Rabinowitz, P. H., Nonlinear Sturm-Liouville problems for second order ordinary differential equations, Comm. Pure Appl. Math., 23, 939-961 (1970).

[7] Rabinowitz, P. H., Some global results for nonlinear eigenvalue problems, J. Funct. Anal., 7, 487-513 (1971).

[8] Rabinowitz, P. H., A global theorem for nonlinear eigenvalue problems and applications, to appear Proc. Symposium on Nonlinear Functional Analysis, University of Wisconsin, Madison, 1971.

[9] Wolkowisky, J., Nonlinear Sturm-Liouville problems, Arch. Rational Mech. Anal. 35, 299-320 (1969).

ON PURSUIT AND EVASION PROBLEMS

Yoshiyuki Sakawa

1. INTRODUCTION

In this paper, we shall discuss pursuit and evasion problems related to a max-min problem first considered by Kelendzheridze [1]. A necessary and sufficient condition for attaining capture will be presented, on the basis of an inclusion relation between two attainable sets of a pursuer and an evader.

The "epsilon technique" which was developed by Balakrishnan [2, 3] for computing optimal control will be applied for solving the pursuit and evasion problems. The advantage of the epsilon technique lies in the fact that the optimization problem containing differential equation constraints can be reduced to a nondynamic optimization problem. We formulate the epsilon problem which can approximate the original problem as closely as desired. Furthermore, a technique for synthesizing the optimal strategies is shown, based on the epsilon problem.

2. STATEMENT OF THE PROBLEM

Let there be two players, the one called pursuer and the other called evader. The states of both players at any time t, $0 \leq t < \infty$, are represented by m-dimensional vectors x(t) and y(t), respectively. The dynamics of the players are given by the following differential equations

$$dx(t)/dt = f(x(t), u(t), t), \quad x(0) = x_0 \tag{1}$$

$$dy(t)/dt = g(y(t), v(t), t), \quad y(0) = y_0, \tag{2}$$

where x_0 and y_0 are initial states.

Let U be a nonempty compact subset of an r-dimensional Euclidean space R^r, and let V be a nonempty compact subset of an s-dimensional Euclidean space R^s. The control $u(\cdot)$ of the pursuer is said to be admissible if $u(\cdot)$ is measurable on $[0, T]$ and for each $t \in [0, T]$

$$u(t) \in U. \tag{3}$$

Let Ω_u denote the set of admissible controls of the pursuer defined on $[0, T]$. Analogously, the set of admissible controls of the evader defined on $[0, T]$ will be denoted by Ω_v.

For the functions f and g, the following assumptions will be made:

Assumption 1. The function $f(x, u, t)$ is continuous on $R^m \times U \times [0, T]$ and continuously differentiable in x. Similarly, the function $g(y, v, t)$ is continuous on $R^m \times V \times [0, T]$ and continuously differentiable in y.

Assumption 2. There exists a positive constant c such that

$$[x, f(x, u, t)] \leq c(1 + \|x\|^2) \tag{4}$$

for all $x \in R^m$, $u \in U$, and $t \in [0, T]$, and that

$$[y, g(y, v, t)] \leq c(1 + \|y\|^2)$$

for all $y \in R^m$, $v \in V$, and $t \in [0, T]$.

Assumption 3. The sets $f(x, U, t)$ and $g(y, V, t)$ defined by

$$\begin{aligned} f(x, U, t) &= \{ f(x, u, t) : u \in U \} \\ g(y, V, t) &= \{ g(y, v, t) : v \in V \} \end{aligned} \tag{5}$$

are convex for every x and $t \in [0, T]$ and for every y and $t \in [0, T]$, respectively.

With Assumptions 1 and 2, for each $u \in \Omega_u$, (1) has a unique solution uniformly bounded on $[0, T]$ which will be denoted by $x(\cdot, u)$, and for each $v \in \Omega_v$, (2) has a unique solution uniformly bounded on $[0, T]$ which will be denoted by $y(\cdot, v)$ [4], [5]. Let $A_x(T, x_0)$, or in short $A_x(T)$, denote the attainable set of the pursuer defined by

$$A_x(T, x_0) = \left\{ x_0 + \int_0^T f(x(t, u), u(t), t)dt : u(\cdot) \in \Omega_u \right\}. \tag{6}$$

In the same manner, the attainable set of the evader is denoted by $A_y(T, y_0)$ or $A_y(T)$. Under Assumption 3, the attainable sets $A_x(T)$ and $A_y(T)$ turn out to be compact [4], [5].

Now, let π be an $n \times m (n \leq m)$ matrix corresponding to the projection from R^m onto an n-dimensional linear subspace. The game is said to be completed from the initial states x_0 and y_0 if, no matter what admissible control may be chosen by the evader, the pursuer can choose an admissible control such that

$$\|\pi x(T) - \pi y(T)\| \leqq \delta \tag{7}$$

for some finite time T, where $\delta \geqq 0$ is a given constant. It is assumed in this paper that both players know the present states of both himself and the opponent and the differential equations (1) and (2). Let us define

$$\begin{aligned} B_x(T) &= \pi A_x(T) = \left\{ \pi x : x \in A_x(T) \right\} \subset R^n \\ B_y(T) &= \pi A_y(T) = \left\{ \pi x : x \in A_y(T) \right\} \subset R^n, \end{aligned} \tag{8}$$

and let S_δ denote a closed sphere defined by

$$S_\delta = \left\{ x \in R^n : \|x\| \leqq \delta \right\}.$$

Then, it is clear that the pursuit and evasion game can be completed if and only if

$$B_x(T) + S_\delta \supset B_y(T). \tag{9}$$

Let η be an arbitrary point of R^n. The distance between a point η and a set $B_x(T)$ is defined by

$$\rho(\eta, B_x(T)) = \inf \left\{ \|\eta - \xi\| : \xi \in B_x(T) \right\}.$$

Further, we define an asymmetrical distance between two sets $B_y(T)$ and $B_x(T)$ as follows:

$$\begin{aligned} \rho^*(B_y(T), B_x(T)) &= \sup \left\{ \rho(\eta, B_x(T)) : \eta \in B_y(T) \right\} \\ &= \sup_{\eta \in B_y(T)} \inf_{\xi \in B_x(T)} \|\eta - \xi\|. \end{aligned} \tag{10}$$

<u>Theorem 1.</u> Under Assumptions 1, 2, and 3, the necessary and sufficient condition for attaining the capture is that the relation

$$\begin{aligned} \rho^*(B_y(T), B_x(T)) &= \sup_{\eta \in B_y(T)} \inf_{\xi \in B_x(T)} \|\eta - \xi\| \\ &= \sup_{y \in A_y(T)} \inf_{x \in A_x(T)} \|\pi y - \pi x\| \leqq \delta \end{aligned} \tag{11}$$

holds for some finite time T.

<u>Proof.</u> To prove the necessity, let us assume that the relation

$$B_x(T) + S_\delta \supset B_y(T)$$

holds for some finite time T. The above relation implies that for all

$\eta \in B_y(T)$ there exists a $\xi \in B_x(T)$ such that

$$\|\eta - \xi\| \leqq \delta.$$

Therefore, it follows that

$$\inf_{\xi \in B_x(T)} \|\eta - \xi\| \leqq \delta \quad \text{for all } \eta \in B_y(T).$$

Hence we obtain

$$\sup_{\eta \in B_y(T)} \inf_{\xi \in B_x(T)} \|\eta - \xi\| \leqq \delta.$$

For proving the sufficiency, let us assume that there is a point $\hat{\eta} \in B_y(T)$ such that $\hat{\eta} \notin B_x(T) + S_\delta$. Since the set $B_x(T)$ is compact, it follows that

$$\rho(\hat{\eta}, B_x(T)) = \inf\left\{\|\hat{\eta} - \xi\| : \xi \in B_x(T)\right\} > \delta. \tag{12}$$

Therefore

$$\rho^*(B_y(T), B_x(T)) \geqq \rho(\hat{\eta}, B_x(T)) > \delta. \tag{13}$$

This is a contradiction. (Q.E.D.)

3. APPLICATION OF THE EPSILON TECHNIQUE

Now, the problem has been reduced to computing (11). The completion of the game, i.e., the existence of T satisfying (11) will not be studied further. In the following, computing the left-hand side of (11) for a fixed time T will be studied. Since the sets $A_x(T)$ and $A_y(T)$ are the attainable sets, most known methods for computing (11) will involve the solution of the dynamic equations (1) and (2) as an essential step. If the epsilon technique is applied, however, (11) can be computed without solving the dynamic equations. Thus we formulate a nondynamic problem for fixed $\varepsilon' > 0$ and $\varepsilon'' > 0$. We seek a sup-inf of the following functional, the time T being fixed,

$$\begin{aligned} h_T(\varepsilon', \varepsilon''; x, u; y, v) &= \|\pi x(T) - \pi y(T)\| \\ &+ \frac{1}{2\varepsilon'} \int_0^T \|\dot{x}(t) - f(x(t), u(t), t)\|^2 dt \\ &- \frac{1}{2\varepsilon''} \int_0^T \|\dot{y}(t) - g(y(t), v(t), t)\|^2 dt, \end{aligned} \tag{14}$$

over the class of absolutely continuous state functions $x(\cdot)$ and $y(\cdot)$ satisfying the given initial conditions, and over the class of admissible control functions $u(\cdot)$ and $v(\cdot)$. We formulate now the epsilon problem more precisely. Let X_1 denote the class of absolutely continuous functions $x(\cdot)$ over $[0, T]$ subject to $x(0) = x_0$, with the derivative square integrable over $[0, T]$. Likewise, let Y_1 denote the class of absolutely continuous functions $y(\cdot)$ over $[0, T]$ subject to $y(0) = y_0$, with the derivative square integrable over $[0, T]$. Let us define product spaces X and Y by

$$X = X_1 x \Omega_u, \quad Y = Y_1 x \Omega_v. \tag{15}$$

By introducing the following notations

$$\phi(\cdot) = (x(\cdot), u(\cdot)), \quad \psi(\cdot) = (y(\cdot), v(\cdot)), \quad \varepsilon = (\varepsilon', \varepsilon''), \tag{16}$$

Equation (14) can be abbreviated as

$$h_T(\varepsilon', \varepsilon''; x, u; y, v) = h_T(\varepsilon; \phi; \psi).$$

It has been proved [6] that under Assumptions 1, 2, and 3, the sup-inf of (14) is attained for each $\varepsilon > 0$; i.e., there exist $\phi^0(\varepsilon) \varepsilon X$ and $\psi^0(\varepsilon) \varepsilon Y$ such that

$$h_T(\varepsilon) = \sup_{\psi \varepsilon Y} \inf_{\phi \varepsilon X} h_T(\varepsilon; \phi; \psi) = h_T(\varepsilon; \phi^0(\varepsilon); \psi^0(\varepsilon)). \tag{17}$$

Now the relation between the solution of the epsilon problem and the original problem is given by the following theorem.

<u>Theorem 2.</u> Suppose there exist $\phi^0(\varepsilon) \varepsilon X$ and $\psi^0(\varepsilon) \varepsilon Y$ such that (17) holds. Then

$$\lim_{\varepsilon \to 0} h_T(\varepsilon) = \lim_{\varepsilon \to 0} \sup_{\psi \varepsilon Y} \inf_{\phi \varepsilon X} h_T(\varepsilon; \phi; \psi)$$

$$= \sup_{y \varepsilon A_y(T)} \inf_{x \varepsilon A_x(T)} \| \pi x - \pi y \|. \tag{18}$$

<u>Proof.</u> Now it is clear that the relation

$$\sup_{\psi \varepsilon Y} \inf_{\phi \varepsilon X} h_T(\varepsilon; \phi; \psi) \geqq \inf_{\phi \varepsilon X} h_T(\varepsilon; \phi; \psi) \tag{19}$$

holds for all $\psi \varepsilon Y$, and $\varepsilon = (\varepsilon', \varepsilon'') > 0$. In particular, (19) holds for a $\psi = (y, v)$ which satisfies the differential Equation (2). Hence,

$$\sup_{\psi \varepsilon Y} \inf_{\phi \varepsilon X} h_T(\varepsilon; \phi; \psi) \geqq \inf_{\phi \varepsilon X} [\| \pi x(T) - \pi y(T) \|$$

$$+ \frac{1}{2\varepsilon'} \int_0^T \| \dot{x}(t) - f(x(t), u(t), t) \|^2 dt] \tag{20}$$

holds for any $\varepsilon = (\varepsilon', \varepsilon'') > 0$ and $y(T) \varepsilon A_y(T)$. By [2, Theorem 3. 1] it follows that

$$\lim_{\varepsilon' \to 0} \inf_{\phi \varepsilon X} [\|\pi x(T) - \pi y(T)\| + \frac{1}{2\varepsilon'}\int_0^T \|\dot{x}(t) - f(x(t), u(t), t)\|^2 dt]$$

$$= \inf_{x \varepsilon A_x(T)} \|\pi x - \pi y(T)\| . \tag{21}$$

Letting $\varepsilon \to 0$ in (20), we obtain

$$\lim_{\varepsilon \to 0} \sup_{\psi \varepsilon Y} \inf_{\phi \varepsilon X} h_T(\varepsilon; \phi; \psi) \geqq \inf_{x \varepsilon A_x(T)} \|\pi x - \pi y(T)\| . \tag{22}$$

Since (22) holds for all $y(T) \varepsilon A_y(T)$, it follows that

$$\lim_{\varepsilon \to 0} \sup_{\psi \varepsilon Y} \inf_{\phi \varepsilon X} h_T(\varepsilon;\phi;\psi) \geqq \sup_{y \varepsilon A_y(T)} \inf_{x \varepsilon A_x(T)} \|\pi x - \pi y\| . \tag{23}$$

On the other hand, if a set $P \subset X$ is defined by

$$P = \left\{ \phi(\cdot) = (x(\cdot, u), u(\cdot)) : u \varepsilon \Omega_u \right\}, \tag{24}$$

it holds that

$$\inf_{\phi \varepsilon X} h_T(\varepsilon; \phi; \psi) \leqq \inf_{\phi \varepsilon P} h_T(\varepsilon; \phi; \psi). \tag{25}$$

Further it is obvious that

$$\inf_{\phi \varepsilon P} \|\pi x(T) - \pi y(T)\| = \inf_{x \varepsilon A_x(T)} \|\pi x - \pi y(T)\|. \tag{26}$$

Therefore, the inequality

$$\inf_{\phi \varepsilon X} h_T(\varepsilon; \phi; \psi) \leqq \inf_{x \varepsilon A_x(T)} \|\pi x - \pi y(T)\|$$

$$- \frac{1}{2\varepsilon''} \int_0^T \|\dot{y}(t) - g(y(t), v(t), t)\|^2 dt \tag{27}$$

holds for all $\varepsilon = (\varepsilon', \varepsilon'') > 0$ and $\psi \varepsilon Y$. From (27) it follows that

$$\sup_{\psi \varepsilon Y} \inf_{\phi \varepsilon X} h_T(\varepsilon; \phi; \psi) \leqq \sup_{\psi \varepsilon Y} [\inf_{x \varepsilon A_x(T)} \|\pi x - \pi y(T)\|$$

$$- \frac{1}{2\varepsilon''} \int_0^T \|\dot{y}(t) - g(y(t), v(t)\|^2 dt]. \tag{28}$$

Applying [2, Theorem 3. 1] again, we obtain

$$\lim_{\varepsilon'' \to 0} \sup_{\psi \varepsilon Y} [\inf_{x \varepsilon A_x(T)} \|\pi x - \pi y(T)\|$$

$$- \frac{1}{2\varepsilon''} \int_0^T \|\dot{y}(t) - g(y(t), v(t), t)\|^2 dt]$$

$$= \sup_{y \varepsilon A_y(T)} \inf_{x \varepsilon A_x(T)} \|\pi x - \pi y\|. \tag{29}$$

Relations (28) and (29) imply that

$$\lim_{\varepsilon \to 0} \sup_{\psi \varepsilon Y} \inf_{\phi \varepsilon X} h_T(\varepsilon; \phi; \psi)$$

$$\leqq \sup_{y \varepsilon A_y(T)} \inf_{x \varepsilon A_x(T)} \|\pi x - \pi y\|. \tag{30}$$

From (23) and (30), we finally obtain

$$\lim_{\varepsilon \to 0} \sup_{\psi \varepsilon Y} \inf_{\phi \varepsilon X} h_T(\varepsilon; \phi; \psi)$$

$$= \sup_{y \varepsilon A_y(T)} \inf_{x \varepsilon A_x(T)} \|\pi x - \pi y\|. \quad \text{(Q.E.D.)} \tag{31}$$

Theorem 2 shows that the epsilon problem approximates the original pursuit and evasion problem as closely as desired and provides an approximating sequence of controls of pursuer and evader that approximates the optimum.

4. SOLVING THE EPSILON PROBLEM

The problem is now to seek

$$\sup_{\psi \varepsilon Y} \inf_{\phi \varepsilon X} h_T(\varepsilon; \phi; \psi)$$

$$= \sup_{y \varepsilon Y_1} \sup_{v \varepsilon \Omega_v} \inf_{x \varepsilon X_1} \inf_{u \varepsilon \Omega_u} h_T(\varepsilon; x, u; y, v), \tag{32}$$

for some fixed time T and for sufficiently small value of $\varepsilon = (\varepsilon', \varepsilon'') > 0$. Because of the point-wise nature of the constraint, it is observed that for any fixed $x(\cdot)$, $y(\cdot)$, and $v(\cdot)$, the $u \varepsilon \Omega_u$ which minimizes (14) must attain

$$\inf_{u \varepsilon U} \| \dot{x}(t) - f(x(t), u, t)\|^2. \tag{33}$$

Therefore, the minimizing u is determined as a (possibly multivalued) function of $x(t)$, $\dot{x}(t)$, and t, which will be denoted by $u^0(x(t), \dot{x}(t), t)$. For fixed choice of $y(\cdot)$ in Y_1, it is clear from (14) that

$$\sup_{v \varepsilon \Omega_v} \inf_{x \varepsilon X_1} h_T(\varepsilon; x, u^0(x, \dot{x}, \cdot); y, v)$$

$$= \inf_{x \varepsilon X_1} \sup_{v \varepsilon \Omega_v} h_T(\varepsilon;\, x,\, u^0(x,\, \dot{x},\, \cdot);\, y,\, v). \tag{34}$$

In the same way, the $v \varepsilon \Omega_v$ which maximizes (14) must attain

$$\inf_{v \varepsilon V} \|\dot{y}(t) - g(y(t),\, v,\, t)\|^2 \tag{35}$$

for fixed choice of $y(\cdot) \varepsilon Y_1$. Thus, the minimizing v to be sought is determined just as a function of y(t), $\dot{y}(t)$, and t, which will be denoted by $v^0(y(t),\, \dot{y}(t),\, t)$. To further the argument, the following assumption will be made.

Assumption 4. $u^0(x(t),\, \dot{x}(t),\, t)$ and $v^0(y(t),\, \dot{y}(t),\, t)$ are uniquely determined, respectively.

Now the problem is to seek the sup-inf of

$$L_T(\varepsilon;\, x;\, y) = \|\pi x(T) - \pi y(T)\|$$

$$+ \frac{1}{2\varepsilon'} \int_0^T \|\dot{x}(t) - f(x(t),\, u^0(x(t),\, \dot{x}(t),\, t),\, t)\|^2 dt$$

$$- \frac{1}{2\varepsilon''} \int_0^T \|\dot{y}(t) - g(y(t),\, v^0(y(t),\, \dot{y}(t),\, t),\, t)\|^2 dt, \tag{36}$$

over the spaces X_1 and Y_1. By a result due to Pshenichniy [7], under Assumption 4, $u^0(x(t),\, \dot{x}(t),\, t)$ and $v^0(y(t),\, \dot{y}(t),\, t)$ are continuous functions of the arguments and (33) and (35) are continuously differentiable with respect to each argument. In this event, it is clear that the problem of minimizing and maximizing (36) is reduced to the classical calculus of variations. Let us define the following functions

$$\mu(x,\, \xi,\, t,\, u) = \frac{1}{2} \|\xi - f(x,\, u,\, t)\|^2, \tag{37}$$

$$\Phi(x,\, \xi,\, t) = \inf_{u \varepsilon U} \mu(x,\, \xi,\, t,\, u). \tag{38}$$

Since the function μ and the gradient vectors μ_x and μ_ξ are continuous in all the variables, by Pshenichniy [7] it follows that

$$\begin{aligned} \Phi_x(x,\, \xi,\, t) &= -f_1^*(x,\, u^0(x,\, \xi,\, t),\, t)(\xi - f(x,\, u^0(x,\, \xi,\, t),\, t)), \\ \Phi_\xi(x,\, \xi,\, t) &= \xi - f(x,\, u^0(x,\, \xi,\, t),\, t), \end{aligned} \tag{39}$$

where $u^0(x,\, \xi,\, t) \varepsilon U$ is a unique element satisfying

$$\mu(x,\, \xi,\, u^0(x,\, \xi,\, t),\, t) = \Phi(x,\, \xi,\, t), \tag{40}$$

$f_1 = \partial f/\partial x$ is an $m \times m$ matrix, and * denotes the transpose.

Let $\theta(t)$ be any m-dimensional vector function which is smooth on $[0, T]$ and satisfies $\theta(0) = 0$. Then, one must have

$$\frac{d}{d\lambda} L_T(\varepsilon; x(\cdot) + \lambda\theta(\cdot); y(\cdot))\Big|_{\lambda=0} = 0, \tag{41}$$

where λ is a real variable. Now by use of (39), it follows that

$$\frac{d}{d\lambda} L_T(\varepsilon; x(\cdot) + \lambda\theta(\cdot); y(\cdot))\Big|_{\lambda=0} = \frac{[\pi^*\pi(x(T) - y(T)), \theta(T)]}{\|\pi x(T) - \pi y(T)\|}$$

$$+ \frac{1}{\varepsilon'} \int_0^T [\dot{x}(t) - f(x(t), u^0(x(t), \dot{x}(t), t), t), \dot{\theta}(t)]dt$$

$$- \frac{1}{\varepsilon'} \int_0^T [f_1^*(x(t), u^0(x(t), \dot{x}(t), t), t)$$

$$\cdot(\dot{x}(t) - f(\dot{x}(t), u^0(x(t), \dot{x}(t), t), t)), \theta(t)]dt, \tag{42}$$

where $[\cdot,\cdot]$ denotes an inner product. Let

$$k(x, \dot{x}, t) = \frac{1}{\varepsilon'} (\dot{x} - f(x, u^0(x, \dot{x}, t), t)). \tag{43}$$

Then, from (41) and (42), it follows that

$$\frac{d}{dt} k(x, \dot{x}, t) + f_1^*(x, u^0(x, \dot{x}, t), t)k(x, \dot{x}, t) = 0, \tag{44}$$

$$k(x(T), \dot{x}(T), T) + \frac{\pi^*\pi(x(T) - y(T))}{\|\pi x(T) - \pi y(T)\|} = 0. \tag{45}$$

In the same way, let

$$\ell(y, \dot{y}, t) = \frac{1}{\varepsilon''} (\dot{y} - g(y, v^0(y, \dot{y}, t), t)). \tag{46}$$

Then from the condition:

$$\frac{d}{d\lambda} L_T(\varepsilon; x(\cdot); y(\cdot) + \lambda\theta(\cdot))\Big|_{\lambda=0} = 0,$$

it follows that

$$\frac{d}{dt} \ell(y, \dot{y}, t) + g_1^*(y, v^0(y, \dot{y}, t), t)\ell(y, \dot{y}, t) = 0, \tag{47}$$

$$\ell(y(T), \dot{y}(T), T) + \frac{\pi^*\pi(x(T) - y(T))}{\|\pi x(T) - \pi y(T)\|} = 0. \tag{48}$$

Equations (43) to (48) are necessary conditions for the x(t) and y(t) which satisfy

$$L_T(\varepsilon;\ x(\cdot);\ y(\cdot)) = \sup_{y\varepsilon Y_1}\ \inf_{x\varepsilon X_1} L_T(\varepsilon;\ x;\ y). \tag{49}$$

If $\varepsilon = (\varepsilon', \varepsilon'')$ goes to zero, then the solution of the original game problem is obtained. In the above, the controls are first obtained in the forms of $u^0(x, \dot{x}, t)$ and $v^0(y, \dot{y}, t)$. It is clear that the optimal trajectories x(t) and y(t) are related to each other through the terminal conditions (45) and (48).

If the differential equations (1) and (2) are linear and of the form:

$$\left.\begin{aligned} dx(t)/dt &= Ax + Bu, \quad x(0) = x_0, \\ dy(t)/dt &= Cy + Dv, \quad y(0) = y_0, \end{aligned}\right\} \tag{50}$$

then (44) is written as

$$\frac{dk(x,\ \dot{x},\ t)}{dt} = -A^{*}k(x,\ \dot{x},\ t). \tag{51}$$

Integrating (51) with terminal condition (45) yields

$$k(x(t),\ \dot{x}(t),\ t) = e^{A^{*}(T-t)}k(x(T),\ \dot{x}(T),\ T) = -e^{A^{*}(T-t)}\pi^{*}\gamma, \tag{52}$$

where γ is a unit n-dimensional vector given by

$$\gamma = \frac{\pi(x(T) - y(T))}{\|\pi(x(T) - y(T))\|}. \tag{53}$$

Let $u^0(\dot{x} - Ax)\ \varepsilon\ U$ be a unique element satisfying

$$\|\dot{x} - Ax - Bu^0(\dot{x} - Ax)\|^2 = \inf_{u\varepsilon U} \|\dot{x} - Ax - Bu\|^2. \tag{54}$$

Then from (43) and (52) it follows that

$$\dot{x} - Ax - Bu^0(\dot{x} - Ax) + \varepsilon' e^{A^{*}(T-t)}\pi^{*}\gamma = 0. \tag{55}$$

In the same fashion, let $v^0(\dot{y} - Cy)\ \varepsilon\ V$ be a unique element satisfying

$$\|\dot{y} - Cy - Dv^0(\dot{y} - Cy)\|^2 = \inf_{v\varepsilon V} \|\dot{y} - Cy - Dv\|^2. \tag{56}$$

Then, it follows that

$$\dot{y} - Cy - Dv^0(\dot{y} - Cy) + \varepsilon'' e^{C^{*}(T-t)}\pi^{*}\gamma = 0. \tag{57}$$

Equations (53) to (57) are necessary conditions for the optimal controls.

5. EXAMPLE

Let us consider a pursuit and evasion game, where the differential equations of a pursuer and an evader are described, respectively, by

$$\left.\begin{aligned} \ddot{x}_1 + \alpha\dot{x}_1 &= \rho u, \\ \ddot{y}_1 + \beta\dot{y}_1 &= \sigma v. \end{aligned}\right\} \qquad (58)$$

In (58), x_1, y_1, u, and v are assumed to be n-dimensional vectors ($n \geqq 2$), α, β, ρ, and σ are positive numbers, and controls are subject to

$$\|u\| \leqq 1, \quad \|v\| \leqq 1. \qquad (59)$$

This example was treated by Pontryagin [8] and Pshenichniy [9]. In this paper, this problem will be solved by the epsilon technique described above. Equation (58) is rewritten as

$$\left.\begin{aligned} \dot{x} &= \begin{pmatrix} 0 & 1 \\ 0 & -\alpha \end{pmatrix} x + \begin{pmatrix} 0 \\ \rho \end{pmatrix} u = Ax + Bu, \\ \dot{y} &= \begin{pmatrix} 0 & 1 \\ 0 & -\beta \end{pmatrix} y + \begin{pmatrix} 0 \\ \sigma \end{pmatrix} v = Cy + Dv, \end{aligned}\right\} \qquad (60)$$

where

$$x = \begin{pmatrix} x_1 \\ x_2 \end{pmatrix}, \qquad y = \begin{pmatrix} y_1 \\ y_2 \end{pmatrix}.$$

It should be noted that the elements of the matrices A, B, C, and D are $n \times n$ diagonal matrices.

Let $\pi x(T) - \pi y(T) = x_1(T) - y_1(T)$. Then it is clear that $\pi = (1 \quad 0)$. From (54) it follows that

$$\left.\begin{aligned} \rho u^0 &= \dot{x}_2 + \alpha x_2, \quad \text{if} \quad \|\dot{x}_2 + \alpha x_2\| \leqq \rho, \\ u^0 &= \frac{\dot{x}_2 + \alpha x_2}{\|\dot{x}_2 + \alpha x_2\|}, \quad \text{if} \quad \|\dot{x}_2 + \alpha x_2\| > \rho. \end{aligned}\right\} \qquad (61)$$

If $x_1(T) \not\equiv y_1(T)$, it is clear that the first case of (61) cannot occur. Because, in the first case, (55) implies $\gamma = 0$. Hence it must hold that

$$\|\dot{x}_2 + \ x_2\| > \rho, \quad \text{and} \quad u^0 = \frac{\dot{x}_2 + \alpha x_2}{\|\dot{x}_2 + \alpha x_2\|} \ . \qquad (62)$$

It is easily seen that

$$e^{At} = \begin{bmatrix} 1 & p_1(t) \\ 0 & p_0(t) \end{bmatrix}, \quad e^{Ct} = \begin{bmatrix} 1 & e_1(t) \\ 0 & e_0(t) \end{bmatrix}, \tag{63}$$

where

$$p_0(t) = e^{-\alpha t}, \qquad p_1(t) = (1 - e^{-\alpha t})/\alpha,$$

$$e_0(t) = e^{-\beta t}, \qquad e_1(t) = (1 - e^{-\beta t})/\beta.$$

From (55) it follows that

$$\left.\begin{aligned} \dot{x}_1 &= x_2 - \varepsilon'\gamma \\ \dot{x}_2 &= -\alpha x_2 + \rho u^0 - \varepsilon' p_1(T - t)\gamma, \end{aligned}\right\} \tag{64}$$

where

$$\gamma = \frac{x_1(T) - y_1(T)}{\|x_1(T) - y_1(T)\|} . \tag{65}$$

Substituting (62) into the second equation of (64) yields

$$(\dot{x}_2 + \alpha x_2)(1 - \frac{\rho}{\|\dot{x}_2 + \alpha x_2\|}) = - \varepsilon' p_1(T - t)\gamma.$$

Since $\|\dot{x}_2 + \alpha x_2\| > \rho$ and $p_1(T - t) > 0$ $(T > t)$, the direction of the vector $\dot{x}_2 + \alpha x_2$ must be opposite to the direction of the unit vector γ. Thus

$$u^0 = \frac{\dot{x}_2 + \alpha x_2}{\|\dot{x}_2 + \alpha x_2\|} = -\gamma. \tag{66}$$

Now, letting ε' go to zero, we obtain

$$\left.\begin{aligned} \dot{x}_1 &= x_2 \\ \dot{x}_2 &= -\alpha x_2 - \rho\gamma. \end{aligned}\right\} \tag{67}$$

In the same way, it follows that

$$v^0 = \frac{\dot{y}_2 + \beta y_2}{\|\dot{y}_2 + \beta y_2\|} = -\gamma, \tag{68}$$

$$\left.\begin{aligned} \dot{y}_1 &= y_2 \\ \dot{y}_2 &= -\beta y_2 - \sigma\gamma. \end{aligned}\right\} \tag{69}$$

Integrating (67) and (69), respectively, yields

$$\left.\begin{aligned} x_1(T) &= x_1(t) + p_1(T - t)x_2(t) - \rho p_2(T - t)\gamma, \\ y_1(T) &= y_1(t) + e_1(T - t)y_2(t) - \sigma e_2(T - t)\gamma, \end{aligned}\right\} \tag{70}$$

where

$$p_2(t) = \int_0^t p_1(\tau)d\tau = \frac{1}{\alpha}\left(t - \frac{1 - e^{-\alpha t}}{\alpha}\right),$$

$$e_2(t) = \int_0^t e_1(\tau)d\tau = \frac{1}{\beta}\left(t - \frac{1 - e^{-\beta t}}{\beta}\right).$$

From (70),

$$\begin{aligned} x_1(T) - y_1(T) &= x_1(0) - y_1(0) + p_1(T)x_2(0) - e_1(T)y_2(0) \\ &\quad - (\rho p_2(T) - \sigma e_2(T))\gamma. \end{aligned} \tag{71}$$

Let us assume that

$$\phi(T) = \rho p_2(T) - \sigma e_2(T) \geqq 0 \quad \text{for all } T \geqq 0. \tag{72}$$

Then, in view of (65) and (71), it follows that

$$\begin{aligned} \|x_1(T) - y_1(T)\| &= \|x_1(0) - y_1(0) + p_1(T)x_2(0) \\ &\quad - e_1(T)y_2(0)\| - \phi(T). \end{aligned}$$

Since $\phi(0) = 0$, if it holds that

$$\left.\begin{aligned} \dot{\phi}(t) = \frac{\rho}{\alpha}(1 - e^{-\alpha t}) - \frac{\sigma}{\beta}(1 - e^{-\beta t}) > 0 \quad \text{for all } t > 0, \\ (\rho/\alpha) - (\sigma/\beta) > 0, \end{aligned}\right\} \tag{73}$$

then (72) is satisfied and it is clear that for any initial conditions there is a finite time T such that

$$x_1(T) - y_1(T) = 0.$$

Since the equation $\dot{\phi}(t) = 0$ has at most two real roots and $\dot{\phi}(0) = 0$, in order that (73) holds, it is sufficient that

$$\dot{\phi}(\infty) = \frac{\rho}{\alpha} - \frac{\sigma}{\beta} > 0, \quad \text{and} \quad \ddot{\phi}(0) = \rho - \sigma \geqq 0. \tag{74}$$

From (65) and (70), it follows that

$$\begin{aligned} x_1(T) - y_1(T) &= x_1(t) - y_1(t) + p_1(T - t)x_2(t) - e_1(T - t)y_2(t) \\ &\quad - \left\{\rho p_2(T - t) - \sigma e_2(T - t)\right\}(x_1(T) - y_1(T))/\|x_1(T) - y_1(T)\|. \end{aligned}$$

In view of (66) and (68), the optimal feedback controls are given by

$$u^0 = v^0 = -\gamma = \frac{x_1(t) - y_1(t) + p_1(T - t)x_2(t) - e_1(T - t)y_2(t)}{\|x_1(T) - y_1(T)\| + \rho p_2(T - t) - \sigma e_2(T - t)}$$

$$= \frac{x_1(t) - y_1(t) + p_1(T - t)x_2(t) - e_1(T - t)y_2(t)}{\|x_1(t) - y_1(t) + p_1(T - t)x_2(t) - e_1(T - t)y_2(t)\|}. \quad (75)$$

These results coincide with those of Pontryagin [8] and Pshenichniy [9].

6. CONCLUDING REMARKS

The Assumption 4 seems pretty strong. Even if the assumption does not hold, the epsilon problem can be solved by the Ritz method. Namely, by applying the Ritz method, the epsilon problem is reduced to a max-min problem with respect to a finite number of parameters.

REFERENCES

1. D. L. Kelendzheridze, "On the Theory of Optimal Pursuit," Soviet Math. Dokl., Vol. 2, pp. 654-656, 1961.
2. A. V. Balakrishnan, "On a New Computing Technique in Optimal Control," SIAM J. Control, Vol. 6, No. 2, pp. 149-173, 1968.
3. A. V. Balakrishnan, "The Epsilon Technique - A Constructive Approach to Optimal Control," in A. V. Balakrishnan (Ed.), Calculus of Variations and Control Theory, Academic Press, New York, 1969.
4. A. F. Filippov, "On Certain Questions in the Theory of Optimal Control," SIAM J. Control, Vol. 1, No. 1, pp. 76-84, 1962.
5. H. Hermes, "On the Closure and Convexity of Attainable Sets in Finite and Infinite Dimensions," SIAM J. Control, Vol. 5, No. 3, pp. 409-417, 1967.
6. Y. Sakawa, "An Application of the Epsilon Technique to the Solution of Pursuit-Evasion Games," J. of Computer and System Sciences, Vol. 4, No. 6, pp.557-569, 1970.
7. B. N. Pshenichniy, "On the Pursuit Problem," Kibernetika, No. 6, pp. 54- 64, 1967.
8. L. S. Pontryagin, "On the Theory of Differential Games," Uspekhi Mat. Nauk, Vol. 21, pp. 219-274, 1966.

9. B. N. Pshenichniy, "Linear Differential Games," Avtomatika i Telemekhanika, No. 1, pp. 65-78, January 1968.

APPENDIX

Theorem (Pshenichniy [7])

Let $\mu(y, u)$ be a continuous real function of $y \in R^n$ and $u \in U \subset R^r$, where U is a compact subset, and let the gradient of $\mu(y, u)$ with respect to y (denoted by $\mu_y(y, u)$) be continuous in y and u. Define the function

$$\Phi(y) = \min_{u \in U} \mu(y, u). \tag{A. 1}$$

If the minimizing element $u^*(y)$ which satisfies

$$\Phi(y) = \mu(y, u^*(y)) \tag{A. 2}$$

is unique in some neighborhood of a point y, then in this neighborhood $u^*(y)$ depends continuously on y, $\Phi(y)$ is continuously differentiable with respect to y, and

$$\text{grad } \Phi(y) = \mu_y(y, u^*(y)). \tag{A. 3}$$

DISCONJUGACY AND WRONSKIANS

Philip Hartman

It seems appropriate that I should talk here on this subject since some of my work on it was an outgrowth of generalizations of the paper [13], presented by J. D. Schuur at the last United States-Japan Seminar on Ordinary Differential and Functional Equations in 1967. Most of my remarks will deal with variants of results of A. Yu. Levin [9] and of Hartman [4], [5]. The problems treated and the results obtained in these papers are similar. Some common arguments were used by Levin and me, but our principal methods are different. The results concern (i) conditions, necessary and/or sufficient, that an n-th order, linear, homogeneous, differential equation be disconjugate; (ii) the notion, existence, and properties of principal solutions of disconjugate equations; and (iii) estimates for solutions of such equations. Below we give new results of the type (i) and (iii), and illustrate the nature of the proofs initiated in [5], [7]. An exposition of these and related matters will be contained in the lecture notes of Coppel [1]. In view of the bibliographies in [1], [5] and [9], it will not be necessary for me to give complete references.

1. DISCONJUGACY

Let I denote a t-interval which, unless otherwise specified, can be bounded or unbounded and can be open, closed, or half-closed. In the equation

$$(*) \qquad P_n(t,D)[x] \equiv x^{(n)} + a_{n-1}(t)x^{(n-1)} + \dots + a_0(t)x = 0,$$

where $D[x] = dx/dt = x'$, the coefficients $a_0(t),\dots,a_{n-1}(t)$ are assumed to be continuous on I. Often we write $P_n(D)$ in place of $P_n(t,D)$.

Definition. Disconjugacy. (*) is said to be disconjugate on I if every solution $x = x(t) \not\equiv 0$ has at most $n - 1$, zeros, counting multiplicities, on I.

This terminology is due to Wintner in the case $n = 2$ where "counting multiplicities" can obviously be omitted. If $n \geq 2$, the omission of "counting multiplicities" does not change the notion of "disconjugate on I" if I is open or half-open, but does change it if I is a compact interval. For the case of an open I, this follows from general results [3], [6] on non-linear interpolating families of functions; for a simpler proof in the linear situation applicable here, see [11]. For I half-open, cf. the Remark 1 following Theorem 1.1 below. See also [14].

There is a large literature giving sufficient conditions for (*) to be disconjugate on a compact interval $I = [a,b]$ in the form of inequalities involving $b - a$ and $||a_0||,\dots,||a_{n-1}||$, where $||\dots||$ denotes some norm (e.g., sup or L^p norm). Here, we shall not be interested directly in this type of theorem. Rather, we are interested in generalizations of the following: If $n = 2$ and $I = [a,b]$ is compact, then (*) is disconjugate on I if and only if there exists a function $u \in C^2([a,b))$ such that

$$u > 0,\ P_2(D)[u] \leqq 0 \quad \text{on} \quad [a,b).$$

Of course, this form of Sturm's comparison theorem had been used by Bôcher, de la Vallée Poussin, and Wintner to derive criteria for disconjugacy which are of the inequality type above; see also [5] and [9] for such derivations in the case $n \geqq 2$.

In order to state our results, it will be necessary to introduce some terminology. The Wronskian determinant of k functions $u_1,\dots,u_k$ will be denoted by

$$W(u_1,\dots,u_k) \equiv \det(u_j^{(i-1)})_{i,j=1,\dots,k}; \text{ so that } W(u_1) \equiv u_1.$$

Definition. $w_n(I)$-system. An ordered set of $n - 1$ functions $u_1,\dots,u_{n-1}$ of class C^{n-2} on a t-interval I is said to be a $w_n(I)$-system if

$$W(u_1,\dots,u_k) > 0 \text{ on } I \text{ for } 1 \leqq k < n. \tag{1.1}$$

Definition. $W_n(I)$-system. An ordered set of $n - 1$ functions $u_1,\dots,u_{n-1}$ of class $C^{n-2}(I)$ is said to be a $W_n(I)$-system if

$$W(u_{i(1)},\dots,u_{i(k)}) > 0 \text{ on } I \text{ for } 1 \leqq i(1) < \dots < i(k) < n \tag{1.2}$$

or equivalently

$$W(u_j,\dots,u_m) > 0 \text{ on } I \text{ for } 1 \leqq j \leqq m < n. \tag{1.3}$$

The equivalence of (1.2) and (1.3) follows from a general theorem of Fekete [2] on minors of a determinant; cf. (3.4) below. The simplest example of a $W_n(I)$-system is

$$(1.5) \qquad u_k(t) = \exp \alpha_k t \quad \text{for } 1 \leqq k < n, \text{ where } \alpha_1 < \ldots < \alpha_{n-1}.$$

Theorem 1.1. Let I be open or compact [or let I be half-open and I^0 its interior]. Then a necessary and sufficient condition for (*) to be disconjugate on I is that there exist a $w_n(I)$-system [or a $w_n(I^0)$-system] of solutions $x_1,\ldots,x_{n-1}$ of (*).

See [5] or [9]. This is a completion of results of Pólya [12] who proved the sufficiency in the case of an open I and the necessity in the case of a half-open I.

Remark 1. This theorem implies, for example, that if $I = [a, \beta)$, $-\infty < a < \beta \leqq \infty$, is half-open, then (*) is disconjugate on I if only if it is disconjugate on the interior $I^0 = (a, \beta)$.

Theorem 1.1 does not give a useful criterion for disconjugacy since it requires a knowledge of the solutions of (*). Our main results in this direction are the following:

Theorem 1.2. Let $I = [a,b]$ be a compact interval. Then a sufficient condition for the disconjugacy of (*) on I is that there exist functions $u_1,\ldots,u_{n-1}$ satisfying the conditions

$$(1.6) \qquad u_1,\ldots,u_{n-1} \in C^n([a,b)),$$

$$(1.7) \qquad (-1)^{n+k} P_n(D)[u_k] \geqq 0 \quad \text{for } 1 \leqq k < n,$$

$$(1.8) \qquad u_1,\ldots,u_{n-1} \text{ is a } w_n([a,b))\text{-system},$$

and the condition

$$(1.9) \qquad u_1,\ldots,u_{n-1} \text{ is a } W_n((a,b))\text{-system},$$

or, more generally, the condition

$$(1.10) \qquad W(u_1,\ldots,\hat{u}_j,\ldots,u_k) \geqq 0 \text{ at } t = a \text{ for } 1 \leqq j < k < n.$$

Furthermore, a necessary condition is that there exist a $W_n(I)$-system of solutions $x_1,\ldots,x_{n-1}$ of (*).

Remark 2. In the first ("sufficiency") part, the conditions (1.10) at $t = a$ cannot be omitted (if $n > 2$). This is seen by the trivial example: $n = 3$, $I = [1,b]$ with large b,

$$P_3(D)[x] \equiv x''' + x' = 0,$$

$u_1 = t$ and $u_2 = -1$.

Theorem 1.2 is a variant of results of [5] and [9]. In the first ("sufficiency") part, the proof of [5] essentially requires the set of conditions involving (1.6)-(1.9), where [a,b) can be replaced by (a,b], while [9] essentially requires (1.6)-(1.8) and the strict form of the inequalities in (1.10) for $a \leqq t < b$ (instead of $t = a$). The proof of "sufficiency" in [9] is by contradiction and depends on Theorem 1.1 and properties of Green's functions. The proof in [5] is direct by induction, and does not use Theorem 1.1. Since these "sufficiency" conditions are "necessary and sufficient" by Theorem 1.2, they are equivalent. One can easily show this directly by reducing the "sufficiency" conditions of Theorem 1.2 to those used in [5], as follows:

Lemma 1.1. Let $u_1,\dots,u_{n-1} \in C^{n-2}([a,b))$ satisfy (1.8), (1.10). Then there exist functions $v_1,\dots,v_{n-1}$ of the form

$$(1.11)\qquad v_k(t) = \sum_{j=1}^{k} (-1)^{j+k} c_{kj}\, u_j(t), \text{ where } c_{kj} \geqq 0 \text{ and } c_{kk} > 0$$

are constants, such that $v_1,\dots,v_{n-1}$ is a $w_n([a,b))$-system and a $W_n((a,b))$-system. (In particular, $(-1)^{n+k} P_n(D)[v_k] \geqq 0$ wherever $u_1,\dots,u_{n-1}$ are of class C^n and satisfy (1.7).)

Since (*) is disconjugate on an interval if and only if it is disconjugate on every compact subinterval, the proof of Theorem 1.2 implies the "sufficiency" in the next theorem (by virtue of Remark 1 above).

Theorem 1.3. Let $I = [\alpha,\beta)$, $-\infty < \alpha < \beta \leqq \infty$, and I^0 is its interior or let $I = (\alpha,\beta)$, $-\infty \leqq \alpha < \beta \leqq \infty$. Then a sufficient condition for (*) to be disconjugate on I is that there exist $u_1,\dots,u_{n-1} \in C^n(I^0)$ which is a $w_n(I^0)$-system satisfying (1.7) and

$$(1.12)\qquad W(u_1,\dots,\hat{u}_j,\dots,u_k) \geqq 0 \text{ on } (\alpha,\gamma) \text{ for } 1 \leqq j < k < n$$

for some γ, $\alpha < \gamma < \beta$. Furthermore, a necessary condition is that, for every

γ, $\alpha < \gamma < \beta$, <u>there exist a</u> $W_n([\gamma,\beta))$-<u>system of solutions</u> $x_1,\dots,x_{n-1}$ <u>of</u> (*).

If $\alpha_1 < \dots < \alpha_{n-1}$, the following result is a consequence of Theorem 1.2 and the choice of the $W_n(I)$-system (1.5). If $\alpha_k = \alpha_{k+1}$ for some k, an additional argument is needed; [5].

Corollary 1.2. <u>Let</u> $\alpha_1 \leqq \dots \leqq \alpha_{n-1}$ <u>be constants satisfying</u>

$$(-1)^{n+k} P_n(t,\alpha_k) \geqq 0 \text{ on } I \tag{1.13}$$

<u>for</u> $1 \leqq k \leqq n-1$ <u>(so that the characteristic equation</u> $P_n(x,\lambda) = 0$ <u>has</u> n <u>real roots</u> $\lambda_1(t),\dots,\lambda_n(t)$ <u>separated by constants</u>

$$\lambda_1(t) \leqq \alpha_1 \leqq \lambda_2(t) \leqq \dots \leqq \lambda_{n-1}(t) \leqq \alpha_{n-1} \leqq \lambda_n(t) \tag{1.14}$$

<u>on</u> I). <u>Then</u> (*) <u>is disconjugate on</u> I.

Lemma 1.1 will be proved in Section 4, and the first part of Theorem 1.2 in Section 5.

2. ESTIMATES FOR SOLUTIONS

The first goal of my work on this subject had been to prove the following generalization of Schuur's result [13].

Corollary 2.1. <u>Let</u> $I = [\gamma,\beta)$, $-\infty < \gamma < \beta < \infty$, <u>be a half-open interval and let there exist constants</u> $\alpha_0 < \alpha_1 < \dots < \alpha_n$ <u>satisfying</u> (1.13). <u>Then</u> (*) <u>has positive, linearly independent solutions</u> $x_1,\dots,x_n$ <u>such that</u>

$$\alpha_0 \leqq x_1'/x_1 \leqq \alpha_1 < x_2'/x_2 < \dots < \alpha_{n-1} < x_n'/x_n < \alpha_n.$$

<u>This result is valid if there is no assumption or assertion involving</u> α_0 <u>and/or</u> α_n.

For $n = 2$, this result is contained in Olech [10]. For $n = 3$ (where no $\alpha_0, \alpha_n = \alpha_3$ occur), the existence of x_2 satisfying $x_2 > 0$ and $\alpha_1 \leqq x_2'/x_2 \leqq \alpha_2$ was proved by Schuur [13]; see also Jackson [8]. Corollary 2.1 is a particular case of the following

Theorem 2.1. <u>Let</u> $I = [\gamma,\beta)$, $-\infty < \gamma < \beta \leqq \infty$. <u>Let</u> $u_0,\dots,u_{n-1} \in C^n(I)$ <u>satisfy</u>

$$(-1)^{n+k} P_n(t,D)[u_k] \geqq 0 \text{ for } 0 \leqq k < n, \tag{2.1}$$

$$u_1,\dots,u_{n-1} \text{ is a } W_n(I)\text{-system,} \tag{2.2}$$

(2.3) $$W(u_0,\ldots,u_k) \geqq 0 \quad \text{for} \quad 0 \leqq k < n.$$

Then (*) has positive linearly independent solutions $x_1,\ldots,x_n$ satisfying

(2.4) $$x_1'/x_1 \leqq u_1'/u_1 < x_2'/x_2 < \ldots < u_{n-1}'/u_{n-1} < x_n'/x_n,$$

(2.5) $$W(x_1,u_0) \leqq 0, \text{ so that } u_0'/u_0 \leqq x_1'/x_1 \text{ where } u_0 > 0,$$

and the inequalities

(2.6)
$$W(x_1,\ldots,x_k) > 0 \quad \text{for} \quad 1 \leqq k \leqq n,$$
$$W(x_1,\ldots,x_k,u_j,\ldots,u_m) > 0 \; [\text{or} \geqq 0] \text{ for } 1 \leqq k < [\text{or} \leqq]\, j \leqq m < n,$$
$$W(x_1,\ldots,x_k,u_{k-1}) \leqq 0 \quad \text{for} \quad 1 \leqq k \leqq n,$$
$$W(x_1,\ldots,x_k,u_{k-1},u_j,\ldots,u_m) \leqq 0 \quad \text{for} \quad 1 \leqq k < j \leqq m < n.$$

If, in addition, $u_n \in C^n(I)$ satisfies

(2.7) $$P_n(D)[u_n] \geqq 0,$$

(2.8) $$W(u_k,\ldots,u_n) > 0 \quad \text{for} \quad 1 \leqq k \leqq n,$$

then the solution $x_n(t)$ can be chosen to satisfy

(2.9) $$W(x_n,u_n) > 0, \text{ i.e., } x_n'/x_n < u_n'/u_n,$$

and $m = n$ is permitted in (2.6).

Theorem 2.1 remains correct if there is no assumption or assertion concerning u_0, but if u_0 is not otherwise specified, we can always make the trivial choice $u_0 \equiv 0$. For $n = 2$, see Olech [10]. Professor Coppel has called my attention to the fact that the version of this theorem stated in [4], [5] was incorrect. There $I = (\alpha,\beta)$ was an open interval $-\infty \leqq \alpha < \beta \leqq \infty$, the inequalities (2.6) did not occur, and the strict inequalities "$>$" in the assumption (2.8) and "$<$" in the conclusions (2.4), (2.9) were replaced by "$\geqq$" and "$\leqq$". It turns out that this incorrect statement becomes valid if the assertion of the "linear independence of $x_1,\ldots,x_n$" is omitted and, in fact, this corrected statement is a corollary of Theorem 2.1. In [9], Levin states an analogous form of Theorem 2.1. He weakens $W_n(I)$ to $w_n(I)$ in (2.2), makes the additional assumption $u_{k-1} = o(u_k)$ as $t \to \beta$ but, in place of the inequalities (2.4),(2.5),(2.6),(2.9) for all t on I, he merely asserts $x_k = O(u_k)$, $u_{k-1} = O(x_k)$ as $t \to \beta$.

The proof of Theorem 2.1 in [7] depends on a refinement of the techniques of [5]. First, we prove the existence of a suitable x_1 and, second, we employ an induction on n (using the known solution x_1 and a variation of constants to reduce (*) to an equation of order $n-1$). The existence of x_1 is proved in the next lemma by replacing the n-th order equation (*) by suitable first order systems.

Lemma 2.1. Let $n > 1$, $I = [\gamma, \beta)$, $P_n(t,D)$, and $u_0,\ldots,u_{n-1}$ as in Theorem 2.1. Then (*) has a solution $x = x_1(t)$ on I satisfying

(2.10) $$x_1 > 0 \text{ and } W(x_1,u_1,\ldots,u_k) \geqq 0 \text{ for } 1 \leqq k < n,$$

(2.11) $$W(x_1,u_0) \leqq 0 \text{ and } W(x,u_0,u_j,\ldots,u_k) \leqq 0 \text{ for } 1 < j \leqq k < n,$$

(2.12) $$x_1,u_2,\ldots,u_{n-1} \text{ is a } W_n(I)\text{-system.}$$

If, in addition, $u_n \in C^n(I)$ satisfies (2.7) and

(2.13) $$W(u_k,\ldots,u_n) \geqq 0 \text{ for } 1 \leqq k \leqq n,$$

then $k = n$ is permitted in (2.10) and (2.11) and, if (2.13) is strengthened to (2.8), then

(2.14) $$W(x_1,u_k,\ldots,u_n) > 0 \text{ for } 2 \leqq k \leqq n.$$

The proofs of Lemma 2.1 and Theorem 2.1 will not be given here, see [7].

3. WRONSKIAN IDENTITIES

Computations below depend on some identities for Wronskians which we state here for easy reference. Here $\sigma_0 = 1, \sigma_k = W(v_1,\ldots,v_k)$, and $\hat{v}_j$ indicates the omission of v_j.

(3.1) $$[W(v_1,\ldots,v_{k-1},x)/\sigma_k]' = W(v_1,\ldots,v_k,x)\sigma_{k-1}/\sigma_k^2;$$

(3.2) $$W(v_0,\ldots,v_k) = v_0^{k+1}W((v_1/v_0)',\ldots,(v_k/v_0)');$$

(3.3) $$v_0^{k-1}W(v_0,\ldots,v_k) = W[W(v_0,v_1),\ldots,W(v_0,v_k)];$$

(3.4) $$\sigma_{m-1}W(xv_1,\ldots,\hat{v}_j,\ldots,v_m) = \sigma_m W(x,v_1,\ldots,\hat{v}_j,\ldots,v_{m-1}) + W(x,v_1,\ldots,v_{m-1})W(v_1,\ldots,\hat{v}_j,\ldots,v_m);$$

$$(3.5)\qquad W(x,v_1,\ldots,\hat{v}_j,\ldots,v_m)$$

$$= \sigma_m \sum_{k=j-1}^{m-1} W(x,v_1,\ldots,v_k)W(v_1,\ldots,\hat{v}_j,\ldots,v_{k+1})/\sigma_k\sigma_{k+1}\,.$$

(For these identities, see, e.g., (4.9),(2.3) and Corollary 2.2 in [5] and Proposition 15.1 in [7].)

4. PROOF OF LEMMA 1.1

Assume that (1.8),(1.10) hold. Put $v_1 = u_1$ and, for $2 \leqq k \leqq n - 1$, let $v_k(t)$ be the $k \times k$ determinant in which the j-th column is

$$(4.1)\qquad (u_j(a),u_j'(a),\ldots,u_j^{(k-2)}(a),u_j(t)).$$

Expanding the determinant along the last row gives

$$(4.2)\qquad v_k(t) = \sum_{j=1}^{k}(-1)^{j+k}\,c_{kj}u_j(t),\ \text{where}\ \ c_{kj} = W(u_1,\ldots,\hat{u}_j,\ldots,u_k)(a) \geqq 0$$

and, for $j = k$,

$$c_{kk} = W(u_1,\ldots,u_{k-1})(a) > 0.$$

Note that $v_1,\ldots,v_{n-1}$ is a $w_n([a,b))$-system, since $v_1 = u_1$ and

$$W(v_1,\ldots,v_k) = c_{22}\ \cdots\ c_{kk}\ W(u_1,\ldots,u_k) > 0.$$

Also, $v_1(a) = u_1(a) > 0$ and

$$v_k(a) = \ldots = v_k^{(k-2)}(a) = 0,\ v_k^{(k-1)}(a) = W(u_1,\ldots,u_k)(a) > 0.$$

We now use a double induction to show that

$$(4.3)\qquad W(v_j,\ldots,v_k) > 0\ \text{ on }\ (a,b)\ \text{ for }\ 1 \leqq j \leqq k < n.$$

The inequalities in (4.3) hold for $j = 1$ and $1 \leqq k < n$. Assume their validity if $j > 1$ and j is replaced by $j - 1$. Note that

$$(v_j/v_{j-1})' = W(v_{j-1},v_j)/v_{j-1}^2 > 0\ \text{ on }\ (a,b),$$

so that $v_j(a) = 0$ implies $v_j > 0$ on (a,b). This gives the inequality in (4.3) if $k = j$. Assume its validity if k is replaced by $k - 1$, where $j \leqq k - 1 < n - 1$. By (3.1),

$$[W(v_j,\ldots,v_k)/W(v_{j-1},\ldots,v_{k-1})]' = W(v_{j-1},\ldots,v_k)W(v_j,\ldots,v_{k-1})/[W(v_{j-1},\ldots,v_{k-1})]^2$$

is positive on (a,b). Thus, since $W(v_j,\ldots,v_k)$ vanishes at $t = a$, it is positive on (a,b). This completes the proof.

5. PROOF OF "SUFFICIENCY" IN THEOREM 1.2

In view of Lemma 1.1, it suffices to prove the sufficiency of the set of conditions (1.6)-(1.9). It will be convenient to replace [a,b) by (a,b] and prove

(A_n) Let $I = [a,b]$, $I' = (a,b]$, $I^0 = (a,b)$. Let $u_1,\ldots,u_{n-1} \in C^n(I')$ satisfy (1.7) on I' and be a $w_n(I')$-system and a $W_n(I^0)$-system. Then (*) is disconjugate on I.

Let $\omega_0 = 1$ and $\omega_k = W(u_1,\ldots,u_k)$ for $1 \leqq k \leqq n$, where $u_n \in C^n(I')$ satisfies $\omega_n > 0$. Let $y = (y_1,\ldots,y_n)$ be the vector with the components

(5.1) $$y_k = W(x,u_1,\ldots,u_{k-1})/\omega_k \quad \text{for } 1 \leqq k \leqq n, \text{ where } y_1 = x/u_1.$$

Then, by (3.1),

$$y_k' = -W(x,u_1,\ldots,u_k)\omega_{k-1}/\omega_k^2 \quad \text{for } 1 \leqq k \leqq n,$$

so that

(5.2) $$y_k' = -y_{k+1}\omega_{k-1}\omega_{k+1}/\omega_k^2 \quad \text{for } 1 \leqq k < n,$$

(5.3) $$y_n' = -W(x,u_1,\ldots,u_n)\omega_{n-1}/\omega_n^2 .$$

To the last row of $W(x,u_1,\ldots,u_n)$ in (5.3), add $a_0(t)$ times the first row, $a_1(t)$ times the second, etc. When x is a solution of (*), the last row becomes $(0,P_n(D)[u_1],\ldots,P_n(D)[u_n])$. Expanding the determinant along the last row, we see that

(5.4) $$W(x,u_1,\ldots,u_n) = \sum_{j=1}^{n} (-1)^{n+j}P_n(D)[u_j].W(x,u_1,\ldots,\hat{u}_j,\ldots,u_n).$$

Hence (3.5) implies the existence of continuous $b^m(t)$ on I' such that (5.3) becomes

(5.5) $$y_n' = -\sum_{m=1}^{n} b^m(t)y_m, \quad \text{where } b^m(t) \geqq 0 \text{ on } I' \text{ for } 1 \leqq m < n;$$

in fact, we have

(5.6) $$b^m(t) = (\omega_{n-1}/\omega_m\omega_n) \sum_{j=1}^{m}(-1)^{n+j}P_n(D)[u_j].W(u_1,\ldots,\hat{u}_j,\ldots,u_m).$$

Thus (*) is equivalent to a first order system $z' = -A(t)z$ for the vector $z = (y_1,\ldots,y_{n-1},y_n \exp(-\int_t^b b^n(s)ds)$, where the entries in the matrix $A(t)$ are non-negative on I'.

Let $x = x(t) = x_0(t)$ be the solution of (*) determined by initial conditions

(5.7) $$z_k = 1 > 0 \quad \text{at} \quad t = b.$$

Then $z_k > 0$, $z_k' \leqq 0$ on I', so that $x_0 > 0$ on I and

(5.8) $$W(x_0,u_1,\ldots,u_k) > 0 \quad \text{on} \quad I' \quad \text{for} \quad 1 \leqq k < n.$$

In particular, $x_0,u_1,\ldots,u_{n-1}$ is a $W_{n+1}(I^0)$-system, so that

(5.9) $$x_0,u_2,\ldots,u_{n-1} \text{ is a } W_n(I^0)\text{-system;}$$

cf. the remark concerning the equivalence of (1.2) and (1.3). Also

(5.10) $$x_0,u_2,\ldots,u_{n-1} \text{ is a } w_n(I')\text{-system.}$$

In fact, $x_0 > 0$, while the inequality $W(x_0,u_2,\ldots,u_m) > 0$ on I' for $2 \leqq m \leqq n - 1$ follows from (3.5) if $j,x,v_1,\ldots,v_m$ are replaced by $1,x_0,u_1,\ldots,u_m$, since all terms on the right are non-negative, while the term for $k = j - 1 = 0$ is $\omega_m x_0/u_1 > 0$ on I'.

Assume $n > 2$ and the validity of (A_{n-1}). Make the variation of constants

(5.11) $$x = x_0(t)v$$

in (*). Since $v = 1$ is a solution of the resulting equation, there is a polynomial $P_{n-1}(t,\lambda)$ of degree $n - 1$ with continuous coefficients and positive leading coefficient such that

(5.12) $$P_{n-1}(t,D)[v'] = P_n(t,D)[x] \quad \text{if} \quad v = x/x_0(t).$$

The equation

(5.13) $$P_{n-1}(t,D)[z] = 0$$

is disconjugate on I by the induction hypothesis. For if $v_k = (u_{k+1}/x_0)'$ for $1 \leqq k \leqq n - 2$, then $v_1,\ldots,v_{n-2}$ is a $w_{n-1}(I')$-system and a $W_{n-1}(I^0)$-system by (3.2) and (5.9)-(5.10), and satisfies

$$(-1^{n-1+k} P_{n-1}(D)[v_k] = (-1)^{n+k+1} P_n(D)[u_{k+1}] \geqq 0$$

on I'. The mean value theorem implies that (*) is disconjugate on I (for otherwise there is a solution $x \not\equiv 0$ of (*) having n zeros on I, in which case,

$z = (x/x_0)' \not\equiv 0$ is a solution of (5.13) having $n - 1$ zeros on I). This proves (A_n).

REFERENCES

[1] W. A. Coppel, Disconjugacy, Australian National University, mimeographed notes (1970); to appear as Lecture Notes, Springer.

[2] M. Fekete, Ueber ein Problem von Laguerre, Rend. Circ. Mat. Palermo 34(1912) 89-100.

[3] P. Hartman, Unrestricted n-parameter families, Rend. Circ. Mat. Palermo (2) 7(1958)123-142.

[4] P. Hartman, Disconjugate n-th order equations and principal solutions, Bull. Amer. Math. Soc. 74(1968)125-129.

[5] P. Hartman, Principal solutions of disconjugate n-th order linear differential equations, Amer. J. Math. 91(1969)306-362.

[6] P. Hartman, On N-parameter families and interpolation problems for non-linear ordinary differential equations, Trans. Amer. Math. Soc. 154(1971) 201-226.

[7] P. Hartman, Corrigendum and addendum: Principal solutions of disconjugate n-th order linear differential equations, Amer. J. Math. 93(1971)439-451.

[8] L. K. Jackson, Disconjugacy conditions for linear third order equations, J. Diff. Equations 4(1968)369-372.

[9] A. Yu. Levin, Non-oscillation of solutions of the equation $x^{(n)} + p_1(t)x^{(n-1)} + \dots + p_n(t)x = 0$, Uspehi Mat. Nauk. 24 (1969) No.2(146)43-96; Russian Math. Surveys 24(1969)No.2,43-99.

[10] C. Olech, Asymptotic behavior of the solutions of second order differential equations, Bull. Acad. Polon. Sci. (Ser. Sci. Math. Astro. Phys.) 7(1959)319-326.

[11] Z. Opial, On a theorem of Aramă, J. Diff. Equations 3(1967)88-91.

[12] G. Pólya, On the mean value theorem corresponding to a given linear homogeneous differential equation, Trans. Amer. Math. Soc. 24(1922)312-324.

[13] J. D. Schuur, The asymptotic behavior of solutions of third order linear differential equations, Proc. Amer. Math. Soc. 18(1967)391-393.

[14] T. L. Sherman, On solutions of n-th order linear differential equations with n zeros, Bull. Amer. Math. Soc. 74(1968)923-925.

ON THE RIEMANN PROBLEM ON RIEMANN SURFACES

Tosihusa Kimura

§ 1. Introduction

Let X be an arbitrary Riemann surface and consider on X the system of linear differential equations

$$(E) \qquad dy = y\Omega,$$

where y is a complex row vector of dimension n and Ω is an $n \times n$ matrix whose entries are meromorphic 1-forms over X. Let a be a point of X, (θ, U) an admissible chart at a and t the local parameter associated to (θ, U) : $t = \theta(x)$. Then system (E) is expressible as

$$(1.1) \qquad \frac{dy}{dt} = yA(t),$$

where $A(t)$ is meromorphic in $\theta(U)$. Since it does not depend on the choice of charts whether or not $A(t)$ is holomorphic at $t = \theta(a)$, we are naturally led to the concept of a singular point of (E). Furthermore, we can give the definition of a regular singular point.

System (E) is called a Fuchsian one of its singularities are all regular singular points. Throughout this paper it is supposed that system (E) is a Fuchsian one.

Denote by S the set of all singular points of (E), which is discrete in X.

We take an arbitrary point x_0 in $X - S$. Let $\Phi(x)$ be a fundamental matrix of solutions of (E) defined in the neighborhood of x_0. $\Phi(x)$ is analytically continuable along any path in $X - S$ and naturally induces a homomorphism κ from the fundamental group $\pi_1(X-S, x_0)$

into $GL(n, \mathbb{C})$, called the <u>monodromy representation</u> associated to $\Phi(x)$. It is well known that the monodromy representation is uniquely determined up to inner automorphism. This situation does not depend on the choice of x_0. Thus system (E) determines an equivalent class of representations K of $\pi_1(X - S)$ on $\mathbb{C}^n$.

We call a triple (X, S, K) a <u>Riemann datum of rank</u> n, if X is a Riemann surface, S is a discrete subset of X and K is an equivalent class of representations of $\pi_1(X - S)$ on $\mathbb{C}^n$. X will be called the base surface, S the set of singularities and K the monodromy.

The above fact is expressible by saying that system (E) determines a Riemann datum of rank n. Roughly speaking, the Riemann problem is that of determining Fuchsian systems whose Riemann datum coincides with a given one.

§2. Classification of regular singular points

Let a be a regular singular point of (E) and (1.1) a representation of (E) by means of a local parameter at a : $t = \theta(x)$. We may suppose that $\theta(a) = 0$.

System (1.1) admits a fundamental matrix of solutions $\Phi(t)$, called a <u>canonical fundamental matrix</u>, of the form

$$\phi(t) = t^E t^N P(t),$$

where

(1) E is a constant matrix such that

$$E = \begin{bmatrix} E_1 & & 0 \\ & \ddots & \\ 0 & & E_s \end{bmatrix}, \qquad E_j = \begin{bmatrix} J_{j1} & & 0 \\ * & \ddots & \\ * & * & J_{je_j} \end{bmatrix}, \qquad (j = 1, \cdots, s)$$

$$J_{jk} = \begin{bmatrix} \lambda_{jk} & & & \\ 1 & \ddots & & \\ & \ddots & \ddots & \\ & & 1 & \lambda_{jk} \end{bmatrix} \quad (k=1,\cdots,e_j) \text{ and } 0 \leq \operatorname{Re} \lambda_j < 1 \ (j=1,\cdots,s);$$

(2) N is a constant matrix with integral entries such that

$$N = \begin{bmatrix} N_1 & & 0 \\ & \ddots & \\ 0 & & N_s \end{bmatrix}, \quad N_j = \begin{bmatrix} \nu_{j1} I_{j1} & & 0 \\ & \ddots & \\ 0 & & \nu_{je_j} I_{je_j} \end{bmatrix} (j=1,\cdots, s),$$

$\nu_{j1} \geq \nu_{j2} \geq \cdots \geq \nu_{je_j}$ $(j = 1,\cdots, s)$, I_{jk} being the identity matrix of the same degree as J_{jk};

(3) $P(t)$ is a matrix-valued function holomorphic at $t = 0$ and $P(0) \neq 0$.

Definition 2.1. The regular singular point a is called

(s) simple if $\det P(0) \neq 0$;

(n) normal if $E_j = \bigoplus_{k=1}^{e_j} J_{jk}$ for all j;

(a) apparent if $E = 0$.

It is well known that:

a is simple $\iff$ $t = 0$ is a simple pole of $A(t)$,

a is normal $\iff$ a is simple and $E+N$ is similar to the residue of $A(t)$ at $t = 0$,

a is apparent $\iff$ all the solutions of (E) is single-valued in the neighborhood of a.

§ 3. Equivalence of Riemann data

An apparent singularity plays only a trivial role in the monodromy. This suggests the following definition.

Definition 3.1. Let (X, S, K) be a Riemann datum. A Riemann datum (X, S_0, K_0) is called the reduced one for (X, S, K) if $S_0 \subset S$ and for any $\kappa \in K$ there exists a $\kappa_0 \in K_0$ such that the diagram

$$\begin{array}{ccc} \pi_1(X-S,\ x_0) & \xrightarrow{i_*} & \pi_1(X-S_0,\ x_0) \\ & \searrow^{\kappa} \quad {}^{\kappa_0}\swarrow & \\ & GL(n,\ \mathbb{C}) & \end{array}$$

commutes and if S_0 is the smallest subset with the above properties, where $x_0 \notin S$ and i_* is the map induced from the inclusion map $i : X-S \longrightarrow X-S_0$. Points in $S-S_0$ are called apparent.

Definition 3.2. A Riemann datum (X, S, K) is said to be equivalent to (X, S', K') if the reduced datum for (X, S, K) coincides with that for (X, S', K').

It is clear that this relation is an equivalence relation.

§ 4. Riemann's problem

We are now in a position to formulate the Riemann problem precisely: To find Fuchsian systems whose Riemann data are equivalent to a prescribed one.

This problem was solved by Röhrl in full generality.

Theorem (Röhrl). Let (X, S, K) be an arbitrary Riemann datum. Then there exist Fuchsian systems whose Riemann data are equivalent to (X, S, K). If X is not compact, then there exist Fuchsian systems whose data are just (X, S, K) and whose singularities are all normal.

In what follows we suppose that X is compact. Then S is a finite set and we put

$$S = \{a_1, \cdots, a_r\}.$$

Historically, the Riemann problem has been studied mainly in the case that X is the complex projective line $\mathbb{P}'$ (the Riemann sphere) by Schlesinger, Plemelji, Garnier, Lappo-Danilevsky and so on. In this case system (E) is written as

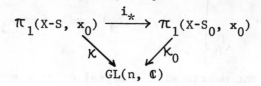

$$\frac{dy}{dx} = yA(x),$$

where $A(x)$ is a matrix whose entries are rational functions.

If the genus of X is ≥ 1, then apparent singularities appear in general outside S.

§ 5. Our problem

Let (E) be a Fuchsian system with Riemann datum equivalent to (X, S, K). Then any transformation

$$y = z\,T(x) \qquad (T(x)\text{: matrix of meromorphic functions})$$

takes (E) into a system whose Riemann datum also is equivalent to (X, S, K) and the converse is true. This shows that there exists in general an infinite number of systems with Riemann data equivalent to (X, S, K).

Our problem is as follows: To find for a given Riemann datum (X, S, K) a Fuchsian system, canonical in some sense, with datum equivalent to (X, S, K). The followings may be considered as criteria for canonicity:

(1) the number of apparent singularities is minimum regardless of the character of singularities;

(2) all singularities have a good character, simple or normal say;

(3) combination of (1) and (2) in an appropriate sense.

§ 6. Complex analytic principal bundle induced from a Riemann datum

Let there be given a Riemann datum (X, S, K). Röhrl's method of proving his theorem consists in constructing a complex analytic principal bundle over X with structure group $GL(n, \mathbb{C})$ from (X, S, K) and then showing the existence of meromorphic cross-sections. Our method is based on his idea.

Let κ be a monodromy representation belonging to K : κ : $\pi_1(X-S, x_0) \longrightarrow GL(n, \mathbb{C})$. We choose a suitable open covering $\mathcal{U} = \{U_j\}$, and, using κ and $\mathcal{U}$, construct a cocycle in $Z^1(\mathcal{U}, \mathcal{GL}(n, \mathbb{C}))$, $\{F_{j\kappa}(x)\}$, $F_{j\kappa} : U_j \cap U_\kappa \longrightarrow GL(n, \mathbb{C})$, where $\mathcal{GL}(n, \mathbb{C})$ is the sheaf of germs of holomorphic maps from X into $GL(n, \mathbb{C})$. The cocycle $\{F_{j\kappa}(x)\}$ defines a complex analytic principal bundle ξ over X with structure group $GL(n, \mathbb{C})$. Our construction of ξ is made so that ξ possesses the following properties:

(1) the restriction of ξ to $X - S$ is flat,

(2) any meromorphic section of ξ over X can be interpreted in a natural way as a fundamental matrix of solutions of a certain Fuchsian system with Riemann datum equivalent to (X, S, K),

(3) for any meromorphic section, a point $a \notin X-S$ is an apparent singularity of the Fuchsian system corresponding to the section if and only if a is a singular point of the section,

(4) if a meromorphic section is holomorphic at $a_j \in S$, then a_j is a normal singularity of the corresponding Fuchsian system.

Here the following convention is made: a is said to be a <u>singular point</u> of a meromorphic section, if a is a pole or a point where the section has zero determinant.

It should be noted that ξ is not unique and the arbitrariness arises from the fact that the logarithmic function of a matrix is many-valued.

By the properties above, our problem is reduced to that of finding suitable meromorphic sections of ξ. In fact, under the assumption that (X, S, K) itself is a reduced datum, the determination of a Fuchsian system satisfying condition (2) stated in § 5 is equivalent to that of a meromorphic section with minimum number of singularities outside S.

§ 7. Case that the genus of X is zero

Suppose that $X = \mathbb{P}^1$. There is no loss in generality in supposing that none of the singularities a_j is the point at infinity and that the point at infinity is contained in only one open set $U_\infty \in \mathcal{U}$. From Grothendieck's result it follows that ξ is isomorphic to the direct sum η of n line bundles, that is, there exists a cocycle $\{G_{jk}(x)\}$ such that each $G_{jk}(x)$ has a diagonal form and is connected to $F_{jk}(x)$ by $G_{jk}(x) = T_j(x)G_{jk}(x)T_k^{-1}(x)$ $(x \in U_j \cap U_k)$, where $T_j(x)$ is a holomorphic map from U_j into $GL(n, \mathbb{C})$. We may suppose that η has a section $\{\Psi_j(x)\}$ defined by

$$\Psi_j(x) = \begin{cases} \begin{bmatrix} x^{-d_1} & & 0 \\ & \ddots & \\ 0 & & x^{-d_n} \end{bmatrix} & \text{in } U_\infty \\ I & \text{in } U_j \neq U_\infty, \end{cases}$$

where the d's are integers. We conclude that the bundle ξ admits the section $\{\Phi_j(x)\}$ defined by

$$\Phi_j(x) = \begin{cases} T_j^{-1}(x) \begin{bmatrix} x^{-d_1} & & 0 \\ & \ddots & \\ 0 & & x^{-d_n} \end{bmatrix} & \text{in } U_\infty \\ T_j^{-1}(x) & \text{in } U_j \neq U_\infty . \end{cases}$$

Lemma. Let $\Phi(x)$ be a matrix-valued function meromorphic in a neighborhood of $x = \infty$ and $\det \Phi(x) \not\equiv 0$. Then there exists a matrix $T(x)$ with the following properties:

(1) the elements of $T(x)$ are all polynomials,

(2) $\det T(x) \equiv \text{const.}\ (\neq 0)$

(3)

$$\Phi(x)T(x) = \begin{bmatrix} x^{-\nu_1} & & 0 \\ & \ddots & \\ 0 & & x^{-\nu_n} \end{bmatrix} P(x),$$

where the ν's are integers and $P(x)$ is holomorphic at $x = \infty$ and $\det P(\infty) \neq 0$.

Combining the fact above and this lemma, we obtain the

Theorem. Let $(\mathbb{P}^1, S, K)$ be a Riemann datum with $S = \{a_1, \cdots, a_r\}$, $a_j \neq \infty$, $r \geqq 3$. Then there exists a Fuchsian system of the form

$$\frac{dy}{dx} = y \sum_{j=1}^{r} \frac{A_j}{x - a_j} \qquad (A_j : \text{ constant matrices}),$$

whose Riemann datum is equivalent to $(\mathbb{P}^1, S, K)$ and whose singularities a_j are normal. The singularity $x = \infty$ is in general an apparent singularity.

If there exists at least one singularity in S to which corresponds a matrix similar to a diagonal one by a monodromy representation in K, we can determine the system so that $x = \infty$ is a regular point, that is,

$$\sum_{j=1}^{r} A_j = 0.$$

§8. Case that the genus of X is > 0

Suppose that the genus of X is positive. We may suppose that (X, S, K) is a reduced datum.

In the easiest case when $S = \phi$ the bundle ξ itself is flat.

Suppose that $S \neq \phi$. Then there exists a point in S, a_1 say, which is not apparent. We may suppose that a_1 is contained in only one open set $U_1 \in \mathcal{U}$. Since any vector bundle over a compact Riemann surface is isomorphic to the direct sum of reducible but indecomposable bundles, there exists a cocycle $\{G_{jk}(x)\}$ in $Z^1(\mathcal{U}, \mathcal{GL}(n, \mathbb{C}))$ and a cocycle $\{T_j(x)\}$ in $Z^0(\mathcal{U}, \mathcal{GL}(n, \mathbb{C}))$ such that

$$G_{jk}(x) = T_j(x)F_{jk}(x)T_k^{-1}(x) \qquad (x \in U_j \cap U_k)$$

and

$$G_{jk}(x) = \begin{bmatrix} G_{jk1}(x) & & 0 \\ & \ddots & \\ 0 & & G_{jks}(x) \end{bmatrix}, \qquad G_{jk\alpha}(x) = \begin{bmatrix} g_{jk\alpha 1}(x) & * & * \\ & \ddots & * \\ 0 & & g_{jk\alpha n_\alpha}(x) \end{bmatrix}.$$

For any fixed α, $\{G_{jk\alpha}(x)\}$ defines a bundle η_α of rank n_α and $\{g_{jk\alpha 1}(x)\}$ does a line bundle $\eta_{\alpha 1}$ which is subbundle of η_α. Let d_α be the degree of η_α and we define a line bundle by

$$\psi_{jk}(x) = \begin{cases} t_1^{-d_\alpha} & \text{in } U_1 \cap U_k \quad (U_k \neq U_1) \\ 1 & \text{in } U_j \cap U_k \quad (U_j, U_k \neq U_1), \end{cases}$$

where t_1 is a local parameter in U_1 taking zero at a_1. We set

$$H_{jk}(x) = \begin{bmatrix} H_{jk}(x) & & 0 \\ & \ddots & \\ 0 & & H_{jks}(x) \end{bmatrix},$$

$$H_{jk\alpha}(x) = \begin{cases} \begin{bmatrix} g_{jk\alpha 1}(x)t^{-d_1} & * & * \\ & g_{jk\alpha 2}(x) & * \\ & & \ddots & \\ 0 & & & g_{jk\alpha n_\alpha}(x) \end{bmatrix} & \text{in } U_1 \cap U_k \ (U_k \neq U_1) \\ G_{jk\alpha}(x) & \text{in } U_j \cap U_k \ (U_j, U_k \neq U_1). \end{cases}$$

Then for every α, $\{H_{jk\alpha}(x)\}$ defines an indecomposable bundle with degree 0, so it is isomorphic to a flat bundle. We can deduce from ξ a flat bundle ζ in this way. We see that if a meromorphic section of ζ is holomorphic at a point different from a_1, then the corresponding section of ξ also is holomorphic at this point.

Thus we arrive at the

Theorem. Suppose that the genus of X is > 0 and $S \neq \emptyset$. Then our problem is reduced to that of obtaining suitable sections of a flat bundle, provided that the existence of a point is permitted where a non simple singularity appears.

ANALYTIC THEORY OF LINEAR DIFFERENTIAL SYSTEMS

W. A. Harris, Jr.

The purpose of this paper is to present some recent results in the analytic theory of linear differential systems. In particular, we are concerned with the classification of singular points and the behavior of solutions near such singular points.

1. Introduction

A natural starting point for the analytic theory of linear differential systems is the fundamental existence and uniqueness theorem.

If the n by n matrix A is holomorphic at $z = z_0$ and w_0 is a constant n-vector, then the linear differential system

$$\frac{dw}{dz} = A(z)\,w$$

has a unique solution holomorphic at $z = z_0$ satisfying the initial condition $w(z_0) = w_0$.

Thus, linear differential systems have the characteristic property that singular behavior of solutions is directly attributable to singularities of the coefficient matrix A and we are led to consider the case in which $z = z_0$ is an isolated singular point of the coefficient matrix A. For such systems we have the following general result.

If the matrix A is holomorphic for $0 < |z - z_0| < \delta$, then the linear differential system

$$\frac{dw}{dz} = A(z)\,w \qquad (1)$$

Supported in part by the United States Army under Contract DA-AR0-D-31-124-71-G14.

has a fundamental matrix of the form

$$W(z) = S(z)\, z^{R}$$

such that $S(z)$ is holomorphic for $0 < |z-z_0| < \delta$ and R is a constant matrix.

The nature of the possible singularity of S at the point $z = z_0$ leads to the following classification of the singular point $z = z_0$.

The point $z = z_0$ is called a regular singular point for the linear differential system (1) if S has at most a pole at the point $z = z_0$, otherwise $z = z_0$ is called an irregular singular point for the linear differential system (1).

This classification, based on knowledge of a fundamental matrix, is not immediately apparent for a given system. In the case that the coefficient matrix A has a pole at the point $z = z_0$ other classification are helpful.

If A has a pole at the point $z = z_0$ we may write

$$A(z) = (z - z_0)^{-p-1}\, \tilde{A}(z)$$

where p is an integer, A is holomorphic at $z = z_0$, and $\tilde{A}(z_0) \neq 0$. The nonnegative integer p is called (after Poincaré) the rank of the singularity. There is a significant difference between the cases $p = 0$, and $p \geq 1$, and accordingly if $p = 0$, the point $z = z_0$ is called a singular point of the first kind (simple singularity), and if $p \geq 1$, the point $z = z_0$ is called a singular point of the second kind.

The classification, regular - irregular singular point, gives more information while the classification, singular point of the first-second kind, is easier to check. These two classifications are not directly comparable(except for the scalar case). If $z = z_0$ is a singular point of the first kind, then the point $z = z_0$ is a regular singular point, whereas if $z = z_0$ is a singular point of the second kind the point $z = z_0$ may or may not be a regular singular point.

Clearly it is an important problem to characterize linear differential systems for which $z = z_0$ is a singular point of the second kind with rank $p \geq 1$

for which $z = z_0$ is a regular singular point. We shall discuss characterizations of linear differential systems with a regular singular point in Section 2.

There exist methods by which linear differential systems of rank $p \geq 1$ may be studied in detail. These methods are complicated and in general will mask the existence of holomorphic solutions. In Section 3 we present a new method, based on the techniques of functional analysis, for determining the existence of holomorphic solutions of linear differential systems. This method provides simple proofs of the basic results for linear differential systems with a singular point of the first kind ($p = 0$, simple singularity) as well as establishing the existence of holomorphic solutions for linear differential systems of rank $p \geq 1$. This method holds great promise of further applications.

2. Characterizations of a regular singular point

Conceptually, a regular singular point is more intrinsic than the rank of a singular point and correspondingly more difficult to ascertain for a given system. This is easily explained by noting that a regular singular point is invariant under changes of dependent variable whereas rank is not, i.e. if T is a nonsingular matrix meromorphic at $z = 0$, the transformation $w = Tu$ carries the linear differential system

$$\frac{dw}{dz} = A(z)\, w$$

into the linear differential system

$$\frac{du}{dz} = B(z)\, u$$

where

$$B = T^{-1} A T - T^{-1} \frac{d}{dz} T \tag{2}$$

and clearly, $z = 0$ is a regular singular point for $\frac{dw}{dz} = A(z)\, w$ if and only if $z = 0$ is a regular singular point for $\frac{du}{dz} = B(z)\, u$. For meromorphic T with $\det T \neq 0$, equation (2) sets up an equivalence relation between the linear differential systems $\frac{dw}{dz} = A(z)\, w$ and $\frac{du}{dz} = B(z)u$ which we denote by $A \sim B$.

Clearly, a necessary and sufficient condition that the linear differential system $\frac{dw}{dz} = A(z)\, w$ have $z = 0$ as a regular singular point is that it be equivalent to a system of rank zero. Even though this condition is necessary and sufficient for the desired structure of a fundamental matrix, it cannot be used to resolve the question for a pre-assigned system. J. Moser [9] has given an algorithm to determine whether a given system of rank $p \geq 1$ has a regular singular point which is based on the following theorems.

Theorem 1.

Let $A(z)$ *and* $B(z)$ *be meromorphic at* $z = 0$,

$$A(z) = z^{-q} \sum_{k=0}^{\infty} A_k z^k,\ A_0 \neq 0\ ,\ B(z) = z^{-q} \sum_{k=0}^{\infty} B_k z^k,\quad |z| < \delta\ .$$

A necessary and sufficient condition that $A \sim B$ *such that* $r =$ *rank* $A_0 >$ *rank* B_0 *is that the polynomial*

$$\beta(\lambda) = z^r \det(\lambda I + z^{q-1} A(z))\,\Big|_{z=0} = \sum_{k=0}^{n-r} \lambda^k \beta_k(A_0, A_1)$$

vanish identically in λ.

(In particular, a necessary condition for a regular singular point is that A_0 is nil potent, i.e. $A_0^n = 0$)

Theorem 2.

Let $A(z)$ *and* $B(z)$ *be meromorphic at* $z = 0$

$$A(z) = z^{-q} \sum_{k=0}^{\infty} A_k z^k\ ,\ A_0 \neq 0,\quad B(z) = z^{-q} \sum_{k=0}^{\infty} B_k z^k,\ |z| < \delta\ .$$

If $A \sim B$ *and rank* $A_0 >$ *rank* B_0, *then* $A \sim C = T^{-1} A T - T^{-1} \frac{d}{dz} T$, *where*

$$T = (P_0 + P_1 z) \text{ diag } (1, 1, \cdots 1, z, \ldots, z)$$

with constant matrices P_0, P_1, det $P_0 \neq 0$.

The proof of Theorem 2 provides a means of computing the matrix T, i.e. the matrices P_0 and P_1. A repeated application of these two theorems will determine whether a given system has $z = 0$ as a regular singular point.

W. B. Jurkat and D. A. Lutz [5, 6] have given the following algorithm to determine whether a system of rank $p \geq 1$ has a regular singular point.

Construct the sequence of matrices

$$\mathcal{a}_0 = I$$

$$\mathcal{a}_{\nu+1} = \mathcal{a}_\nu A + \frac{d}{dz} \mathcal{a}_\nu \qquad \nu = 0, 1, \cdots$$

and let $\rho(\mathcal{a}_\nu)$ denote the order of the meromorphic matrix $\mathcal{a}_\nu$.

Theorem 3.

Let the n by n matrix A be meromorphic at $z=0$, $A(z) = z^{-q} \sum_{k=0}^{\infty} A_k z^k$, $A_0 \neq 0$, $|z| < \delta$. The linear differential system $\frac{dw}{dz} = A(z)w$ has $z = 0$ as a regular singular point if and only if

$$\rho[\mathcal{a}_\nu] \leq \nu + (n-1)(q-1)$$

for $\nu = n, n+1, \ldots$, $(n-1)(2n(q-1)-1)$.

The application of these algorithms is lengthy and it is desirable to have alternate criteria. W. A. Harris, Jr and D. A. Lutz [2, 7] have given further criteria in the form of necessary conditions which are contained in the following theorems.

Theorem 4.

Let A be meromorphic at $z=0$,

$$A(z) = z^{-q} \sum_{k=0}^{\infty} A_k z^k, \; A_0 \neq 0, \; |z| < \delta, \; q > 0,$$

and

$$\underline{\det}\ (\lambda I - A) = \lambda^n + a_1 \lambda^{n-1} + \cdots + a_n .$$

If $\frac{dw}{dz} = A(z)w$ has $z = 0$ as a regular singular point, then

(i) $a_k(z) = 0(z^{-(k-1)q} + z^{-k}),\ k=1,\ldots,n$,

(ii) $A_0^k = 0$ for some $k \leq n$,

(iii) $\underline{\mathrm{trace}}\ (A_0^k A_1) = 0 \begin{cases} k=0,1,\ldots,n-1, & \text{if } q \geq 3 \\ k=1,\ldots\ n-1, & \text{if } q=2 \end{cases}$.

Theorem 5.

Let A be as in Theorem 1 and let

$$a_k(z) = 0\ (z^{-(k-1)q} + z^{-k}),\quad k=1,\ldots,n \quad .$$

Then

(i) $A_0^n = 0$;

and if $A_0^{n-1} \neq 0$,

(ii) $\underline{\mathrm{trace}}\ (A_0^k A_1) = 0 \begin{cases} k=0,1,\ldots,n-1, & \text{if } q \geq 3 \\ k=1,\ldots,n-1, & \text{if } q=2 \end{cases}$.

D.A. Lutz [8] has shown that H.L. Turrittin's [10] rank reduction process preserves a regular singular point and utilized the results of Theorems 4, 5 to obtain further necessary conditions.

Theorem 6.

Let the n by n matrix A be meromophic at $z = 0$,

$$A(z) = z^{-q} \sum_{k=0}^{\infty} A_k z^k ,\quad A_0 \neq 0,\ |z| < \delta .$$

If the linear differential system $\frac{dw}{dz} = A(z)w$ has $z=0$ as a regular singular point, then

$$\underline{\text{trace}} \left(\sum_{j=1}^{q-1} \sum_{m=0}^{j-1} \left(\sum_{i_1+\cdots+i_k=m} A_{i_1} A_{i_2} \cdots A_{i_k} \right) A_{q-1-m} \right) = 0$$

for $1 \leq k \leq rn - 1$.

These conditions are not sufficient for a regular singular point since they involve only $A_0, \ldots, A_{q-1}$, and it is known that necessary and sufficient conditions may depend non-trivially on the first $n(q-1)$ coefficients in the expansion for $A(z)$.

Clearly, additional research is needed on the characterization of linear differential systems with a regular singular point.

3. A New Method.

W. A. Harris, Jr., Y. Sibuya, and L. Weinberg [3] have presented a new method for determining the existence of holomorphic solutions of linear differential systems. This method is based on the techniques of functional analysis and reduces the existence problem to the solution of a determining (bifurcation) equation. This method is embodied in the following theorem.

Theorem 7.

Let $A(z)$ be an n by n matrix which is holomorphic at $z = 0$, and let $D = \mathrm{diag}(d_1, d_2, \ldots, d_n)$ with non-negative integers d_i. For every N sufficiently large and every polynomial $\varphi(z)$ in z with $z^D \varphi(z)$ of degree N, there exists a polynomial $f(z; A, \varphi, N)$ in z of degree $N-1$ with coefficients that depend upon A, N, and φ such that the linear differential system

$$(1) \qquad z^D \frac{dy}{dz} = A(z)\, y + f(z; A, \varphi, N)$$

has a solution holomorphic at $z = 0$. Further, f and y are linear and homogeneous in φ and $z^D(y - \varphi) = O(z^{N+1})$.

The proof of this theorem is a simple application of the Banach fixed point theorem and hence of the same mathematical level as the method of successive approximations. Hence, it is reasonable that it be used as a tool in the study of linear differential systems at a singular point. For example, the following basic results are simple corollaries.

Corollary 1.

Let d=trace D and $n-d \geqq 0$. Then the system

(2) $$z^D \frac{dy}{dz} = A(z)\, y$$

has at least n-d linearly independent solutions holomorphic at z=0.

Corollary 2.

Let $A(z) = \sum_{k=0}^{\infty} A_k z^k$ be convergent for $|z| < \delta$, and let $y(z) = \sum_{k=0}^{\infty} y_k z^k$

be a formal solution of

(3) $$z \frac{dy}{dz} = A(z)\, y$$

in the sense that $n y_n = \sum_{k=0}^{n} A_{n-k} y_k$, $n=0, 1, \ldots$. Then y(z) is convergent for

$|z| < \delta$.

Corollary 3.

Let λ be a fixed complex constant and n_λ ($\geqq 0$) the number of linearly independent vectors y satisfying $A_0 y = \lambda y$. Then the number N_λ($\geqq 0$) of linearly independent solutions of the differential systems $z \frac{dy}{dz} = A(z)y$ of the form $y = z^\lambda \sum_{k=0}^{\infty} y_k z^k$ satisfies

$$\max (n_\lambda, n_{\lambda+1}, \ldots) \leqq N_\lambda \leqq n_\lambda + n_{\lambda+1} + \cdots .$$

This method hold great promise for future applications.

REFERENCES

1. E. A. Coddington and N. Levinson, Theory of ordinary differential equations, McGraw-Hill, New York, 1955.

2. W. A. Harris, Jr., Characterizations of linear differential systems with a regular singular point, Proc. Edinburgh Math. Soc. (to appear).

3. W. A. Harris, Jr., Y. Sibuya, and L. Weinberg, Holomorphic solutions of linear differential systems at singular points, Arch, Rational Mech. Anal. 35, 245-248 (1969).

4. P. Hartman, Ordinary differential equation, J. Wiley and Sons, New York, 1964.

5. W. B. Jurkat and D. A. Lutz, On the order of solutions of analytic differential equation, Proc. London Math. Soc. 22, 465-482 (1971).

6. D. A. Lutz, Some characterizations of systems of linear differential equations having regular singular solutions, Trans. Amer. Math. Soc. 126, 427-441 (1967).

7. D. A. Lutz, On systems of linear differential equations having regular singular solutions, J. of Differential Equations 3, 311-322 (1967).

8. D. A. Lutz, On the reduction of rank of linear differential system, Pacific J. Math. (to appear).

9. J. Moser, The order of a singularity in Fuchs' theory, Math. Z. 72, 379-398 (1960).

10. H. L. Turrittin, Reducing the rank of ordinary differential equations, Duke Math. J. 30, 271-274 (1963).

11. W. Wasow, Asymptotic expansions for ordinary differential equations, J. Wiley and Sons, New York, 1965.

CONNECTION PROBLEMS FOR SYSTEMS OF LINEAR DIFFERENTIAL EQUATIONS

Kenjiro Okubo

1. Introduction. We consider a system of linear ordinary differential equations

(1.1) $dx/dt = A(t)x$

where x is an n-vector, and A(t) is an n by n matrix with elements rational functions of the complex variable t. If $t=t_0$ is not a pole of A(t), there is a fundamenta l set X(t) of solutions such that $X(t_0)=I$. If we continue this set of solutions along a simple closed path P which encircles one and only one pole t= of A(t), we have another set of solutions $X(t,\lambda)$ which is linearly related to the original set X(t) by

(1.2) $$X(t,\lambda) = X(t)C_\lambda .$$

We call the matrix C_λ, the circuit matrix at $t=\lambda$ of the system (1.1) with respect to the solution X(t). If R is an n by n constant matrix such that

(1.3) $$\exp(2\pi i R_\lambda) = C_\lambda ,$$

Then clearly $Y(t,\lambda)$ defined by

(1.4) $$X(t) = Y(t,\lambda)(t-\lambda)^{R_\lambda}$$

is single-valued analytic near $t=\lambda$, and hence it has Laurent expansion:

(1.5) $$Y(t,\lambda) = \sum_{s=-\infty}^{\infty} G(s)(t-\lambda)^s.$$

We call the singular point $t=\lambda$ a regular singular point if the principal part of the expansion is finite, if not, we call λ an irregular singular point.

A sufficient condition that a pole of A(t) is a regular singular point is that it is a pole of order one. Even if a pole may have an order greater than one there is a rational transformation which transforms the given system into a new system which has a pole of order one at that point as long as the point is known to be a regular singular point.

Let λ be a regular singular point. We write the solution matrix $X(t,\lambda)$ in the form

$$(1.6) \qquad X(t,\lambda) = \sum_{s=0}^{\infty} G(s)(t-\lambda)^s(t-\lambda)^{R_\lambda}$$

by replacing R_λ in (L.4) by $(R_\lambda - kI)$ for some positive integer k, if necessary. For the sake of convenience, we call a solution $X(t,\lambda)$ of the form (1.6) a local canonical set at $t=\lambda$ if R_λ is in Jordan canonical form. A local canonical set X is transformed into another local canonical set $X^*=XT$ by a non-singular constant matrix T is and only if T and R_λ commute.

A system (1.1) is called a Fuchsian system if all the singular points of the system are regular singular points. A fundamental problem about a Fuchsian system is to compute its monodromy group. Let P be a closed path in $\overline{C}-S$, where $\overline{C}$ is the closed complex plane, and S is the set of singular points of (1.1). The resultant X(t,P) of the analytic continuation along P, of a fundamental set X(t) takes the

$$X(t,P) = X(t)C_P$$

The set G(X) of all such matrices C_P constitutes a group, the monodromy group of the system (1.1) relative to the set X(t). If Y is another fundamental Y=XT, then G(Y) is the set of the matrices of the form $T^{-1}C_PT$. Given a fundamental set X(t) of (1.1), to compute the group G(X), it is sufficient to know the connection formulae :

$$(1.7) \qquad X(t) = X(t,\lambda)T_\lambda$$

for local canonical systems $X(t,\lambda)$ at each point λ in S. Then we see, the group is generated by the set

$$\{ T_\lambda^{-1}CT \quad : \lambda \in S \}$$

2. Hypergeometric system

We call a system

$$(2.1) \qquad (t-B)dx/dt = Ax$$

a hypergeometric system of equations. Here x is an n vector, A is a constant n by n matrix and B is a constant n by n diagonal matrix. We denote the k-th diagonal elements of B by λ_k, and (j-k)-th element of A by a_{jk}.

The system (2.1) is a Fuchsian system, $t=\lambda_1,\ldots,\lambda_n$ and ∞ are regular singular points of the system and characteristic exponents are

(2.2) $\quad 0,0,\ldots,a_{kk},\ldots,0$

at $t=\lambda_k$, and

$$p_1,\ldots\ldots,p_n \qquad\qquad (\det A-p_jI = 0)$$

at $t=\infty$.

<u>Proposition 1</u>. Let

$$L[y] = \sum_{r=0}^{n} Q_r(t)D^{n-r}y$$

be a single n-th order differential equation where $Q_r(t)$ is a polynomial of degree n-r. Then it is equivalent to a hypergeometric system.

<u>Proof</u>. Omitted.

<u>Proposition 2.</u> In particular, for

$$t(1-t)y''+ c-(a+b+1)t\ y'-aby = 0$$

we set

$$y = x_1, \quad ty'-(1-c)y = x_2.$$

Then for the vector $x=(x_1,x_2)$, we have

$$(t-B)dx/dt = Ax$$

where

$$A = \begin{pmatrix} 1-c & 1 \\ -(c-1-a)(c-1-b) & c-a-b-1 \end{pmatrix} \qquad B = \begin{pmatrix} 0 & 0 \\ 0 & 1 \end{pmatrix}$$

<u>Proposition 2</u>. (Fuchs' relation)

$$\sum_{k=1}^{n} a_{kk} = \sum_{j=1}^{n} \rho_j \tag{2.4}$$

<u>Proof</u>. This is just the invariance of the trace of the matrix A.

<u>Proposition 3</u>. The most general constant non-singular linear transformation Tx=y which leaves B invariant, is a diagonal transformation.
<u>Proof</u>. Such a transformation must commute with B. If B is diagonal, any diagonal matrix T commutes with B.

<u>Proposition 4</u>. By a sutable choice of a diagonal matrix T, we can always make (n-1) off diagonal non-zero elements of A to take some specified values.
any preassigned

<u>Proposition 5</u>.(Number of Accessary Parameters) Given B and 2n constants $a_{11},\ldots\ldots,a_{nn}$; $\rho_1,\ldots,\rho_n$ ($\sum a_{kk} = \sum \rho_j$), we need (n-2)(n-1) more constants to determine a hypergeometric system. These constants are called accessary parameters.
<u>Proof</u>. A contains n^2 constants. But n-1 of which can be assinged any numerical values. We have to determine n^2-n+1 constants from the given (2n-1) data.

<u>Corollary</u>. Given B = diag(λ_1,λ_2), $a_{11},a_{22},\rho_1,\rho_2$, we can determine a hypergeometric system except for the case $\rho_j - a_{kk} = 0$.
<u>Proof</u>. We only have to determine a_{12} and a_{21}. Take $a_{21}=1$, then we have

$$a_{21} = a_{11}a_{22} - \rho_1\rho_2$$

In the exceptional case, we have, say, $\rho_1 = a_{11}$. Then by Fuchs relation $\rho_2 = a_{22}$. We have three possibilities.

$$\begin{pmatrix} \rho_1 & 0 \\ 0 & \rho_2 \end{pmatrix} \qquad \begin{pmatrix} \rho_1 & 1 \\ 0 & \rho_2 \end{pmatrix} \qquad \begin{pmatrix} \rho_1 & 0 \\ 1 & \rho_2 \end{pmatrix}.$$

We find it convenient to introduce a parameter μ in the following way:

$$(t-B)dx/dt = (A+\mu)x \tag{2.1.μ}$$

<u>Proposition 6</u>. (Truesdell's F-equations) Let $X(t,\mu)$ be a solution matrix of the system (2.1. μ), then for some solution matrix $X(t,\mu-1)$ of (2.1. μ-1), we have

$$(2.5) \qquad (d/dt)X(t,\mu) = X(t,\mu-1)$$

<u>Proof</u>. Trivial.

<u>Proposition 7</u>. (Invariance under Euler's Transformation) We have

$$(2.7) \qquad X(t,\mu) = \int_C (s-t)^{\mu-1} X(s,0)\,ds$$

for some path C such that

$$\left[(s-t)^{\mu-1}(s-B)X(s,o)\right]_C = 0$$

<u>Proof</u>. If μ is a non-negative integer, we may consider the right hand side of (2.7) as the Riemann-Liouville Integral of order μ. Then the recurrence relation (2.5) repeated μ times give the desired results. For complex μ we may differentiate under the integral sign, or we may take the finite part of divergent integrals in the sense of J. Hadamard.

<u>Corollary</u>. If for some value of μ, (2.1. μ) is integrable, then we have integral representation of the solution matrix of (2.1.0).

<u>Example</u>. We take the case n=2 of the classical hypergeometric equation. Let $(\mu+A)$ has eigenvalue -1 for the choice $\mu=-\rho_1-1$: $|\rho_1 I-A|=o$. We consider two cases separately.

<u>Case I</u>. A has distinct eigenvalues. There is a non-singular transformation T such that x=Ty transforms the system (2.1) into

$$(2.7) \qquad (t-\overline{B})dy/dt = \overline{A}y \quad , \quad \overline{B} = (\,\overline{b}_{jk}\,)\ ,\ \overline{A} = \begin{pmatrix} -1 & 0 \\ 0 & * \end{pmatrix}$$

If we write down the first equation, we have

$$(d/dt)\ (t-\overline{b}_{11})y_1 + \overline{b}_{12}y_2 \quad = 0$$

<u>Case II</u>. If A has multiple eigenvalue, then integrate the second equation first. In this case, the Jordan canonical form $\overline{A}$ of the matrix A may have a non-zero element at (1.2) position, but the second equation is now integrable as above.

Remark. Rouhgly speaking, a solution of F-equation is representable as Euler integral, and any function representable as an Euler integral is a solution of F-equation.

Definition. $x_k(t,\mu)$ be the singular solution at $t=\lambda_k$ corresponding to the non-zero characteristic exponent $\mu+a_{kk}$ with the normalization condition:

$$x_k(t,\mu) = (t-\lambda_k)^{\mu+a_{kk}} \sum_{s=0}^{\infty} g_k(s,\mu)(t-\lambda_k)^s \tag{2.8}$$

$$g_k(o,\mu) = e_k = (0,0,\dots,1,\dots,0) \quad \text{(1 at the k-th place).} \tag{2.9}$$

Proposition 8. If we write

$$g_k(s,\mu) = \Gamma(a_{kk}+s+1)/\Gamma(a_{kk}+\mu+s+1)\, h_k(s) \tag{2.10}$$

then $h_k(s)$ is independent of μ and is determined recursively by

$$(B-\lambda_k)h_k(s+1) = (a_{kk}-A)h_k(s) + sh_k(s) \tag{2.11}$$

and

$$h_k(o) = e_k \tag{2.12}$$

Proof. By direct computation.

Definition. Let $X(t,\mu)$ be the matrix solution whose k-th vertical vector is the solution $x_k(t,\mu)$ defined above.

Proposition 9. (Modified Truesdell's Equation) We have

$$(d/dt)X(t,\mu) = X(t,\mu-1)(D+\mu) \tag{2.13}$$

where D is the diagonal matrix $D=\mathrm{diag}(a_{11},a_{22},\dots,a_{nn})$.

Proof. $(d/dt)X(t,\mu)$ is a solution matrix of (2.1. μ-1) whose k-th vertical vector behaves like

$$(a_{kk}+\mu)(t-\lambda_k)^{a_{kk}+\mu-1}e_k$$

but there is only one such solution, namely,

$$(d/dt)x_k(t,\mu) = (a_{kk}+\mu)x_k(t,\mu-1) \tag{2.14}$$

Just put these vector solutions into a matrix, we have the proposition.

Proposition 10. (Recurrence formula)

$$(2.15) \qquad (\mu+A)X(t,\mu) = (t-B)X(t,\mu-1)(\mu+D)$$

Proof. Omitted.

In the following analysis, it is convenient to make the following assumption upon the locations of singular points.

$$(2.16) \qquad |\lambda_j-\lambda_k| > |\lambda_k| > 0 \qquad (j\neq k).$$

We shall refer to this condition by "pentagonal condition". It is easy to deduce from this condition that the origin is in the intersection of the circles of convergence of local solutions.

Theorem. For any simply connected domain in $\overline{C} - S$, we have

$$(2.17) \qquad W(t) = \det X(t,0) = \prod_{k=1}^{n} (t-\lambda_k)^{a_{kk}} \frac{\Gamma(a_{kk}+1)}{\Gamma(\rho_k+1)}$$

Proof. Write $W(t,\mu)=\det X(t,\mu)$. Then from (2.16) we have

$$W(t,\mu) = W(t,0)\cdot \prod_{k=1}^{n}(t-\lambda_k)^{\mu}\, \Gamma(\mu+a_{kk}+1)\Gamma(\rho_k+1)/\Gamma(\mu+\rho_k+1)\Gamma(a_{kk}+1)$$

And the power series expansions of $x_k(t,\mu)$ $(k=1,2,\ldots,n)$ are inverse factorial series expansion in μ be Prop.8, hence we have

$$W(t,\ \ ) = \prod_{k=1}^{n} (t-\lambda_k)^{\mu+a_{kk}}\cdot\{1+O(i/\mu)\}$$

in a domain D_0 which is contained in the intersection of the circles of convergence of local solutions at $\lambda_1,\ldots,\lambda_n$. Combine these two expressions and take the limit as μ goes to $+\infty$. Then we get (2.17) with the use of Fuchs' relation (2.4). On the other hand W(t) is a solution of

$$W'/W = \prod_{k=1}^{n} a_{kk}/(t-\lambda_k)$$

Consequently, we can continue W(t) to any domain which is simply connected in $\overline{C}$-S.

3. Computing the Group. We now return to the system (2.1) which is free from the parameter μ. We denote by X(t), the matrix solution whose k-th vertical vector $x_k(t)$ is the normalized solution corresponding to the characteristic exponent a_{kk} at $t=\lambda_k$. To compute the group

of the system (2.1), we have to fix local canonical systems at each singular points. Let us express the solution x_j near $t=\lambda_k$ by

$$(3.1) \qquad x_j(t) = \sigma_k^{\,j} x_k(t) + y_k^{\,j}(t)$$

where $y_k^{\,j}(t)$ be holomorphic at $t=\lambda_k$. This completely specifies the solution $y_k^{\,j}(t)$ in a simply connected domain D in $\overline{C}$-S. We take the matrix solution $X(t,\lambda_k)$:

$$(3.2) \qquad X(t,\lambda_k) = (y_k^{\,1}, y_k^{\,2}, \ldots, x_k, \ldots\ldots, y_k^{\,n})$$

as the local canonical system at $t=\lambda_k$. This is clearly a linearly independent set. Because if we substitute (3.1) into (3.2) we easily have

$$\det[X(t,\lambda_k)] = \det[X(t)] \neq 0.$$

Clearly, the circuit matrix C_k at $t=\lambda_k$ with respect to $X(t,\lambda_k)$ is given by the diagonal matrix:

$$(3.3) \qquad C_k = \operatorname{diag}(1,1,\ldots,\pi_k,\ldots,1) \qquad \pi_k = \exp(2\pi i a_{kk})$$

and the connection formula becomes

$$(3.4) \qquad X(t) = X(t,\lambda_k)T_k = X(t,\lambda_k)(I+\Sigma_k)$$

where Σ_k is given by

$$(3.5) \qquad \Sigma_k = \begin{pmatrix} 0\ ,\ 0 & \ldots\ldots & 0 \\ & \ldots\ldots\ldots & \\ & \ldots\ldots\ldots\ldots & \\ \sigma_k^{\,1}\ ,\ \sigma_k^{\,2}, & \ldots, 0, \ldots\ldots\ldots, & \sigma_k^{\,n} \\ & \ldots\ldots\ldots\ldots & \\ & \ldots\ldots\ldots\ldots & \\ 0\ ,\ 0 & \ldots\ldots\ldots & 0 \end{pmatrix}$$

The group is generated by

$$(3.6) \qquad M_k = T_k^{-1}C_kT_k = C_k - \Sigma_k C_k + C_k\Sigma_k - \Sigma_k C_k \Sigma_k = C_k + (\Sigma_k - 1)\Sigma_k.$$

To determine n(n-1) constants $\sigma_k^{\,j}$: $k\neq j$, $j,k=1,2,\ldots,n$, we make use of $X(t,\infty)$, a local canonical system at the point $t=\infty$, for which the circuit matrix C_∞ at ∞ becomes

(3.7) $$C_\infty = \mathrm{diag}(\ \exp(-2\pi i \rho_1),\ldots\ldots,\exp(-2\pi i \rho_n)\ ).$$

Although, we do not yet know the connection formula

(3.8) $$X(t) = X(t,\infty)T_\infty$$

the circuit matrix at ∞ with respect to our reference system $X(t)$ is given by

(3.9) $$M_\infty = T_\infty^{-1} C_\infty T_\infty$$

gives (n-1) more relations for $\sigma_k^{\ j}$'s because of the identity:

(3.10) $$M_1 M_2 \cdots\cdots M_n M_\infty = I$$

We write this identity in the form:

(3.10) $$T_\infty (M_1 M_2 \cdots\cdots M_n) T_\infty^{-1} = C_\infty^{-1}.$$

That is to say, n eigenvalues of the matrix $M_1 M_2 \cdots M_n$ are given by n eigenvalues of the matrix C_∞^{-1}: $\exp(2\pi i \rho_j)$ $(j=1,2,\ldots,n)$. This gives n relations, but one of these relations is already known. Because,

(3.11) $$\det(M_1 M_2 \cdots M_n) = \det(C_\infty^{-1})$$

is nothing but the Fuchs' relation (2.4).

We finally remark that (n-1) of $\sigma_k^{\ j}$ may be given any preassined values. Because we started from the fixed set $X(t)$, but we may replace each of the vector solutions $x_k(t)$ $(k=1,2,\ldots,n)$ by its non-zero constant maultiple.

These in total, give 2(n-1) relations among σ's. Thus we need $n^2-3n+2 = (n-2)(n-1)$ more relations to determine the group, the number is exactly the number of accessary parameters in Prop.5 of the section 2.

Again the classical case n=2 is the special case where we can compute the group. Given $\rho_1, \rho_2, a_{11}, a_{22}, \lambda_1\lambda_2$ such that $\sum a_{kk} = \sum \rho_j$, we compute :

$$\Sigma_1 = \begin{pmatrix} 0 & \sigma_1^{\ 2} \\ 0 & 0 \end{pmatrix},\quad C_1 = \begin{pmatrix} e_1 & 0 \\ 0 & 1 \end{pmatrix},\quad M_1 = \begin{pmatrix} e_1 & \sigma_1^{\ 2}(e_2-1) \\ 0 & 1 \end{pmatrix}$$

$$\Sigma_2 = \begin{pmatrix} 0 & 0 \\ \sigma_2^1 & 0 \end{pmatrix}, \; C_2 = \begin{pmatrix} 1 & 0 \\ 0 & e_2 \end{pmatrix}, \; M_2 = \begin{pmatrix} 1 & 0 \\ \sigma_2^1(e_2-1) & e_2 \end{pmatrix},$$

where we abbreviated $\exp(2\pi i a_{kk})$ by e_k, $k=1,2$. Since the equation

$$\det(M_1M_2 - \theta I) = \theta^2 - \theta[e_1+e_2-\sigma_1^2\sigma_2^1(e_1-1)(e_2-1)] + e_1e_2 \tag{3.12}$$

whould have roots

$$\theta = \exp(2\pi i\rho_1),\ \exp(2\pi i\rho_2)$$

we have explicit value for the product $\sigma_1^2\sigma_2^1$. Then take a new fundamental set

$$Y(t) = X(t)\begin{pmatrix} (\sigma_1^2)^{1/2} & 0 \\ 0 & (\sigma_2^1)^{1/2} \end{pmatrix}$$

Then we easily see that the group is generated by

$$M_1^* = \begin{pmatrix} e_1 & p(e_1-1) \\ 0 & 1 \end{pmatrix} \text{ and } M_2^* = \begin{pmatrix} 1 & 0 \\ p(e_2-1) & e_2 \end{pmatrix} \tag{3.13}$$

where $p^2 = \sigma_1^2\sigma_2^1$.

4. Further Comments. We list some of the results without proof.

A. When the diagonal matrix B contains multiple eigenvalues, we may apply further linear transformations in lower dimensional subspaces to reduce the matrix A into blocks some of which are in Jordan canonical forms. This procedure decrease the number of accessary parameters. In particular, when there is only two finite singular points with the multiplicities m and n-m respectively, we can prove that the group can be computed when $(m-1)(m-n+1)=0$. This is the case of the so-called generalized hypergeometric functions.

B. Our theorem gives, in the case n=2, the cerebrated identity of Gauss

$$F(a,b,c,1) = \frac{\Gamma(c)\cdot\Gamma(c-a-b)}{\Gamma(c-a)\cdot\Gamma(c-b)}$$

after a suitable choice of constants.

C. The case where some of A a_{kk} are negative integers, there may appear the logarithmic functions. We must use di-gamma function in our statement of the theorem.

D. By the invariance under Euler's transformation, if we know the group of (2.1.0), then we know the group for all (2.1. μ). The proof is easy if μ is a positive integer, but for complex values of the parameter, we must use the arguments in Jordan, and we need some computations.

Bibliography

- G.D. Birkhoff Collected Mathematical Works.
- M. Kohno; The convergence condition of a series appearing in connection problems and the determination of Stokes multiplierw, Publ. R.I.M.S. Kyoto Univ. 3 (1968),337-350.
- C. Jordan; Cours de Analyse de L'ecole polytechnique, 1882.
- K. Okubo; A global representation of a fundamental set of solusions and a Stokes phenomenon for a system of linear ordinary differential equation, J. Math. Soc. Japan, 15(1963),268-288.
- K. Okubo; An extension of Gauss' Formula for hypergeometric series. Suriken-Kobyuroku 117(1971), R.I.M.S. Kyoto Univ.
- H. Poincare; Sur les group des equations lineaires.
- B. Riemann; Beitrage zur Theorie der durch die Gausschen Reihe F(a,b,c,x) darstellbaren Functionen" 1857, Collected Works, 69-87.

S E C T I O N II

DIRICHLET SERIES SOLUTIONS FOR CERTAIN FUNCTIONAL DIFFERENTIAL EQUATIONS

Paul O. Frederickson*

The object of this paper is to display some particularly well behaved solutions -- analytic, or almost periodic, or $\mathcal{L}_p$, or totally monotone -- to certain functional differential equations of advanced type. The first equation considered is

1) $$y'(t) = a\, y(\lambda t) + b\, y(t)$$

in which $\lambda > 1$, an equation discussed most recently by T. Kato and J. B. McLeod [5], with high speed train applications in mind. Related equations were studied earlier by S. Izumi [4], P. Flamant [1], F. Gross [3], and R. Oberg [7], among others, but primarily in the retarded case. (See also the current work of G. R. Morris, A. Feldstein, and E. W. Bowen [6].)

The same methods, appropriately modified, yield solutions to the equation

2) $$y'(t) = b\, y(t) + \sum_{s=0}^{\infty} a_s y(\lambda_s t)$$

in which all $\lambda_s > 1$, provided $b \neq 0$ and provided that the condition

3) $$\sum_{s=0}^{\infty} (1 + (\lambda_s - 1)^{-1})|a_s| < \infty$$

is satisfied. These solutions are then used to construct solutions, equally well behaved, to the related integral equation

4) $$y'(t) = b\, y(t) + \int_1^{\infty} y(st)\, d\alpha(s)$$

in which we also assume that $b \neq 0$, and in which the measure α satisfies

5) $$\int_1^{\infty} (1 + (s - 1)^{-1})|d\alpha|(s) < \infty.$$

---

* Lakehead University, Thunder Bay, Ontario, Canada. Support of this research by the National Research Council of Canada is gratefully acknowledged.

Theorem A: *Equations 1), 2), and 4) have solutions* $\phi(t)$ *which are continuous in the closed half plane* $Re(bt) \le 0$, *analytic in its interior, and also*

(i) in $\mathcal{L}_p[0, \infty)$ for $1 \le p \le \infty$ if $Re(b) < 0$,

(ii) Bohr almost periodic on $(-\infty, \infty)$ if $Re(b) = 0$ and if either 1) or 2) is satisfied,

(iii) totally monotone on $[0, \infty)$ if the coefficients are all non positive.

In the proof we will consider equation 1) by itself, to start with, since this offers a quick outline of the rest of the proof. We can say more, however, in the case of equation 1), and we may as well prove this at the same time.

Theorem B: *If* $|b| < |a|$ *equation 1) has a one parameter family of solutions* $\phi(t) = \phi(t, \beta)$, *analytic in the parameter* β, *each one of which is continuous in the closed half plane* $Re(t\beta) \le 0$, *analytic in its interior, and also*

(i) in $\mathcal{L}_p[0, \infty)$ for $\infty \ge p > max(1, log(\lambda)/log|a/b|)$ if $Re(\beta) < 0$

(ii) Bohr almost periodic on $(-\infty, \infty)$ if $Re(\beta) = 0$

(iii) totally monotone if β, a, and b are non positive.

The whole idea is to consider only those solutions to equation 1) which are represented in a half-plane by a Dirichlet series. This seems, in retrospect, like a very natural way to represent a function analytic in a half plane. Any function totally monotone on the positive reals can be extended to a function analytic on the right half plane. Totally monotone solutions to equation 1) can be constructed, when $b = 0$, using nonlinear operators analogous to those used by the author [2] in the construction of bounded oscillatory solutions to the same equation. This, then, is how the present research began.

The simplest Dirichlet series solutions to equation 1) take the form

$$\phi(t, \beta) = \sum_n c_n e^{\beta\lambda^n t}. \tag{6}$$

If we impose the additional constraint

$$\sum_n (1 + \lambda^n)|c_n| < \infty \tag{7}$$

on the coefficients $c_n = c_n(\beta)$ the series converges absolutely on the closed half plane $Re(\beta t) \leq 0$ to an analytic function $\phi(t)$, for each fixed β, and we can differentiate term by term to obtain the equations

$$8) \qquad (\lambda^n\beta - b)c_n = a\, c_{n-1}$$

which the coefficients c_n must satisfy. This reduces to the recurrence relation

$$9) \qquad c_n = \frac{a/b}{\lambda^n - 1}\, c_{n-1}, \qquad n > 0$$

if we set $\beta = b$, provided $b \neq 0$, and the solutions to 9) are easily seen to satisfy 7). An equivalent recurrence recurrence relation results when $\beta = \lambda^{-k}b$ for any integer k, and the resulting $\phi(t)$ is the same within a multiplicative constant. If β avoids the discrete set $\lambda^{-k}b$ and zero we set $c_0 = 1$ and observe that the resulting sequence $c_n = c_n(\beta)$ satisfies 9) if and only if $|b| < |a|$. In fact, if μ satisfies $1 > \mu > |b|/|a|$ we find that $|c_{n-1}| < \mu\,|c_n|$ for all large negative n, and it follows that there is a constant K such that $|c_n| < K\lambda^{|n|}$ if $n < 0$. Convergence to the right is, as before, no problem. Observe that 7) is satisfied uniformly in β as β varies over any compact set B which contains no point $\lambda^{-k}b$, (and does not contain zero). Thus the Dirichlet series 6) converges uniformly on $T \times B$ to a function $\phi(t, \beta)$, analytic in two variables, provided the set T satisfies $Re(t\beta) \leq 0$ for $t \in T$, $\beta \in B$. We will show later that the line $Re(t\beta) = 0$ is a natural boundary for $\phi(t, \beta)$ if $b = 0$.

Conclusion (ii) of either theorem now follows from the uniform convergence of 6) and the fact that every term is periodic on $(-\infty, \infty)$ if $Re(\beta) = 0$. The conditions on (iii) make all the terms of 6) positive and totally monotone, and hence their sum is totally monotone. We observe, finally, that $e^{\beta\lambda^n t}$ is in $\mathcal{L}_p[0, \infty)$ only if $Re(\beta) < 0$, and $\| e^{\beta\lambda^n t}\|_p = (-Re(\beta)\lambda^n p)^{-1/p}$ in this case. Since these norms decrease as n increases, (i) in Theorem A is immediate. To prove (i) in Theorem B now requires the observation that $\sum_{n=-\infty}^{0} \rho^n c_n$ converges if $1 > \rho > |b/a|$, together with the fact that $\rho = \lambda^{-1/p} > |b/a|$ only if p satisfies the given condition. The proof of Theorem B is now complete.

The second stage in the proof of Theorem A begins with the observation that the Dirichlet series 6) can also solve equation 2) if we interpret it as a multiple series. Thus $n = (n_0, n_1, n_2, \ldots)$ is now a multi-index, a sequence of integers of which only a finite number differ from zero, and $\lambda^n = \prod_s (\lambda_s)^{n_s}$. We continue to assume that the coefficients c_n satisfy 7), so that we can still differentiate equation 6) term by term. When we do so we find that $\phi(t)$ satisfies equation 2) exactly when the coefficients c_n satisfy

10) $$(\lambda^n \beta - b)c_n = \sum_{s=0}^{\infty} a_s c_{n-e_s}$$

(Where the basic multi-indices e^s are defined by $n = \sum_s e^s n_s$.) An easy solution of 10) results if we take $\beta = b$ and impose the additional constraint

11) $$c_n \neq 0 \Rightarrow n \geq 0$$

on the coefficients, for 10) then becomes a recurrence relation and has a unique solution, given c_0. It is trivial to show, by induction, that

12) $$\lambda^n |c_n| \leq \prod_s (K_s^{n_s}/n_s!),$$

where $K_s = \lambda_s |a_s/b|/(\lambda_s - 1)$, and it follows that

13) $$\sum \lambda^n |c_n| \leq e^K < \infty,$$

since $K = \sum_s K_s$ is finite by 3). We now use the absolute convergence of the Dirichlet series to verify conclusions (i), (ii), and (iii), exactly as before.

The third phase of the proof requires a somewhat more complicated argument. The idea is simple enough: we wish to approximate the given measure α by a sequence of discrete measures α^r in such a way that the corresponding solutions $\phi^r(t)$ to equation 2) converge to a solution $\phi(t)$ to equation 4). We will do this in such a way that

14) $$|\phi^r(t) - \phi^{r-1}(t)| < 2^{-r}$$

holds for any t in the closed strip $\mathrm{Re}(tb) \leq 0$, $|\mathrm{Im}(tb)| \leq 2^r$. To achieve these goals the measures α^r must converge to α in a rather particular way. Denote by λ_s^r, $s = 0, 1, 2, \ldots$, the points which support the measure α^r, and let $a_s^r = \alpha_s^r(\lambda_s^r)$. We will construct α^r so that it is possible for the λ_s^r to satisfy $\lambda_{s+1}^r < \lambda_s^r$, and also $\{\lambda_s^r\}_{s=0}^{\infty} \subset \{\lambda_s^{r+1}\}_{s=0}^{\infty}$. We begin by choosing λ_0^r to satisfy $\int_{\lambda_0^r}^{\infty} |b^{-1}|(t-1)^{-1}|d\alpha|(t) < 2^{-r-2}\, e^{-2K}$, where $K = \int_1^{\infty} (1 + (s-1)^{-1})|d\alpha|(s)$. We then require a fine enough mesh that the numbers

$$\tilde{K}_s^r = \int_{\lambda_s^r}^{\lambda_{s-1}^r} |b^{-1}|(1 + (\lambda_s^r - 1)^{-1})|d\alpha|(t)$$

satisfy $\tilde{K}^r = \sum_{s=1}^{\infty} \tilde{K}_s^r < 2K$, and also

$$\sum_{s=1}^{\infty} (\lambda_{s-1}^r - \lambda_s^r)\, \tilde{K}_s^r < |b|^{-1} \exp(-\tilde{K}^r) 2^{-2r-3} \tag{15}$$

We then define $a_s^r = \alpha[\lambda_s^r, \lambda_{s-1}^r)$, $s = 1, s, \ldots$, and $a_0^r = \sum \{a_s^{r+1} : \lambda_s^{r+1} \geq \lambda_0^r\}$. Since $K_s^r = \lambda_s^r |a_s^r/b|/(\lambda_s^r - 1) < \tilde{K}_s^r$, condition 3) is satisfied and we can construct the solutions $\phi^r(t)$ to equation 2).

To prove 14) we consider the solution $\bar{\phi}(t) = \bar{\phi}(t, \lambda)$ which interpolates, in a sense, between $\phi^{r-1}(t)$ and $\phi^r(t)$. Define $\bar{a}_s = a_s^r$ and let $\lambda = (\lambda_0, \lambda_1, \ldots)$ vary between λ^r and $\bar{\lambda}^r$, where $\bar{\lambda}_s^r = \lambda_\sigma^{r-1}$, $\sigma = \sigma(s)$ being chosen so that $\lambda_\sigma^{r-1} \leq \lambda_s^r < \lambda_{\sigma+1}^{r-1}$. Since $a_\sigma^{r-1} = \sum\{\bar{a}_s : \bar{\lambda}_s^r = \lambda_\sigma^{r-1}\}$, one can show that $\bar{\phi}(t, \lambda) = \phi^{r-1}(t)$ when $\lambda = \bar{\lambda}^r$, and clearly $\bar{\phi}(t, \lambda^r) = \phi^r(t)$. Observe that $\dfrac{\partial\phi(t, \lambda)}{\partial\lambda_s} = \sum_n (n_s \lambda^{n-e^s}\, b\, c_n) t\, e^{\lambda^n \beta t} + \sum_n \dfrac{\partial c_n}{\partial\lambda_s} e^{\lambda^n \beta t}$. Now $\sum_n |n_s \lambda^{n-e^s}\, b\, c_n||t\, e^{\lambda^n \beta t}| \leq |bt|\tilde{K}_s^r e^{\tilde{K}^r}$ and $\sum_n |\dfrac{\partial c_n}{\partial\lambda_s}| \leq \tilde{K}_s^r e^{2\tilde{K}^r}$ for any value of λ between λ^r and $\bar{\lambda}^r$. It follows from these inequalities that $|\phi^{r-1}(t) - \phi^r(t)| \leq \sum_s (\lambda_s^r - \bar{\lambda}_s^r)\tilde{K}_s^r e^{\tilde{K}^r}|bte^{bt}| + \sum_s (\lambda_s^r - \bar{\lambda}_s^r)\tilde{K}_s^r e^{2K^r}$. We see, using 15) that either term on the right is bounded by 2^{-r-1}, provided that $|\mathrm{Im}(t)| \leq 2^r$, and we conclude that 14) is valid. The limit $\phi(t)$ of the sequence $\phi^r(t)$ is therefore continuous on the closed half plane $\mathrm{Re}(\beta t) \leq 0$ and analytic in its interior. It is an academic matter to prove that $\phi(t)$ is a solution to 4) on the closed half plane. Conclusion (i) requires the fact that the $\phi^r(t)$ are uniformly bounded in $\mathcal{L}_p$-norm, but (iii) is immediate.

A little more can be said if equation 1) is simplified further.

Theorem C: If $b = 0$, then for any fixed β the solution $\phi(t) = \phi(t, \beta)$ constructed in Theorem B has the line $Re(t\beta) = 0$ as a natural boundary.

To prove this, assume the contrary. If $\phi(t)$ were analytic at the point t_0, $Re(t_0\beta) = 0$, then there would be a $\delta > 0$ and a constant $M > 0$ such that $|\phi^{(r)}(t)| \leq r!M^r$ for $|t - t_0| < \delta$. The fact that ϕ is a solution to 1) means that $|\phi(\lambda^r t)| < r!M^r|a^{-r}|\lambda^{-n(n-1)/2}$ for $|t - t_0| < \delta$. But this contradicts the fact that $\phi(t)$ is almost periodic on $Re(t\beta) = 0$.

The hypothesis $b = 0$ in this theorem seems quite unnatural, (although it is essential to this proof). Are any of the solutions we have generated continuable beyond the half plane given? Since the proof of the following is fairly straightforward, we omit it.

Theorem D: If $|b| > |a|$ the solution $\phi(t)$ given is the only solution to equation 1) represented by a Dirichlet series.

But are there other $\mathcal{A}_p$ solutions, for example? And does this sort of uniqueness extend to equation 2) if $|b| > \sum_s |a_s|$? Finally, we observe that linear combinations of the solutions $\phi(t, \beta)$ interpolate on finite sets when $|b| < |a|$. Do they form a basis for the space of all analytic solutions? All bounded solutions?

[1] Flamant, P., Sur une équation différentielle fonctionnelle lineaire, Rend. Circ. Mat. Palermo 48 (1924), 135-208.

[2] Frederickson, P.O., Global Solutions to Certain Nonlinear Functional Differential Equations, J. Math. Anal. Appl., 33 (1971) pp 355-358).

[3] Gross, F., On a Remark of Utz., Am. Math. Monthly 74 (1967), 1107-08.

[4] Izumi, S., On the theory of linear functional differential equations, Tohoku Math. J. 30 (1929), 10-18.

[5] Kato, T., and J.B. McLeod, The Functional Differential Equation $y'(x) = ay(\lambda x) + by(x)$, (to appear).

[6] Morris, G.R., A. Feldstein, and E.W. Bowen, The Phragmeń-Lindenlöf Principle and a Class of Functional Differential Euqations, Proceedings of the NRL-MRC Conference on Ordinary Differential Euqations, Academic Press, 1972.

[7] Oberg, R.J., Local Theory of Complex Functional Differential Euqations, Trans. Amer. Math. Society, Nov. 1971, pp 302-327.

CONTINUOUS DEPENDENCE FOR SOME FUNCTIONAL DIFFERENTIAL EQUATIONS

Yoshiyuki HINO

In discussing asymptotic behaviors of solutions of functional differential equations with infinite retardation, Coleman and Mizel [1] introduced a class of Banach spaces as the phase spaces and Hino [3] also considered a class of Banach spaces which contains the class considered by Coleman and Mizel. Recently, Hale [2] introduced a more general class of Banach spaces as the phase spaces, and global behaviors of trajectories in the phase space considered by Hale have been discussed by Hale [2], Hino [3], [4], Naito [6] and others. In this paper, we shall discuss the continuous dependence of solutions on initial conditions for functional differential equations with the phase space introduced by Hale.

Let x be any vector in R^n and let $|x|$ be any norm of x. Let B be a Banach space of functions mapping $(-\infty,0]$ into R^n with norm $\|\cdot\|$. For any ϕ in B and any σ in $[0,\infty)$, let ϕ^σ be the restriction of ϕ to the interval $(-\infty,-\sigma]$. This is a function mapping $(-\infty,-\sigma]$ into R^n. We shall denote by B^σ the space of such functions ϕ^σ. For any $\eta \varepsilon B^\sigma$, we define the semi-norm $\|\eta\|_{B^\sigma}$ of η by

$$\|\eta\|_{B^\sigma} = \inf_{\phi} \{\|\phi\|,\ \phi^\sigma = \eta\}.$$

Then we can regard the space B^σ as a Banach space with norm $\|\ \|_{B^\sigma}$. If x is a function defined on $(-\infty,a)$, then for each t in $(-\infty,a)$ we define the function x_t by the relation $x_t(s) = x(t+s)$, $-\infty < s \leqq 0$. For numbers a and τ, $a>\tau$, we denote by $A_\tau{}^a$ the class of functions x mapping $(-\infty,a)$ into R^n such that x is a continuous function on $[\tau,a)$ and $x_\tau \varepsilon B$. The space B is assumed to have the following properties:

(I) If x is in $A_\tau{}^a$, then x_t is in B for all t in $[\tau,a)$ and x_t is a continuous function of t, where a and τ are constants such that $\tau < a \leqq \infty$.

(II) All bounded continuous functions mapping $(-\infty,0]$ into R^n are in B.

(III) If a sequence $\{\phi_k\}$, $\phi_k \varepsilon B$, is uniformly bounded on $(-\infty,0]$ with respect to $|\cdot|$ and converges to ϕ uniformly on any compact subset of $(-\infty,0]$, then $\|\phi_k - \phi\| \to 0$ as $k \to \infty$.

(IV) There are continuous, nondecreasing and nonnegative functions $b(r)$, $c(r)$ defined on $[0,\infty)$, $b(0) = c(0) = 0$, such that

$$\|\phi\| \leqq b(\sup_{-\sigma \leqq s \leqq 0} |\phi(s)|) + c(\|\phi^\sigma\|_{B^\sigma})$$

for any ϕ in B and any $\sigma \geqq 0$.

(V) If σ is a nonnegative number and ϕ is an element in B, then $T_\sigma\phi$ defined by $T_\sigma\phi(s) = \phi(s+\sigma)$, $s \in (-\infty,-\sigma]$, is an element in B^σ.

(VI) $\|T_\sigma\phi\|_{B^\sigma} \to 0$ as $\sigma \to \infty$.

Remark 1. To discuss the invariance principle for autonomous functional differential equations with the phase space B, Hale has assumed properties (I) ∿ (VI) on the space B, but in this paper the property (VI), which is called the fading memory, is not required.

In addition, we shall assume that the space B has the following properties:

(VII) $|\phi(0)| \leqq M_1\|\phi\|$ for some constant $M_1 > 0$.

(VIII) $\|T_\sigma\phi\|_{B^\sigma} \leqq M_2\|\phi\|$ for all $\sigma_0 \geqq \sigma \geqq 0$ and for some constant $M_2>0$, where σ_0 is some positive constant.

Remark 2. When I discussed uniform stability and uniform asymptotic stability of solutions in the phase space B, I needed the following property on the space B instead of the property (VIII):

(VIII)' $\|T_\sigma\phi\|_{B^\sigma} \leqq M_2'\|\phi\|$ for all $\sigma \geqq 0$ and for some constant $M_2'>0$.

Remark 3. The classes of phase spaces considered by Hino and by Coleman and Mizel have properties (I) ∿ (VIII), and hence the result in this paper holds good for functional differential equations with these phase spaces.

Consider the functional differential equation

(1) $$\dot{x}(t) = f(t, x_t).$$

The superposed dot denotes the right-hand derivative and $f(t,\phi)$ is a continuous function of (t,ϕ) which is defined on $I \times B^*$ and takes values in R^n, where I and B^* are open subset of $[0,\infty)$ and B, respectively. We shall denote by $x(t_0,\phi)$ a solution of (1) such that $x_{t_0}(t_0,\phi) = \phi$ and denote by $x(t,t_0,\phi)$ the value at t of $x(t_0,\phi)$.

Theorem. Suppose that a solution $x(t,t_0,\phi^0)$, $(t_0,\phi^0) \in I \times B^*$, of (1) defined on $[t_0,t_0+a]$ for some $a>0$ is unique for initial value problem. Then for any $\varepsilon>0$, there exists a $\delta(\varepsilon) > 0$ such that if $(s,\psi) \in I \times B^*$, $|s - t_0| < \delta$ and $\|\psi - \psi^0\| < \delta$, then $\|x_t(s,\psi) - x_t(t_0,\phi^0)\| < \varepsilon$ for all $t \in [\max\{t_0,s\}, t_0 + a]$.

Sketch of the proof. Since $C = \{x_t(t_o,\phi^o);\ t_o \leqq t \leqq t_o + a\}$ is compact and $I \times B^*$ is open, we can choose a neighborhood D of C and a continuous function $g(t,\phi)$ such that

$$|g(t,\phi)| \leqq M \text{ for a constant } M > 0$$

and that

$$f(t,\phi) = g(t,\phi) \quad \text{on} \quad D.$$

Now, consider the system

$$\dot{y}(t) = g(t,y_t) \tag{2}$$

instead of the system (1). Clearly, $x(t_o,\phi^o)$ is a solution of (2) which is unique for initial value problem, and it is sufficient to prove the theorem for the system (2).

To do this, by using the properties (VII) and (VIII) we verify that the set

$$\{y_s(t_m,\phi^m);\ m = 1,\ 2,\ \ldots,\ s \in [t_o,\tau]\}$$

is a relatively compact subset of B for any $\tau > t_o$, if (t_m,ϕ^m) converges to (t_o,ϕ^o) as $m \to \infty$, where $y(t_m,\phi^m)$ is a solution of (2) through (t_m, ϕ^m). The remained parts of the proof will be given by the standard arguments. For the detailes, see [5].

References

[1] B.D. Coleman and V.J. Mizel, On the stability of solutions of functional differential equations, Arch. Rational Mech. Anal., 30(1968), 174-196.

[2] J.K. Hale, Dynamical systems and stability, J. Math. Anal. Appl., 26(1969), 39-59.

[3] Y. Hino, Asymptotic behavior of solutions of some functional differential equations, Tôhoku Math. J., 22(1970), 98-108.

[4] _____ , On stability of the solution of some functional differential equations, Funkcial. Ekvac., 14(1971), 47-60.

[5] _____ , Continuous dependence for some functional differential equations, to appear in Tôhoku Math. J.

[6] T. Naito, Integral manifold for linear functional differential equations on some Banach space, Funkcial. Ekvac., 13(1970), 199-213.

ON AN ASYMPTOTIC BEHAVIOR OF SOLUTIONS OF DIFFERENTIAL-DIFFERENCE EQUATIONS

Mitsunobu Kurihara

R. Bellman and K. L. Cooke [1] studied a linear differential-difference equation of retarded type of the form

(1) $u'(t)+(a_0+a(t))u(t)+(b_0+b(t))u(t-\omega)=0,$

where both $a(t)$ and $b(t)$ are continuous functions of t for the interval $t_0 \leq t < +\infty$ and satisfy either integrability conditions

$$\int^{+\infty} |a(t)|\,dt < +\infty, \qquad \int^{+\infty} |b(t)|\,dt < +\infty$$

or order conditions

(2) $a(t)=o(1), \quad b(t)=o(1) \quad$ as $t \to +\infty$.

Let λ be any simple root of the equation

(3) $h(s)=s+a_0+b_0 e^{-\omega s}=0.$

They concluded that the equation (1) has a solution $u(t)$ satisfying

(4) $u(t)=(1+o(1))\exp[\lambda t - c_1 \int_{t_0}^{t} (a(\tau)+e^{-\lambda\omega} b(\tau))\,d\tau] \quad$ as $t \to +\infty$,

where $c_1=(1-b_0 \omega e^{-\lambda\omega})^{-1}$ and t_0 is a sufficiently large number.

In the case of the order conditions (2) they, however, imposed further additional conditions on the functions $a(t)$ and $b(t)$ to prove the existence of such a solution. When a root λ is double one, they also assumed many additional conditions other than the order conditions

(5) $a(t)=o(t^{-1}), \quad b(t)=o(t^{-1}) \quad$ as $t \to +\infty$

to establish the existence of a solution with the same asymptotic property.

Since it follows from the order conditions (2) or (5) that

$$\int_{t_0}^{t} (a(\tau)+e^{-\lambda\omega}b(\tau))d\tau=o(t) \quad \text{as } t\to+\infty,$$

a solution satisfying (4), which was studied by R. Bellman and K. L. Cooke, satisfies the relation

(6) $|u(t)|\leq \exp[\mu t+o(t)], \quad \mu=\mathrm{Re}\lambda, \quad \text{as } t\to+\infty.$

Obviously the relation (6) does not imply the relation (4). Therefore it is expected that there exists a solution of the equation (1) satisfying the relation (6) under the conditions weaker than those of R. Bellman and K. L. Cooke. Actually we can prove the existence of such a solution under the conditions (2) or (5) alone.

We consider a system of differential-difference equations of retarded type written in the vectorial form

(7) $u'(t)+(A_0+A(t))u(t)+(B_0+B(t))u(t-\omega)=0,$

where $u(t)$ is an unknown vector of dimension N and $A(t)$ and $B(t)$ are both N by N matrices whose components $a_{ij}(t)$ and $b_{ij}(t)$ are continuous in the interval $t_0\leq t<+\infty$ for all i and j. Corresponding to the equation (3), we cosider the equation

$$h(s)=\det(sI+A_0+e^{-\omega s}B_0)=0, \quad \text{(I is the N by N unit matrix)}$$

which is called characteristic equation and the roots of the equation above which are called characteristic roots. We can prove the following theorem:

<u>Theorem.</u> Let λ be any characteristic root. Denote by μ the real part of λ. Let m+1 be the maximum multiplicity of the roots with real parts μ. Suppose that the components $a_{ij}(t)$ and $b_{ij}(t)$ satisfy

$$a_{ij}(t)=o(t^{-m}), \quad b_{ij}(t)=o(t^{-m}) \quad \text{as } t\to+\infty, \ (i,j=1,\cdots,N).$$

Then the equation (7) has a non-trivial solution $u(t)$ whose components

$u_j(t)$ $(j=1,\cdots\cdots,N)$ satisfy

(8) $\max_j|u_j(t)|\leq\exp[\mu t+o(t)],\quad \mu=\mathrm{Re}\lambda,\quad$ as $t\to+\infty$.

The proof of the theorem I gave consists of two parts. First we convert the problem of solving the equation (7) to that of solving an integral equation of the form

$$(9)\quad u(t)=e^{\lambda t}C+\int_{t_0}^{t}K_1(t-\tau)[A(\tau)u(\tau)+B(\tau)u(\tau-\omega)]d\tau$$

$$+\int_{t}^{+\infty}K_2(t-\tau)[A(\tau)u(\tau)+B(\tau)u(\tau-\omega)]d\tau\quad \text{for } t>t_0.$$

Here C is a constant N-dimensional vector and the kernel matrices $K_1(t)=(k^1_{ij}(t))$ and $K_2(t)=(k^2_{ij}(t))$ satisfy the inequalities respectively

$$|k^1_{ij}(t)|\leq q(t)e^{\mu t},\qquad \text{for } t\geqq 0,\quad (i,j=1,\cdots\cdots,N)$$

and

$$|k^2_{ij}(t)|\leq c_2e^{\nu t},\quad (\nu>\mu)\qquad \text{for } t\leqq 0,\quad (i,j=1,\cdots\cdots,N),$$

where $q(t)$ is a polynomial of degree less than $m+1$ with nonnegative coefficients and c_2 is a constant. A solution of the integral equation (9) satisfying the relation (8) is a desired solution of the equation (7).

Second we establish the existence of such a solution making use of a fixed point theorem due to M. Hukuhara [2]. In this case we consider the family F consisting of the continuous functions $u(t)=(u_j(t))$ for the interval $t_0-\omega\leqq t<+\infty$ which satisfy the inequalities

$$|u_j(t)|\ ,\ |u_j(t-\omega)|\ e^{\mu\omega}$$

$$\leq\eta\exp[\mu t+\delta^{-1}|\int_{t_0}^{t}\tau^m\phi(\tau)d\tau|],\quad \text{for } t\geqq t_0,\quad (j=1,\cdots\cdots,N),$$

where η and δ are constants and

$$\phi(t)=\max_{i}\sum_{j}|a_{ij}(t)| +\max_{i}\sum_{j}|b_{ij}(t)|\, e^{-\mu\omega}.$$

Moreover we cosider a transformation $T[u(t)]$ defined for the functions $u(t)$ belonging to the family F. The transformation $T[u(t)]$ is equal to the right member of the equality (9) for $t>t_0$ and is defined suitably for $t_0-\omega\leq t\leq t_0$.

We can find a member $u(t)$ of the family F which is invariant with respect to the transformation T, that is, a function $u(t)$ such that $u(t)=T[u(t)]$ by the fixed point theorem. It is a solution of our eqation (7) and satisfies the relation (8).

REFERENCES

[1] R. Bellman and K. L. Cooke, Asymptotic behavior of solutions of differential-difference equations, Memoirs of the Amer. Math. Soc. No.35 (1959).

[2] M. Hukuhara, Sur les points singuliers des équations différentielles linéaires; domaine réel, Jour. of the Faculty of Science Hokkaido Imp. Univ. Ser.I Vol.II Nos.1-2 (1934).

ON A NONLINEAR DIFFERENCE EQUATION
OF BRIOT-BOUQUET TYPE

Kyoichi Takano

In this paper, we study a nonlinear difference equation of the form

(E) $$y(x-1) = f(x, y(x))$$

where x, y are complex variables and $f(x, y)$ is developable into a uniformly convergent double power series of x^{-1} and y

$$f(x, y) = y + x^{-1} \sum_{j+k\geqq 1} a_{j,k}\, x^{-j} y^{k}$$

for $|x| > \delta^{-1}$, $|y| < \Delta$.

If we write equation (E) in the form

$$x\, \frac{y(x-1) - y(x)}{1} = \sum_{j+k\geqq 1} a_{j,k}\, x^{-j}\, y(x)^{k}$$

and if we replace the difference quotient $\frac{y(x-1) - y(x)}{1}$ by the differential quotient $-\frac{dy}{dx}$, we get the well known differential equation of Briot-Bouquet type. So we naturally expect to obtain a theory which is analogous to that for the differential equation

$$-x\, \frac{dy}{dx} = \sum_{j+k\geqq 1} a_{j,k}\, x^{-j} y^{k} .$$

Such a theory was studied by Henri Poincaré and followed by many other mathematicians. As is easily seen, the leading coefficient $a_{0,1}$ plays an important role in the classification of the various possible situations. Professor Hukuhara [1] obtained an analytic expression for a general solution of the differential equation in each case. We follow him in the study of the difference equation (E).

By applying a formal transformation of the form

$$y \sim \sum_{j+k\geq 1} p_{j,k}\, x^{-j}\, y^{k} \qquad p_{0,1} = 1,$$

we can reduce equation (E) to an equation which has as simple a form as possible.

The simplified equation depends on the value of $\lambda = a_{0,1}$; In the case λ is neither a positive integer, nor zero, nor a negative rational, if we choose the coefficients $\{p_{j,k}\}$ in a suitable way, equation (E) can be transformed to the equation of the form

$$(E' - 1) \qquad z(z - 1) = (1 + \lambda x^{-1})z(x),$$

in the case λ is a positive integer, into

$$(E' - 2) \qquad z(x - 1) = (1 + \lambda x^{-1})z(x) + b\, x^{-\lambda - 1},$$

in the case $\lambda = 0$, into

$$(E' - 3) \qquad z(x - 1) = z(x)(1 - m c\, x^{-1} z(x)^{m} - m d\, x^{-1} z(x)^{2m})^{-\frac{1}{m}},$$

The general solutions of (E' - 1) and (E' - 2) are given respectively by

$$c(x)\,\frac{\Gamma(x+1)}{\Gamma(x+\lambda+1)}$$

$$(-b\Psi(x+1) + \sum_{k=2}^{\lambda} b_k\, \zeta(k,\ x+1) + c(x))\,\frac{\Gamma(x+1)}{\Gamma(x+\lambda+1)},$$

where $c(x)$ is an arbitrary periodic function of period 1, $\Psi(x)$ is psi function and $\zeta(s,\ a)$ is Hurwitz zeta function defined by

$$\zeta(s,\ a) = \sum_{n=0}^{\infty} \frac{1}{(n+a)^{s}}\ .$$

The general solution of (E' - 3) can be obtained in the special case $d = 0$. In this case it is given by

$$\left(mc \, \frac{\Gamma'(x+1)}{\Gamma(x+1)} + c(x) \right)^{-\frac{1}{m}} .$$

In the following theorem we give an analytic meaning to the formal transformation as in the first case.

Theorem. Suppose that λ is not a positive integer and that $\mathrm{Re}\,\lambda > 0$. Then there exists a function $\varphi(x, z)$ defined in a domain D

$$D = \left\{ (x,z) \in \mathbb{C}^2;\ |x| \geq \frac{1}{\delta'},\ |z| \leq \Delta',\ |\arg x| \leq \frac{\pi}{2} - \varepsilon,\ |\arg x - \arg \lambda| \leq \frac{\pi}{2} - \varepsilon \right\}$$

with the following properties

(1) $\varphi(x, z)$ is holomorphic in D,

(2) the transformation

$$y = \varphi(x, z)$$

takes equation (E) into (E'-1),

(3) $\varphi(x, z)$ is represented by the convergent series

$$\varphi(x, z) = \sum_{k=0}^{\infty} \varphi_k(x)\, z^k$$

in D, where $\varphi_k(x)$ are holomorphic in D_x, the projection of the domain D on the x-plane, and have asymptotic expansions

$$\varphi_k(x) \sim \sum_j p_{j,k}\, x^{-j} ,$$

as x tends to infinity through D_x.

In the other cases λ is a positive integer or λ is zero, similar theorems can be obtained.

References

[1] Hukuhara, M., Kimura, T., et Mme Matuda, T. Équations Differéntielles Ordinaires du Premier Ordre dans le Champ Complexe. Publ. Math. Soc. Japan. (1951).

[2] Takano, K. General solution of a nonlinear difference equation of Briot-Bouquet type. Funkc. Ekvac. 13, 179-198 (1971).

INVARIANT MEASURE IN THE PLANE

Jirō EGAWA

§1. Introduction

The purpose of this paper is to give some necessary and sufficient conditions for the existence of positive invariant measures for a given local dynamical system on R^2, S^2 or its open subset. For the existence of invariant measures of dynamical systems the following are well known.

(1) Every dynamical system on a compact metric space admits an invariant measure ([3]).

(2) Let π be a dynamical system on a complete separable metric space X. π admits a finite invariant measure if and only if there exist a compact set $K \subset X$ and a point $x \in X$ such that

$$\overline{\lim_{\tau\to\infty}} \frac{1}{\tau}\int_0^\tau \phi_K(\pi(x, t))\, dt > 0,$$

where ϕ_K is the characteristic function of K ([5]).

However, the measure considered in these papers are not necessarily positive on a non-empty open set. In the present paper, we are interested in a positive measure invariant for a given local dynamical system π, finite for a compact set. More precisely; we consider only an invariant Lebesque measure μ on X satisfying the following two conditions, where X denotes the phase space of π:

(i) If $G \subset X$ is a non-empty open set, then $\mu(G) > 0$.

(ii) If $K \subset X$ is a compact set, then $\mu(K) < \infty$.

[3] and [5] do not give any results concerning invariant measure having the properties *(i)* and *(ii)*. In the sequel. we shall understand that the term "*measure*" means "*Lebesque measure satisfying (i) and (ii)*". For flows on a torus, J.C.Oxtoby showed an example of a Stepanov flow which admits an invariant measure ([4]). However, it is still unknown, as far as the auther knows, whether there exists an invariant measure for every Stepanov flow.

First we extend the notion of an invariant measure to a local (dynamical) system as follows (for local (dynamical) systems, see [8] or [9]).

Definition. Let π be a local system on X and μ a measure on X. μ is said to be *invariant for* π (or π is said to *admit an invariant measure* μ) if for every measurable set $A \subset X$ and for every real number t, we have

$$\mu(\pi(A, t)) = \mu(A),$$

whenever $\pi(x, t)$ is defined for all $x \in A$. If further the total measure $\mu(X) < \infty$, we say that π *is finite*.

In the following, π denotes a given local system on X which is either R^2, S^2 or its non-empty open subset, unless otherwise stated. In §2, we introduce the notion of an invariant measure locally at a point, and discuss the existence of invariant measures locally at a regular point and an isolated singular point. The results in this section give the foundation of the next section. We assume in §3 that π is a local system on R^2 or S^2, and that π admits only a finite number of singular points. We obtain necessary and sufficient conditions for the global existence of invariant measures for π, finite and infinite respectively, which is the main purpose of this paper.

We shall state only the results without providing proofs, which together with further detailed discussions will appear elsewhere.

§2. Local Existence

Let $U \subset X$ be an open set. We have a local system on U induced by π (see [9]), which we shall denote by $\pi \| U$.

Definition 2.1. Let $x \in X$. We say that π *admits an invariant measure locally at* x if there exists an open neighborhood U of x such that $\pi \| U$ admits an invariant measure.

For regular points in X, using the results in [1] (cf. [8], [9]), we obtain the following.

Theorem 2.2. *Let* $x \in X$ *be a regular point. Then* π *admits an invariant measure locally at* x.

For isolated singular points, the following is known.

Proposition 2.3. (See [7]). *Let* $x \in X$ *be an isolated singular point. If* π *admits an invariant measure locally at* x, *then* x *is a Poincaré center or a generalized saddle.*

Here, a *generalized saddle point* means an isolated singular point x such that only a finite number of orbits approach to x as $t \uparrow \infty$ or $t \downarrow -\infty$.

The converse problem, which is our first purpose, is solved affirmatively as follows:

Theorem 2.4. *If* x *is a Poincaré center or a generalized saddle, then* π *admits an invariant measure locally at* x.

For the proof of Theorem 2.4, we use the results in [1] and [2] (cf. [8], [9]).

§ 3. Global Existence

Theorem 3.1. *Let* π *be a local system on* R^2 *and assume that there are only a finite number of singular points. Let* C *be the subset of* X *defined by*

$$C = \{\text{non-periodic point } x \in X \; ; \; L^+(x) \neq \phi \text{ or } L^-(x) \neq \phi\} .$$

Then a necessary and sufficient condition for the existence of an invariant measure for π *is that* C *consists of only a finite number of orbits, which is equivalent to the conjoint of the following:*

(1) Every singular point is either a Poincaré center or a generalized saddle.

(2) Each of $L^+(x)$ *and* $L^-(x)$ *of a non-periodic point* $x \in X$ *contains at most one point.*

The proof of necessity is rather easy (cf. Poincaré's Recurrence Theorem and its generalization by E.Hopf). To prove the sufficiency, we use the results obtained in §2 and some more global properties of local systems in the plane ([6]).

For the existence of finite invariant measures, we obtain the following:

Theorem 3.2. *Let* π *be a (global) dynamical system on* R^2 *or* S^2 *and assume that there are only a finite number of singular points. Then* π *admits a finite invariant measure if and only if the number of non-periodic orbits is finite.*

References

[1] Egawa, J., Global Parallelizability of Local Dynamical Systems, Math. Syst. Theory, 6, No.1 (1972) (to appear).

[2] McCann, Roger C., A Classification of Centers, Pacific J. Math., 30 (1969), 733-746.

[3] Nemytskii, V.V. and Stepanov, V.V., Qualitative Theory of Differen-

tial Equations, Princeton Univ. Press, Princeton, N.J., 1960.

[4] Oxtoby, J. C., Stepanoff Flows on the Torus, Proc. Amer. Math. Soc., 4 (1953), 982-987.

[5] Oxtoby, J. C. and Ulam, S. M., On the Existence of a Measure Invariant under a Transformation, Ann. of Math., 40 (1939), 560-566.

[6] Hajek, O., Dynamical Systems in the Plane, Academic Press, London and New York, 1968.

[7] Ura, T. and Hirasawa, Y., Sur les Points Singuliers des Equations Différentielles Admettant un Invariant Intégral, Proc. Japan Acad., 30 (1954), 726-730.

[8] Ura, T., Isomorphism and Local Characterization of Local Dynamical Systems, Funkc. Ekvac., 12 (1969), 99-122.

[9] Ura, T., Local Dynamical Systems and Their Isomorphisms, in this Proceeding.

ON THE SWIRLING FLOW PROBLEM

Philip Hartman

We prove [2] an existence theorem for the boundary value problem on $0 \leq t < \infty$ given by

$$\text{(DE)} \qquad \begin{cases} f''' + ff'' + a(g^2 - w^2 - f'^2) = 0, \\ g'' + fg' - 2bf'g = 0, \end{cases}$$

$$(C_0) \qquad f(0) = f_0, \; f'(0) = f_0', \; g(0) = g_0,$$

$$(C_\infty) \qquad f'(\infty) = 0 \quad \text{and} \quad g(\infty) = w,$$

where a, w, f_0, f_0', g_0 are constants satisfying

$$(*) \qquad 0 < a < 2b \quad \text{and} \quad g_0 > 0, \; w > 0.$$

McLeod [3] has proved existence for the case $f_0' = 0$, $a = b = \frac{1}{2}$ which corresponds to an axially symmetric solution of the Navier-Stokes equations for a flow above an infinite rotating disc, where the angular velocity of the disc is g_0 and of the fluid at ∞ is w. The question of existence when $(*)$ is not assumed is left open.

Our proof (which is quite different from McLeod's) depends on three steps:

(1) Let $T > 0$. There exists a solution (f_T, g_T) of the boundary value problem on $[0,T]$ consisting of (DE), (C_0) and

$$(C_T) \qquad f'(T) = 0 \quad \text{and} \quad g(t) = w$$

(2) $C^3 - \lim (f_T, g_T) = (f,g)$ as $T \to \infty$ exists on all bounded intervals of $t > 0$ and is a solution of (DE) and (C_0) such that f', g, $1/g$ are bounded. Here, $T \to \infty$ through a suitable sequence.

(3) (f,g) in step (2) satisfies (C_∞).

The treatment of the boundary value problem on $[0,T]$ in step (1) depends on a device of McLeod [3], an estimate of Nagumo [4], and the methods of Hartman [1]. The _a priori_ bounds in step (1) imply (2). Step (3) is non-trivial.

REFERENCES

[1] P. Hartman, On boundary value problems for systems of ordinary non-linear, second order differential equations, Trans. Amer. Math. Soc. 96 (1960) 493-509.

[2] P. Hartman, On the swirling flow problem, to appear.

[3] J. B. McLeod, The existence of axially symmetric flow above a rotating disc, Proc. Royal Soc. (Ser. A), to appear.

[4] M. Nagumo, Ueber die Differentialgleichung $y'' = f(x,y,y')$, Proc. Phys. Math. Soc. Japan (3) 10 (1937) 861-886.

ON AN EXTENSION THEOREM AND ITS APPLICATION FOR TURNING POINT PROBLEMS

Toshihiko NISHIMOTO

1. Introduction

We consider here the following 2-nd order ordinary differential equation containing small positive parameter ε ,

(1.1) $$\varepsilon^{2h}\frac{d^2y}{dx^2} = p(x,\varepsilon)y.$$

Here h is a positive integer, x is the complex variable, and we suppose the function $p(x,\varepsilon)$ has an asymptotic expansion

$$p(x,\varepsilon) \sim \sum_{\nu=0}^{\infty} p_\nu(x)\varepsilon^\nu .$$

The points a_k $(k=1,2,\cdots,s)$ where $p_0(x)$ vanishes are called turning points and at these points asymptotic expansion of solution becomes complicated. There has been a lot of contributions about this type of equation, and one of the most well-known form of asymptotic expansion is so-called W-K-B-J approximation:

(1.2)
$$Y(x,\varepsilon) \sim p_0(x)^{-1/4}\left\{\sum_{\nu=0}^{\infty} Y_\nu(x)\varepsilon^\nu\right\} \exp\begin{bmatrix}\xi_h(x,x_0,\varepsilon), & 0\\ 0, & -\xi_h(x,x_0,\varepsilon)\end{bmatrix},$$
$$\xi_h(x,x_0,\varepsilon) = \int_{x_0}^{x}\frac{\sqrt{p_0}}{\varepsilon^h}\left\{1+\frac{p_1}{2p_0}\varepsilon+\cdots+\frac{p_h}{2p_0}\varepsilon^h\right\}dx,$$

where $Y_\nu(x)$ are 2-2 matrices. The purpose of this talk is to specify the region of existence of this approximation and its application to the local turning point problems. Our results presented here are of interest in two respects. At first, this is concerned with the Kaplun's Extension Theorem which is used quite often in the perturbation methods in fluid mechanics. Roughly speaking, this asserts that if an asymptotic approximation is uniformly valid in an interval of a variable x, then it is uniformly valid in a wider interval depending on the parameter . According to the known results, the formal solution (1.2) is uniformly asymptotic approximation in a certain sectorial region which is arbitrarily large but bounded and deleted arbitrarily small neighborhood of turning points, then we can expect that it is correct in a wider region depending on the parameter ε . Secondly our results can be applied to the analyses of the local turning point problems of large order, that is, even if the Iwano's characteristic polygon consists of two segments, it is possible to solve the central connection problem by using matching methods.

2. The domain of influence and the canonical region

We consider the differential equation (1.1) in vector form

(2.1) $$\varepsilon^h\frac{dy}{dx} = \begin{bmatrix}0 & 1\\ p(x,\varepsilon) & 0\end{bmatrix}y \quad \text{in } D=\{|x|<\infty, |x^\alpha\varepsilon|\leq\delta_0, 0<\varepsilon\leq\varepsilon_0\},$$

where δ_0 and ε_0 are sufficiently small positive constants, and we assume that the function $p(x,\varepsilon)$ has an asymptotic expansion in power series of ε such that for every m there exists a positive constant M ;

$$\left\|p(x,\varepsilon) - \sum_{\nu=0}^{m} p_\nu(x)\varepsilon^\nu\right\| \leq M(1+|x|^{(m+1)\alpha+\beta})\,\varepsilon^{m+1} ,$$

where $p_\nu(x)$ are polynomials of x of degree at most $\alpha\nu+\beta$ for $\nu \geq 1$. Here α and β are some positive numbers which may be zero. In particular

$$p_o(x) = x^q + p_{o,q-1}x^{q-1} + \cdots + p_{o,o}.$$

To each turning point a_i it corresponds a rational number ρ_{a_i} which is determined from the characteristic polygon Γ_{a_i} associated with a_i, that is, $\rho_{a_i} = -\mu_{a_i}^{-1}$, here μ_{a_i} denotes the tangent of the segment of Γ_{a_i} which is situated at the highest position. For characteristic polygon of second order equations we refer for example to Nakano and Nishimoto[2]. The equation (2.1) is changed by the transformation

$$y = \begin{bmatrix} 1 & 1 \\ \sqrt{p_0} & -\sqrt{p_0} \end{bmatrix} (E+\varepsilon Q_1)(E+\varepsilon^2 Q_2)\cdots\cdots(E+\varepsilon^{m+h} Q_{m+h})\, z_m$$

into

$$\text{(2.2)} \qquad \varepsilon^h \frac{dz_m}{dx} = \sqrt{p_0}\left\{\sum_{\nu=0}^{m+h} G_\nu(x)\varepsilon^\nu + R_{m+h+1}(x,\varepsilon)\,\varepsilon^{m+h+1}\right\} z_m .$$

Here $G_\nu(x)$ and $Q_\nu(x)$ are diagonal and antidiagonal 2-2 matrices respectively, and their elements are determined successively.

Lemma2.1. The grouth order of elements of the matrices $G_\nu(x)$ and $Q_\nu(x)$ as x tends to infinity is at most $\nu(\alpha+\beta-q)$ if $\beta \geq q$, and $\alpha\nu+\beta-q$ if $\beta<q$. The order of pole of them at a turning point a_i is at most ν/ρ_{a_i} .

Lemma 2.2. For the remainder term $\varepsilon^{m+h+1}\sqrt{p_o(x)}\,R_{m+h+1}(x,\varepsilon)$, we have

$$\varepsilon^{m+h+1}\sqrt{p_o}R_{m+h+1} = \begin{cases} O[(x^{\alpha+\beta-q}\varepsilon)^{m+h+1}\cdot x^{q/2}] & \beta \geq q, \\ O[(x^{\alpha}\varepsilon)^{m+h+1}\cdot x^{\beta-q/2}] & \beta < q \end{cases}$$

as x tends to infinity under the restriction $|x^{\alpha_1}\varepsilon| \ll 1$, here α_1 denotes the number $\alpha+\beta-q$ if $\beta \geq q$, and α if $\beta<q$. and we have

$$\varepsilon^{m+h+1}\sqrt{p_o}R_{m+h+1} = O[((x-a_i)^{-1/\rho_{a_i}}\varepsilon)^{m+h+1}\cdot (x-a_i)^{r_i/2}]$$

as x tends to a turning point a_i of order r_i under the restriction $|(x-a_i)^{-1/\rho_{a_i}}\varepsilon| \ll 1.$

For each turning point a_i, the domain of influence N_{a_i} is defined by

$$N_{a_i} : \quad |x-a_i| \leq N\varepsilon^{\lambda_{a_i}},\ \lambda_{a_i} = [\frac{h+1}{\rho_{a_i}} - \frac{r}{2}i - 1]^{-1},$$

where N is an appropriately chosen positive constant.

Next, we use the notion of the canonical region with respect to $\xi_o(x,a) = \int_a^x \sqrt{p_o(x)}dx$ which was firstly introduced by Evgrafov and Fedoryuk[1]. I can not go into the details of this region, but it is simply connected unbounded region, bounded by Stokes curves, and contains no turning

point in its interior.
Let X be the complex x-plane, $D[\xi_0]$ be a canonical region with respect to $\xi_0(x,a)$ and $A=\{a_i\}$ be a set of turning points that are at the boundary of $D[\xi_0]$. Here we define a few types of region successively which depend on ε.

$$D[\xi_0,N] = D[\xi_0] \cap \{X - \bigcup_{a_i\in A} N_{a_i}\}$$

where N_{a_i} is the domain of influence at a_i.

$$D[\xi_0,N,\delta_2,\varepsilon] = D[\xi_0,N] \cap \{x: |x^{\alpha_2}\varepsilon| \leqq \delta_2\} \text{ for } 0<\varepsilon\leqq\varepsilon_0 \quad ,$$

where δ_2 is a sufficiently small positive constant and α_2 is a positive number defined by $(h+1)(\alpha+\beta-q)+q/2+1$ if $\beta\geqq q$, $(h+1)\alpha+\beta-q/2+1$ if $q/2-1-\alpha h\leqq\beta<q$, and α if $\beta<q/2-1-h$.

Lemma2.3. There exists a region $D[\xi_h,N,\delta_2,\varepsilon]$ which we call region of admissibility with respect to $\xi_h(x,a,\varepsilon)$ and is obtained by a slight deformation of $D[\xi_0,N,\delta_2,\varepsilon]$ with the conditions: for each x in $D[\xi_h] = D[\xi_h,N,\delta_2,\varepsilon]$, there exist two points $x_0^+(\varepsilon), x_0^-(\varepsilon)$, and two curves $c^+(s,x,x_0^+)$, $c^-(s,x,x_0^-)$ defined for $0\leqq s\leqq s_{x_0^\pm}$ and connecting x and $x_0^+(\varepsilon)$, $x_0^-(\varepsilon)$ respectively such that

(1) $x_0^\pm(\varepsilon)$ are on the boundary of $D[\xi_h]$ and $c^\pm(s,x,x_0^\pm)$ are contained in $D[\xi_h]$,
(2) $c^\pm(0,x,x_0^\pm) = x$, $c^\pm(s_{x_0}^\pm, x, x_0^\pm) = x_0^\pm(\varepsilon)$,
(3) $\mathrm{Re}\,\xi_h(x,a,\varepsilon)$ is monotone increasing along $c^+(s,x,x_0^+)$, monotone decreasing along $c^-(s,x,x_0^-)$, and $\lim \mathrm{Re}\,\xi_h(x,a,\varepsilon) = \pm\infty$ as x tends to infinity along $c^\pm(s,x,x_0^\pm)$.

3. Existence theorem

To state the main theorem precisely, it is convenient to divide the region $D[\xi_h]$ into a finite number of subregions such that

$$D^{(a_i)}[\xi_h] = D[\xi_h] \cap \{x: N\varepsilon^{\lambda a_i} \leqq |x-a_i| \leqq \delta_3\} \quad ,$$

for $a_i\in A$, $0<\varepsilon\leqq\varepsilon_0$, where δ_3 and ε_0 are taken sufficiently small.

$$D^{(\infty)}[\xi_h] = D[\xi_h] - \bigcup_{a_i\in A} D^{(a_i)}[\xi_h].$$

Theorem3.1. There exists a fundamental system of solutions $z_m(x,\varepsilon)$ of the equation (2.2) such that

$$\left\| \{z_m(x,\varepsilon) - w_m(x,\varepsilon)\} p_0^{1/4} \exp^{-1}\begin{bmatrix} \xi_h(x,x_0,\varepsilon), & 0 \\ 0 & , \xi_h(x,x_0,\varepsilon)\end{bmatrix} \right\|$$

$$\leqq \begin{cases} K[1+|x|^{(m+1)\alpha_2}]\varepsilon^{m+1} & \text{for } x \text{ in } D^{(\infty)}[\xi_h], \\ K[|x-a_i|^{-1/\lambda a_i}\varepsilon]^{m+1} & \text{for } x \text{ in } D^{(a_i)}[\xi_h], \end{cases}$$

where $w_m(x,\varepsilon)$ is given by

$$p_0^{-1/4}\exp\int^x \sum_{j=1}^m \sqrt{p_0}G_{h+j}\varepsilon^j dx \cdot \exp\begin{bmatrix} \xi_h(x,x_0,\varepsilon), & 0 \\ 0 & , \xi_h(x,x_0,\varepsilon) \end{bmatrix}.$$

Theorem3.2. The differential equation (2.1) has a fundamental system of solutions of the form

$$y(x,\varepsilon) = \begin{bmatrix} 1 & 1 \\ \sqrt{p_0} & \sqrt{p_0} \end{bmatrix} p_0^{-1/4}\left\{E+\varepsilon Y_1(x)+\varepsilon^2 Y_2(x)+ \quad +\varepsilon^m Y_m(x)+Y_{m+1}(x,\varepsilon)\right\}$$
$$\times \exp\begin{bmatrix} \xi_h(x,x_0,\varepsilon), & 0 \\ 0 & , \xi_h(x,x_0,\varepsilon) \end{bmatrix},$$

where E is the 2-2 unit matrix and the remainder term $Y_{m+1}(x,\varepsilon)$ satisfies the inequalities

$$\left\| Y_{m+1}(x,\varepsilon) \right\| \leqq \begin{cases} K[1+|x|^{(m+1)\alpha_2}]\varepsilon^{m+1} & \text{for } x \text{ in } D^{(\infty)}[\xi_h], \\ K[|x-a_i|^{-1/\lambda a_i}\varepsilon]^{m+1} & \text{for } x \text{ in } D^{(a_i)}[\xi_h], \end{cases}$$

for some positive constant K.

4. Application

We consider an asymptotic expansion of solution of equation

$$(4.1) \qquad \varepsilon\frac{dy}{dx} = \begin{bmatrix} 0 & , & 1 \\ p(x,\varepsilon), & & 0 \end{bmatrix} y,$$

where the function $p(x,\varepsilon)$ has an asymptotic expansion with holomorphic coefficients:

$$p(x,\varepsilon) \sim x^q + \sum_{\nu=1}^{\infty} p_\nu(x)\varepsilon^\nu, \quad p_\nu(x) = \sum_{\mu=m_\nu}^{\infty} p_{\nu\mu}x^\mu,$$

in the region D:

$$D: \quad |x| \leqq \delta_o, \quad 0 < \varepsilon \leqq \varepsilon_o$$

Thus the origin is a turning point of order q. If the characteristic polygon consists of two segments, the region D is divided into four types of subregions D_i (i=1,2,3,4) in each of which the equation (4.1) has different principal parts. We introduce the following constants.

$$\rho_1 = \frac{1}{q-m_1}, \qquad \rho_2 = \frac{1}{m_1+2}, \qquad \sigma_1 = \frac{q}{2(q-m_1)}.$$

The subregions D_i are as follow: D_1; $M\varepsilon^{\rho_1} \leqq |x| \leqq \delta_0$, D_2; $m\varepsilon^{\rho_1} \leqq |x| \leqq M\varepsilon^{\rho_1}$, D_3; $M\varepsilon^{\rho_2} \leqq |x| \leqq m\varepsilon^{\rho_1}$, D_4; $|x| \leqq M\varepsilon^{\rho_2}$ for $0 < \varepsilon \leqq \varepsilon_0$ and $m < M$.

In D_2 the equation (4.1) is transformed into the equation which has just the same form of the equation (2.1) after a trivial modification:

$$\varepsilon^{1-\rho_1-\gamma_1}\frac{dz}{ds} = A_2(s,\varepsilon)z, \quad x = \varepsilon^{\rho_1} s, \quad y = \begin{bmatrix} 1 & 0 \\ 0 & \varepsilon^{\gamma_1} \end{bmatrix} z,$$

$$A_2(s,\varepsilon) \sim \begin{bmatrix} 0 & 1 \\ s^q + p_{1,m_1} s^m, & 0 \end{bmatrix} + \sum_{k=1}^{\infty} A_{2k}(s)\varepsilon^{k/(q-m_1)},$$

where $A_{2k}(s)$ are 2-2 matrices of polynomials of s of degree at most $(k+q)/(q-m_1)$. If we apply the previous results, there exists an asymptotic approximation of the above equation in a wider region so that it is overlapped with D_1 and also contains D_3 in absolute value x, and from these facts it is possible to obtain the asymptotic expansion at the origin of the solution defined in the subregion of D_1 by using two times of matching procedure.

REFERENCES

[1] Evgrafov,M.A., and Fedoryuk,M.B. Uspehi Mat. Nauk. 21,3-50(1966)

[2] Nakano,M., and Nishimoto,T. Kodai Math. Sem. Rep. 22, 355-384 (1970).

ON FUCHS' PROBLEM ON A TORUS

Kazuo Okamoto

We consider the Fuchs problem on a complex torus of complex dimension 1. This problem is deeply related to the Riemann problem, which was solved by H. Röhrl [3]. In this paper, however, we do not discuss Fuchs' problem in relation with Riemann's problem

Let

(1) $$y'' + p(x)y' + q(x)y = 0$$

be an Fuchsian differential equation of second order, where

(2) $$p(x) = k_1+c_0\zeta(x)+c_1\zeta(x-t)-\zeta(x-\lambda)-\zeta(x-\mu),$$

(3) $$q(x) = k_2+b_0\zeta(x)+b_1\zeta(x-t)+b_2\zeta(x-\lambda)+b_3\zeta(x-\mu)+a_0\wp(x)+a_0\wp(x-t),$$

$\zeta(x)$ being the Weierstrass ζ-function corresponding to the Weierstrass elliptic function $\wp(x)$. As is well-known, $\zeta(x)$ is not elliptic function, so that we have

(4) $$c_0 + c_1 = 2,$$

(5) $$b_0 + b_1 + b_2 + b_3 = 0.$$

We denote by $2\omega_1$, $2\omega_3$ $(\mathrm{Im}\,\frac{\omega_3}{\omega_1} > 0)$ the primitive periods of $\wp(x)$ and by Ω the set of all period. The set of singularities of (1) is $\Omega\cup\{t+\Omega\}\cup\{\lambda+\Omega\}\cup\{\mu+\Omega\}$, then Riemannian scheme of equation (1) takes following form:

$$\left\{\begin{matrix} \Omega & t+\Omega & \lambda+\Omega & \mu+\Omega \\ \sigma_0 & \sigma_1 & 0 & 0 \\ \sigma_0' & \sigma_1' & 2 & 2 \end{matrix}\right\},$$

where

(6) $$\sigma_0 + \sigma_0' + \sigma_1 + \sigma_1' = 0.$$

This equality (6) is Fuchs' relation of differential equation (1) and it will be easily seen that equality (4) implies (5).

We make the following assumption:

(A) Neither λ nor μ is a logarithmic singularity.

Now, let $\mathcal{Y}(x) = \begin{pmatrix} y_1(x) \\ y_2(x) \end{pmatrix}$ be a fundamental system of solutions of (1). By (A), we see that the circuit matrices of $\mathcal{Y}(x)$ around $x = \lambda$ and $x = \mu$ are both I, where I is the identity matrix in $GL(2, \mathbb{C})$. On the other hand, the circuit matrices C_1, C_2 of $\mathcal{Y}(x)$ around $x = 0$ and $x = t$ may depend on t, λ, μ, k_1, k_2, b_0, b_1, b_2, b_3, c_0, c_1, a_0, a_1 and $\mathcal{Y}(x)$ itself. The same is true for the period matrices D_1, D_3 defined by

$$\mathcal{Y}(x+2\omega_1) = D_1\mathcal{Y}(x)$$
$$\mathcal{Y}(x+2\omega_3) = D_3\mathcal{Y}(x).$$

We propose the following problem:

(P) Considering t as a variable point varying within a simply connected domain U not containing 0, then can we determine λ, μ, k_1, k_2, b_0, b_1, b_2, b_3, c_0, c_1, a_0 and a_1 as functions of t so that equation (1) has a fundamental system of solutions whose circuit matrices and period matrices are independent of t?

This kind of problem was first considered and formulated by L. Fuchs and was called after his name by L. Schlesinger. L. Schlesinger himself, R. Garnier and R. Fuchs, a son of famous L. Fuchs, studied this problem on a Riemann sphere $\mathbb{P}^1$ and gave some interesting result.

Now, it is clear that c_0, c_1, a_0 and a_1 should remain invariant.

Some appropriate consideration and calculations show the following result:

(I) $\lambda = \lambda(t)$ and $\mu = \mu(t)$ satisfy the system of equations

$$(E_a)\quad \frac{1}{2M}\frac{d^2\lambda}{dt^2} = -\frac{1}{4}(\wp(\lambda)-\wp(\lambda-t))\left(\left(\frac{d\lambda}{dt}\right)^2+\left(\frac{d\mu}{dt}\right)^2\right)+\frac{1}{2}(\wp(\mu)-\wp(\mu-t))\frac{d\lambda}{dt}\frac{d\mu}{dt}$$
$$-\frac{1}{2}(\wp(\lambda-t)-\wp(\mu-t))\frac{d\lambda}{dt}+\frac{1}{2}(\wp(\lambda)-\wp(\mu))\frac{d\mu}{dt}$$
$$-M^2(\wp(\lambda)-\wp(\mu))(\wp(\lambda)-\wp(\lambda-t))(\alpha_0-\alpha_1)$$
$$+M\wp'(\lambda)(\alpha_0-\alpha_1)+(\wp(\lambda)-\wp(\lambda-t))\alpha_1 ,$$

$$(E_b)\quad \frac{1}{2M}\frac{d^2\mu}{dt^2} = \frac{1}{4}(\wp(\mu)-\wp(\mu-t))\left(\left(\frac{d\lambda}{dt}\right)^2+\left(\frac{d\mu}{dt}\right)^2\right)-\frac{1}{2}(\wp(\lambda)-\wp(\lambda-t))\frac{d\lambda}{dt}\frac{d\mu}{dt}$$
$$+\frac{1}{2}(\wp(\lambda)-\wp(\mu))\frac{d\lambda}{dt}-\frac{1}{2}(\wp(\lambda-t)-\wp(\mu-t))\frac{d\mu}{dt}$$
$$-M^2(\wp(\lambda)-\wp(\mu))(\wp(\mu)-\wp(\mu-t))(\alpha_0-\alpha_1)$$
$$+M\wp'(\mu)(\alpha_0-\alpha_1)-(\wp(\mu)-\wp(\mu-t))\alpha_1 ,$$

where $M = \dfrac{1}{\zeta(\mu)-\zeta(\lambda)+\zeta(\lambda-t)-\zeta(\mu-t)}$, $\alpha_j = -a_j+\frac{1}{4}c_j^2-\frac{1}{2}c_j$ $(j=0,1)$.

(II) $k_2(t)$, $b_0(t)$, $b_1(t)$, $b_2(t)$, $b_3(t)$ are rational functions in $\frac{d\lambda}{dt}$, $\frac{d\mu}{dt}$, $\wp(\lambda)$, $\wp(\lambda-t)$, $\wp(\mu)$, $\wp(\mu-t)$, $\zeta(\lambda)$, $\zeta(\lambda-t)$, $\zeta(\mu)$ and $\zeta(\ -t)$.

(III) k_1 is independent of t.

Conversely, if the functions $\lambda(t)$, $\mu(t)$, $k_1(t)$, $k_2(t)$, $b_0(t)$, $b_1(t)$, $b_2(t)$ and $b_3(t)$ satisfy the above conditions, then equation (1) admits a fundamental system of solutions for which all circuit matrices and period matrices are independent of t.

This result is an extended one that the author discusses in [2]. We remark that the main difficulty lies in the proof of the first assertion. Once this has been done, the proof of second assertion will be immediate.

References

[1] Fuchs, R. Über lineare homogene Differentialgleichungen zweiter ordnung mit drei im Endlichen gelegene wesentlich singulären Stellen. Math. Ann. 63, 301-321 (1907).

[2] Okamoto, K. On Fuchs' problem on a torus, I. (to appear)

[3] Röhrl, H. Das Riemann-Hilbertsche Problem der Theorie der lineare Differentialgleichungen. Math. Ann. 133, 1-25 (1957).

DISCONTINUOUS ENERGY FUNCTIONS

James S. Muldowney

In this paper sufficient conditions are given for all solutions of the system

(1) $$x' = f(t,x)$$

to satisfy an inequality of the form

(2) $$V(t,x(t)) - V(t_0,x_0) \leq \int_{t_0}^{t} W(s,x(s))ds \ , \ t \geq t_0 \ ,$$

where V and W are prescribed real valued functions.

Sufficient conditions for inequalities of this form to hold have been given by LaSalle and Lefschetz [1], Yoshizawa [3], and Yorke [2]. Indeed Yorke gives a necessary and sufficient condition that (2) hold for some solution through each point (t_0,x_0) provided only that f be continuous, V lower semicontinuous and W continuous. However, his results are not applicable in the case that f satisfies the Carathéodory conditions; their main importance seems to lie in the realm of converse theorems for various types of stability.

For a specific problem of the type (1) where $f(t,x)$ satisfies the Carathéodory conditions one expects a 'natural' choice of energy function $V(t,x)$ to have reasonably smooth behaviour in the variable x, while perhaps having pathologies, such as discontinuities, in t. The following theorem is applicable in many such cases.

Theorem:

Suppose $f(t,x)$ and $W(t,x)$ satisfy the Caratheodory conditions. Assume also that $V(t,x)$ is nonincreasing in t, continuously differentiable in x with

$$\left|\frac{\partial V}{\partial x}(t,x)\right| \leq M \qquad \left(\frac{\partial}{\partial x} \equiv \text{grad}\right)$$

uniformly in (t,x), and

$$\left(\frac{\partial V}{\partial x},f\right)(t,x) \leq W(t,x) .$$

Then each solution of (1) satisfies (2).

Example 1:

Suppose $A(t),H(t)$ are $n\times n$ complex matrix valued functions on $[a,\infty)$ such that $A(t)$ is locally Lebesgue integrable and $H(t)$ is Hermitian, bounded locally and Lebesgue measurable. Then, for each solution $x(t)$ of the system

$$x' = A(t)x ,$$

the function

$$E(t) = x^*Hx(t)$$

is nonincreasing provided

$$d\{H(t) + \int_0^t (A^*H + HA)\} \leq 0 .$$

In the case that $H(t)$ is locally absolutely continuous this reduces to the familiar condition

$$H' + A^*H + HA \leq 0 .$$

The most simple proof is by an integration by parts procedure. However, it also follows from the present theorem by consideration of

$$V(t,x) = x^*\{H(t) + \int_0^t (A^*H + HA)\}x .$$

Example 2:

If $u(t)$ is a solution of the second order scalar equation

$$u'' + f(t,u) = 0$$

then

$$\frac{1}{2}(u')^2 + \int_0^u f(t,s)ds + \mu(t)$$

is nonincreasing provided

$$\int_0^x f(t,s)ds + \mu(t)$$

is nonincreasing in t for each x and $f(t,x)$ is locally bounded, as well as satisfying Carathéodory's conditions.

A proof of the Theorem is based on the following lemma.

Lemma:

Let ϕ and ψ be real valued functions on $[a,b] \times [a,b]$ and $[a,b]$ respectively. Suppose

(i) $\phi(t,s)$ is nonincreasing in t for each s and absolutely continuous in s for each t,

(ii) $\frac{\partial\phi}{\partial s}(t,s) \leq \mu(s)$ a.e. uniformly in t, where $\mu \in L_1[a,b]$,

(iii) $\psi \in L_1[a,b]$,

(iv) $\frac{\partial\phi}{\partial s}(t,t)$ exists a.e. on $[a,b]$ and $\frac{\partial\phi}{\partial s}(t,t) \leq \psi(t)$.

Then

$$\phi(b,b) - \phi(a,a) \leq \int_a^b \psi$$

A proof of this lemma will appear elsewhere.

Acknowledgement

The research reported here was supported by the Defence Research Board of Canada under grant 9540-28.

References

[1] LaSalle, J.P., and Lefschetz, S. 'Stability by Liapunov's direct method'. Academic Press, New York, 1961.

[2] Yorke, J.A., 'Extending Liapunov's second method to non-lipschitz Liapunov functions'. Lecture notes in mathematics 60, 31-36 Springer-Verlag New York 1968.

[3] Yoshizawa, T. 'Stability Theory by Liapunov's second method' The Mathematical Society of Japan 1966.

ON THE ABSOLUTE STABILITY OF A CLASS OF HYPERBOLIC SYSTEMS

Toshiki NAITO

We consider a system with two independent variables

$$(1)\quad \begin{cases} u_t = Cu - Au_x + b\xi \\ \xi_t = f(\sigma) \\ \sigma = \nu' u - \mu' u_x - r\xi, \end{cases}$$

where b, u, μ and ν are n-vectors, r and ξ are scalars and A and C are constant n×n matrices. We assume that $f(\sigma)$ is of class C^2 and $\sigma f(\sigma)>0$ for $\sigma\varepsilon R-\{0\}$ and that the fundamental system $u_t=Cu-Au_x$ is a linear hyperbolic system, that is, the matrix $\lambda I-A$, where λ is a parameter and I is the unit matrix, has simple elementary divisors $\lambda-a_i$, $a_i\varepsilon R$, $i=1, \ldots, n$. For a function $g(t,x)=(g^1(t,x), \ldots, g^m(t,x))$ of class C^k defined for $(t,x)\varepsilon[0,T]\times R \equiv R_T$, we define the norm $\|g(t,\cdot)\|_k$ by the relation

$$\|g(t,\cdot)\|_k = \sum_{j=0}^{k} \sup_{x\varepsilon R} \max_{i=1}^{m} \left|\frac{\partial^j}{\partial x^j} g^i(t,x)\right|,$$

if it exists. $\|g(t,\cdot)\|_k$ is a function of t.

Consider a semilinear hyperbolic system

$$(2)\quad w_t = \mathbb{C}w - \mathbb{A}w_x + h(w),$$

where w is an m-vector, $\mathbb{A}$ and $\mathbb{C}$ are constant m×m matrices and $\mathbb{A}$ satisfies the same condition as the one on A. $h(w)$ is defined for $|w|=\max_{i=1}^{m}|w^i|<\Omega$ and is of class C^2, and $h(0)=0$, $h_w(0)=0$. Concerning existence, uniqueness and stability, we have the following theorems for the system (2).

Theorem 1 ([1] and [2]). Suppose that the system (2) is hyperbolic and that the initial function $\bar{w}(x)$ is of class C^2 and $\|\bar{w}(\cdot)\|_2 < \infty$.

Then there exists a number $T>0$ such that a bounded solution $w(t,x)$ of the system (2) exists on R_T assuming the initial value $\bar{w}(x)$. $w(t,x)$ is of class C^2 and $w_{x,t}$ are bounded. T depends on the bound for $|\bar{w}(x)|$, i.e., $\|\bar{w}(x)\|_0$.

Let $\mathbb{B}$ be a positive definite m×m matrix. Define a Liapunov functional $V_{\rho,y}$, $\rho>0$ and $y\varepsilon R$, by the relation

(3) $$V_{\rho,y} = \int_{-\infty}^{+\infty} [\langle w, Bw\rangle + \langle w_x, Bw_x\rangle] e^{-\rho|x-y|} dx,$$

where w is of class C^2, $\|w(\cdot)\|_1 < \infty$ and $\langle\cdot,\cdot\rangle$ is an inner product. Set

(4) $$V_\rho = \sup_{y \in R} V_{\rho,y}.$$

If $\rho \leq 1$, we have the following inequality (see [2])

(5) $$\|w(\cdot)\|_0^2 \leq KV_\rho \text{ for some } K > 0.$$

Theorem 2 ([2]). *Suppose that there exist positive definite matrices B and D such that*

(6) $$\begin{cases} BC + C'B = -D \\ BA - A'B = 0. \end{cases}$$

Then there exists a constant $\rho_0 > 0$ such that the zero solution of (2) *is asymptotically stable in the sense of V_ρ for any $\rho \in (0, \rho_0]$, that is, the following conditions* (i) *and* (ii) *hold*:

(i) *For any $\varepsilon > 0$, there exists a $\delta > 0$ such that if $V_\rho(\bar{w}(\cdot)) < \delta$, then the solution $w(t,x)$ exists in the future assuming the initial value $\bar{w}(x)$, and $V_\rho(w(t,\cdot)) < \varepsilon$ for $t \in [0,\infty)$.*

(ii) *There exists a constant $\gamma > 0$ such that if $V_\rho(\bar{w}(\cdot)) < \gamma$, then $V_\rho(w(t,\cdot)) \to 0$ as $t \to \infty$.*

Now we consider the absolute stability for the system (1). Under a change of variables

(7) $$L: \begin{cases} v = Cu - Au_x + b\xi \\ \sigma = \nu' u - \mu' u_x - r\xi, \end{cases}$$

(1) becomes

(8) $$\begin{cases} v_t = Cv - Av_x + bf(\sigma) \\ \sigma_t = \nu' v - \mu' v_x - rf(\sigma). \end{cases}$$

Let $X_{p,q}$ be the space of functions $(u(x), \xi(x))$, $u \in C^p$ and $\xi \in C^q$, such that

$$\|u(\cdot)\|_p < \infty, \quad \|\xi(\cdot)\|_q < \infty.$$

Proposition 3. *Suppose that any characteristic root of $A^{-1}C$ has non-zero real part.*

Then the transformation $L: (u,\xi) \mapsto (v,\sigma)$ is a bicontinuous linear transformation from $X_{p+1,p}$ onto $X_{p,p}$, $p \geq 0$, if $|r|$ is sufficiently large.

Proof. It is clear that L is a linear and continuous map from

$X_{p+1,p}$ to $X_{p,p}$. We shall show that L^{-1} exists, i.e., (7) has a unique solution (u,ξ) for any given (v,σ). Eliminating ξ from the system (7), we have an equation for u

$$(A + r^{-1}b\mu')u_x = (C + r^{-1}b\nu')u - (v + r^{-1}b\sigma).$$

Set $A_r = A+r^{-1}b\mu'$ and $C_r = C+r^{-1}b\nu'$. If $A_r^{-1}C_r$ has no eigenvalue with zero real part, which holds if $|r|$ is large enough, this equation has one and only one bounded solution $u(x)$ on R for a bounded continuous function $(v(x),\sigma(x))$ on R. The uniquely existence of ξ is immediate, and clearly $(u,\xi)\in X_{p+1,p}$ if $(v,\sigma)\in X_{p,p}$. Q.E.D.

We have the following proposition concerning the relationship between the solutions of (1) and those of (8). It can be easily proved, so that we omit the proof.

Proposition 4. *Let $X_{p,q}(i)$ be a set of solutions of (i), i=1 or 8, which are defined on R_T and belong to $X_{p,q}$ for each $t\in[0,T]$. Suppose that $A^{-1}C$ has no eigenvalue with zero real part.*

Then L is a bicontinuous transformation from $X_{p+1,p}(1)$ onto $X_{p,p}(8)$, $p\geq 1$, for any fixed $T>0$, if $|r|$ is sufficiently large.

From now on, we consider the system (8). We shall prove the following theorems for the absolute stability of (8).

Theorem 5. *Suppose that $\mu=0$ in (8) and that there exist positive definite matrices B and D such that*

$$(9)\quad \begin{cases} BC + C'B = -D \\ BA - A'B = 0. \end{cases}$$

Then there exist constants $r_0>0$ and $\rho_0>0$ such that if $r>r_0$, the zero solution of (8) is asymptotically stable in the sense of V_ρ, $0<\rho\leq\rho_0$, for any choice of $f(\sigma)$ satisfying the conditions

(i) $f(\sigma)$ is of class C^2,

(ii) the derivative $f'(\sigma)$ of $f(\sigma)$ is positive at $\sigma=0$.

Proof. Set

$$\mathbb{A} = \begin{pmatrix} A & 0 \\ \mu' & 0 \end{pmatrix}, \quad \mathbb{B} = \begin{pmatrix} B & 0 \\ 0 & f'(0)/2 \end{pmatrix}, \quad \mathbb{C} = \begin{pmatrix} C & bf'(0) \\ \nu' & -rf'(0) \end{pmatrix}.$$

Then

$$-(\mathbb{B}\mathbb{C} + \mathbb{C}'\mathbb{B}) = \begin{pmatrix} D & -(Bbf'(0) + \nu f'(0)/2) \\ -(Bbf'(0) + \nu f'(0)/2)' & rf'(0)^2 \end{pmatrix}.$$

This is positive definite if and only if

$$r > (Bb + \nu/2)'D^{-1}(Bb + \nu/2).$$

On the other hand, we have

$$\mathbb{B}\mathbb{A} - \mathbb{A}'\mathbb{B} = \begin{pmatrix} 0 & \mu f'(0)/2 \\ \mu' f'(0)/2 & 0 \end{pmatrix}.$$

Since $f'(0)>0$, $\mathbb{B}\mathbb{A}-\mathbb{A}'\mathbb{B}=0$ if and only if $\mu=0$. Therefore the condition (6) in Theorem 2 is satisfied for B. This proves the theorem.

Theorem 6. Suppose that $\mu=0$ and that $f(\sigma)$ is of class C^2 and satisfies $f_1 \leq f'(\sigma) \leq f_2$ for any $\sigma\varepsilon R$ and for some positive numbers f_1 and f_2. Moreover we assume that there exist positive definite matrices B and D as those in Theorem 5.

Then there exist numbers $r_0>0$ and $\rho_0>0$ such that if $r>r_0$, the zero solution of (8) is asymptotically stable in the large in the sense of V_ρ, $0<\rho\leq\rho_0$.

Proof. Set

$$\mathbb{B} = \begin{pmatrix} B & 0 \\ 0 & \gamma \end{pmatrix}, \quad \gamma > 0, \quad \text{and} \quad w = (v,\sigma).$$

The time derivative of $V_{\rho,y}$ along the solutions of (8) satisfies

$$\frac{d}{dt} V_{\rho,y}(w(t,\cdot)) = \int_{-\infty}^{+\infty}[\langle w, -\mathbb{D}_o w\rangle + \langle w_x, -\mathbb{D}_1 w_x\rangle]e^{-\rho|x-y|}dx,$$

where

$$\mathbb{D}_i = \begin{pmatrix} D + \rho\,\mathrm{sign}(x-y)A'B & -(Bbf'(\sigma_i) + \gamma\nu) \\ -(Bbf'(\sigma_i) + \gamma\nu)' & 2\gamma r f'(\sigma_i) \end{pmatrix}, \quad i = 0,1,$$

for some σ_i, $|\sigma_i|\leq|\sigma|$. Because D is positive definite and $A'B=BA$, there exists a $\rho_0>0$ such that $D+\rho\,\mathrm{sign}(x-y)A'B$ is positive definite for $\rho\varepsilon(0,\rho_0]$. Therefore $\mathbb{D}_i$ is positive definite for $\rho\varepsilon(0,\rho_0]$, if

$$(10) \qquad 2\gamma r f'(\sigma_i) > (Bbf'(\sigma_i)+\gamma\nu)'(D+\rho\,\mathrm{sign}(x-y)A'B)^{-1}(Bbf'(\sigma_i)+\gamma\nu).$$

The right-hand side of (10) is bounded, because $f'(\sigma_i)$ is in $[f_1, f_2]$. If r is sufficiently large, (10) is satisfied and $\mathbb{D}_i$ are positive definite for $x, y\varepsilon R$, $\rho\varepsilon(0,\rho_0]$ and $f_1\leq f'(\sigma_i)\leq f_2$. Hence there exist positive numbers $\underline{d}$ and $\overline{d}$ such that $\underline{d}I<\mathbb{D}_i<\overline{d}I$, $i=0,1$, and we have

$$(11) \qquad \frac{d}{dt} V_{\rho,y}(w(t,\cdot)) \leq -\underline{d} \int_{-\infty}^{+\infty} [\langle w,w\rangle + \langle w_x,w_x\rangle]e^{-\rho|x-y|}dx$$
$$\leq c_0 V_{\rho,y}(w(t,\cdot))$$

for some $c_0<0$ independent of $y\varepsilon R$. The relation (11) holds as long as

$V_{\rho,y}$ exists. Therefore we have

$$V_\rho(w(t,\cdot)) \leq V_\rho(w(0,\cdot))\exp(c_0 t) \quad \text{for} \quad t \geq 0,$$

whatever values are assumed by $V_\rho(w(0,\cdot))$. Q.E.D.

References

[1] K.O.Friedrichs, Nonlinear hyperbolic differential equations for functions of two independent variables, Amer. J. Math., 70(1948), 555-589.

[2] A.Jeffrey and Y.Kato, Liapunov's direct method in stability problems for semilinear and quasilinear hyperbolic systems, J. Math. and Mech., 18(1969), 659-682.

OSCILLATIONS IN FUNCTIONAL DIFFERENTIAL EQUATIONS

Lynn Erbe

The purpose of this note is to show how the theory of second order differential inequalities may be applied to the study of the oscillation of solutions of the functional differential equation

$$x'' + F(t,x(\cdot)) = 0 \tag{1}$$

where $F(t,\psi(\cdot))$ is a continuous real-valued functional defined for $t \geq t_o \geq 0$ and ψ in C_t (real-valued) and $C_t \equiv \{\psi \in C[g(t)-t,0]: t \geq t_o\}$. Here we assume $g(t)$ is a real-valued continuous function with $g(t) \leq t,\ t \geq t_o$ and $g(t) \to \infty$ as $t \to \infty$. Furthermore, we assume $F(t,\psi(\cdot))$ satisfies

$$(\operatorname{sgn} \psi)F(t,\psi(\cdot)) > 0 \text{ if } |\psi| \text{ is positive}, \quad \psi \in C_t$$

and that for each $\tau > t_o$ there exist continuous functions $f_\tau(t,u_1,u_2)$, $h_\tau(v_1,v_2)$ such that

$$f_\tau(t,u_1,u_2) \begin{cases} > 0 & \text{if } u_1,u_2 > 0 \\ < 0 & \text{if } u_1,u_2 < 0 \end{cases} \quad \text{and} \quad h_\tau(v_1,v_2) > 0, \text{ all } v_1,v_2,$$

and such that for $\psi \in C_t^{(1)}$ with ψ positive and increasing, we have

$$F(t,\psi(\cdot)) \geq f_\tau(t,\psi(t),\psi(g(t))h_\tau(\psi'(t),\psi'(g(t)) > 0, \quad t \geq \tau. \tag{2}$$

The reverse inequality is assumed to hold if $\psi \in C_t^{(1)}$ is negative and decreasing. Special cases of (1) are

$$x'' + p(t)x(g(t))^\gamma = 0, \quad p(t) \geq 0, \quad \gamma > 0, \tag{3}$$

where γ is the quotient of odd positive integers, and

$$x'' + f(t,x(t),x(g(t)))h(x'(t),x'(g(t))) = 0 \tag{4}$$

(For a survey of oscillation criteria for (3) and (4) in the case $g(t) = t$, see [7]; recent references for the case $g(t) \leq t$ are [1], [2], [4] and [6]).

Our results show, in general, that the oscillatory nature of solutions of (1) is determined by the asymptotic ratio of $\frac{g(t)}{t}$ rather than the difference $t - g(t)$. In particular, if $g(t) = t-\tau(t)$, where $0 \leq \tau(t) < M$, then oscillation criteria for (3) and (4) are "essentially the same" as for the ordinary differential equation which results by replacing $g(t)$ by t.

Our technique will involve an application of the following existence and comparison theorem for solutions of second order differential equations: [3]

<u>Theorem 1</u>: Let $\alpha(t), \beta(t) \in C^{(2)}[a,b]$ with $\alpha''(t) \geq f(t,\alpha,\alpha')$, $\beta'' \leq f(t,\beta,\beta')$ and $\alpha(t) < \beta(t)$ on $[a,b]$. Assume $f \in C[a,b] \times R^2$ is $O(|x'|^2)$ as $|x'| \to \infty$ for $(t,x) \in K$, K a compact subset of $[a,b] \times R$. Then the differential equation

$$x'' = f(t,x,x')$$

has a solution $x_o(t) \in C^{(2)}[a,b]$ with $\alpha(t) \leq x_o(t) \leq \beta(t)$, $a \leq t \leq b$. In addition, if the partial derivative functions $f_x(t,x,x')$, $f_{x'}(t,x,x')$ are continuous, $x_o(t)$ may be chosen so that the variational equation

$$z'' = f_{x'}(t,x_o,x_o')z' + f_x(t,x_o,x_o')z$$

is disconjugate on $[a,b]$.

To apply the above theorem we need a preliminary lemma.

<u>Lemma 2</u>: Let $g(t) \to \infty$ and assume $y(t) \in C^{(2)}[t_o,\infty)$ satisfies $y(t) > 0$, $y'(t) > 0$, $y''(t) \leq 0$, $t \geq T \geq t_o$. Then for each $0 < k < 1$ there is a $T_k \geq T$ such that

$$y(g(t)) \geq ky(t)\frac{g(t)}{t}, \quad t \geq T_k .$$

The above Lemma allows one to convert solutions of the FDE (1) into solutions of ordinary differential inequalities to which the existence Theorem and subsequent necessary conditions for non-oscillation may be applied. As examples we state the following

Theorem 3: All solutions of the equation (3) with $\gamma = 1$ are oscillatory in case $y'' + \lambda\mu(t)p(t)y = 0$ is oscillatory for some $0 < \lambda < 1$, where $\mu(t) = \frac{g(t)}{t}$.

Theorem 4: All solutions of equation (3) are oscillatory

a) iff $0 < \gamma < 1$ and $\int^{\infty} (g(t))^{\gamma} p(t)dt = +\infty$

b) in case $\gamma > 1$ and $\int^{\infty} t(\mu(t))^{\gamma} p(t)dt = +\infty$

The converse of b) is true if $\liminf_{t \to \infty} \mu(t) > 0$.

If $g(t) \equiv t$ and $\gamma > 1$, then non-oscillatory solutions of (3) are either bounded or grow as a power of t (cf. [5]). For $g(t) < t$ this is no longer true (i.e., $g(t) = y(t) = t^{1/2}$, $p(t) = t^{\alpha}/4$, $\alpha = -3/2 - \gamma/4$). This particular example also has solutions which grow as a power of t, if $\gamma < 2$, as a consequence of

Theorem 5: Let $\gamma > 1$ in equation (3). Then (3) has a solution $y(t)$ with $\frac{y(t)}{t} \to \alpha > 0$ iff $\int^{\infty} (g(t))^{\gamma} p(t)dt < +\infty$

As a final example we state a theorem for the more general equation (1). For convenience, we assume inequality (2) holds with $h_{\tau} \equiv 1$ and $f_{\tau} \equiv f(t,u,v)$ for all $\tau \geq t_o$ and that f has continuous partial derivatives with respect to u,v. Furthermore, we assume $f(t,u,v) = -f(t,-u,-v)$, all t,u,v and that for each fixed t and $u > 0$ $f_v(t,u,v) \geq 0$ and is non-decreasing in v for $v > 0$ and for each fixed t and $v > 0$ $f_u(t,u,v) \geq 0$ and is non-decreasing in u for $u > 0$.

Theorem 6: All solutions of (1) are oscillatory in case the linear equation

$$z'' + [f_u(t,\alpha,k\alpha\mu(t)) + k\mu(t)f_v(t,\alpha,k\alpha\mu(t))]z = 0$$

is oscillatory for all $\alpha \neq 0$ and some $0 < k < 1$.

We conclude by pointing out that the right hand side of inequality (2) may be generalized to involve several retardations and that one may also use this technique to establish asymptotic growth estimates for solutions of (1).

References

[1] J.S. Bradley, Oscillation theorems for a second order delay equation, J. Diff. Eqns. 8 (1970), 397-403.

[2] F. Burkowski, Oscillation theorems for a second order nonlinear functional differential equation, J. Math. Anal. Appl. 33 (1971), 258-262.

[3] L. Erbe, Nonlinear boundary value problems for second order differential equations, J. Diff. Eqns. 7 (1970), 459-472.

[4] H.E. Gollwitzer, On nonlinear oscillations for a second order delay equation, J. Math. Anal. Appl. 26 (1969), 385-389.

[5] R.A. Moore and Z. Nehari, Nonoscillation theorems for a class of nonlinear differential equations, Trans. Amer. Math. Soc. 93 (1959), 30-52.

[6] Odaric, O.M., Sevelo, V.M., Conditions for oscillation of solutions of second order nonlinear differential equations with lag. Dopovidi Akad. Nauk. Ukrain. RSR Ser. A. 1967, 1027-1031.

[7] J.S.W. Wong, On second order nonlinear oscillation, Funk. Ekv. 11 (1968), 207-234.

EXISTENCE OF ALMOST PERIODIC SOLUTIONS BY LIAPUNOV FUNCTIONS

Fumio NAKAJIMA

We shall consider the systems

$$(1) \quad \frac{dx}{dt} = F(t,x)$$

and

$$(2) \quad \frac{dx}{dt} = G(t,x) ,$$

where $F(t,x) \in C(R \times U : R^n)$ (U: open subset of R^n) is almost periodic in t uniformly for $x \in U$ and $G(t,x)$ is an element of the hull of $F(t,x)$.

For the existence of almost periodic solutions of the system (1), we have the following proposition due to Amerio [1].

Proposition. If for each $G(t,x)$ in the hull of $F(t,x)$, the system (2) has one and only one solution which remains in a compact subset S of U for all $t \in R$, then the system (1) has one and only one almost periodic solution in S whose module is contained in the module of $F(t,x)$.

As a sufficient condition in order that for each $G(t,x)$ in the hull of $F(t,x)$, the system (2) has one and only one solution which remains in S, Seifert [2] has assumed some kind of global stability on a bounded solution, and Fink and Seifert [3] have assumed the existence of some Liapunov function.

The purpose of this paper is to weaken the conditions in Fink and Seifert's theorem.

From another point of view, Yoshizawa [4] has obtained the existence theorem for almost periodic solutions. His theorem can be obtained from our corollary.

We have the following theorem.

Theorem. Suppose that the system (1) has a solution $\phi(t)$ which remains in a compact subset S in U for all $t \geqq 0$, and that there exists a function $V(t,x) \in C([0,\infty) \times U : R)$ which satisfies the following conditions;

(i) $V(t,\phi(t))$ is bounded on $[0,\infty)$,

(ii) $|V(t,x) - V(t,y)| \leqq L\|x-y\|$ for $x \in K$, $y \in K$, where K is a compact subset of U and L is a constant which may depend on K,

(iii) $\dot{V}(t,x) \geqq a(\|x - \phi(t)\|)$, where $a(r)$ is continuous and positive definite on $[0,\infty)$ and

$$\dot{V}(t,x) = \overline{\lim_{h\to+0}} \frac{1}{h}\{V(t+h,x+hF(t,x)) - V(t,x)\}.$$

Then the system (1) has an almost periodic solution in U whose module is contained in the module of $F(t,x)$.

In this theorem, we have to know what $\phi(t)$ is. However there is often a case that we know only the existence of a bounded solution of (1). For such a case, the following corollary is useful.

Corollary. Suppose that there exists a function $V(t,x,y)\varepsilon C([0,\infty)\times U\times U:R)$ which satisfies the following conditions

(i) $V(t,x,x)$ is bounded for $t\varepsilon[0,\infty)$, $x\varepsilon U$,

(ii) $|V(t,x,y) - V(t,x',y')| \leqq L\{\|x-x'\| + \|y-y'\|\}$ for $x,x',y,y'\varepsilon K$, where K is a compact subset in U and $L>0$ is constant which may depend on K,

(iii) $\dot{V}(t,x,y) \geqq a(\|x - y\|)$, where $a(r)$ is continuous and positive definite on $[0,\infty)$ and

$$\dot{V}(t,x,y) = \overline{\lim_{h\to+0}} \frac{1}{h}\{V(t+h,x+hF(t,x),y+hF(t,y)) - V(t,x,y)\}.$$

Moreover suppose that the system (1) has a bounded solution which remains in a compact subset S of U.

Then the system (1) has an almost periodic solution in U whose module is contained in the module of $F(t,x)$.

Let $\phi(t)$ be a given bounded solution and consider $V(t,x,\phi(t))$ as $V(t,x)$ in the theorem. Then this corollary follows immediately from the theorem.

We shall show the outline of the proof of the theorem.

For each $G(t,x)$ in the hull of $F(t,x)$, we have a sequence $\{t_k\}$ such that

$$t_k \to +\infty \quad \text{as} \quad k \to \infty$$

and

$$F(t+t_k,x) \to G(t,x) \text{ uniformly on } R\times S \text{ as } k \to \infty.$$

We can also assume that for some function $\psi(t)$, $\phi(t+t_k) \to \psi(t)$ uniformly on any compact set in R as $k \to \infty$. Then $\psi(t)$ is a solution of (2) which remains in S on R. It is sufficient to show that, if $x(t)$ is a solution of (2) such that $x(t) \in S$ on R, then $x(t) = \psi(t)$ on R. If we define

$$v_k(t) = V(t+t_k, x(t)) \quad \text{for} \quad t > -t_k ,$$

we have

$$D^+ v_k(t) = \overline{\lim_{h\to+0}} \frac{1}{h}\{v_k(t+h) - v_k(t)\}$$
$$\geqq a(\|x(t) - \phi(t+t_k)\|) - A_k(t) ,$$

where

$$A_k(t) = L\|G(t,x(t)) - F(t+t_k, x(t))\|.$$

On any interval $[s_1, s_2]$ in R, we have

$$(4) \qquad v_k(s_2) - v_k(s_1) \geqq \int_{s_1}^{s_2} a(\|x(s) - \phi(s+t_k)\|)ds - \int_{s_1}^{s_2} A_k(s)ds ,$$

where k is sufficiently large so that $s_1 + t_k \geqq 0$. Since $|V(t,x)| < B$ for $t \geqq 0$, $x \in S$ and for some $B > 0$, we have

$$2B > \int_{-\infty}^{+\infty} a(\|x(s) - \psi(s)\|)ds ,$$

because $A_k(t) \to 0$ uniformly on R as $k \to \infty$. Therefore there are sequences $\{\tau_\ell^1\}$ and $\{\tau_\ell^2\}$ such that

$$\tau_\ell^1 \to -\infty , \quad \tau_\ell^2 \to +\infty \qquad \text{as } \ell \to \infty$$

and that

$$\|x(\tau_\ell^i) - \psi(\tau_\ell^i)\| \to 0 \quad \text{as} \quad \ell \to \infty, \ i=1,2.$$

We put $s_i = \tau^i$ $(i=1,2)$ in (4), and subtract $W(\ell,k)$ from the both side of (4), where

$$W(\ell,k) = V(\tau_\ell^2 + t_k, \phi(\tau_\ell^2 + t_k)) - V(\tau_\ell^1 + t_k, \phi(\tau_\ell^1 + t_k)).$$

Then we have

$$L\left\{\|x(\tau_\ell^2) - \phi(\tau_\ell^2 + t_k)\| + \|x(\tau_\ell^1) - \phi(\tau_\ell^1 + t_k)\|\right\}$$
$$\geqq \int_{\tau_\ell^1}^{\tau_\ell^2} a(\|x(s) - \phi(s+t_k)\|)ds - \int_{\tau_\ell^1}^{\tau_\ell^2} A_k(s)ds - W(\ell,k).$$

Thus, for any $\varepsilon > 0$, if ℓ is sufficiently large,

$$2\varepsilon L > \int_{\tau_\ell^1}^{\tau_\ell^2} a(\|x(s) - \psi(s)\|)ds - \lim_{k\to\infty} W(\ell,k).$$

Since $\lim_{t\to\infty} V(t,\phi(t))$ exists, we have

$$\lim_{k\to\infty} W(\ell,k) = 0,$$

which implies that

$$2\varepsilon L > \int_{\tau_\ell^1}^{\tau_\ell^2} a(\|x(s) - \psi(s)\|)ds.$$

Letting $\ell \to\infty$, we have

$$\int_{-\infty}^{+\infty} a(\|x(s) - \psi(s)\|)ds = 0.$$

This implies that $x(s)=\psi(s)$ on R. Hence, by the proposition we have almost periodic solutions for (1) in U.

Remark 1. We can prove that there exists only one almost periodic solution in U.

Remark 2. If we have a Liapunov function for the system (1) which is defined on $R\times U$ and satisfies the conditions (i), (ii) and (iii) in our theorem, the system (1) has one and only one solution which remains in S on R. Fink and Seifert have constructed a Liapunov function satisfying the conditions (i), (ii) and (iii) for each system (2) by using a Liapunov function assumed for (1). To do this, in addition to the conditions in our theorem, they have actually assumed the conditions that

(i) $\phi(t)$ and $V(t,x)$ are defined on R and $R\times U$, respectively,

(ii) $V(t,\phi(t)) = 0$ on R,

(iii) $V(t,x)$ is continuous in t uniformly for $(t,x)\in R\times K$, where K is any compact subset in U.

References

[1] Amerio, L., Soluzioni quasi-periodiche, 0 limitati di sistemi differenziali non lineari quasi-periodici, 0 limitati. Ann. Math. Pura Appl., 39(1955), 97-119.

[2] Seifert, G., Stability conditions for the existence of almost-periodic solutions of almost-periodic systems. J. Math. Analysis

and Appl., 10(2)(1965), 409-418.

[3] Fink, A.M. and Seifert, G., Liapunov functions and almost periodic solutions for almost periodic systems. J. Differential Equations, 5(1969), 307-313.

[4] Yoshizawa, T., Stability theory by Liapunov's second method Math. Soc., Japan, (1965).

MEAN STABILITY OF QUASI-LINEAR SYSTEMS WITH RANDOM COEFFICIENTS

Tetsuo FURUMOCHI

There are many results on the mean stability of solutions of linear systems with random coefficients. Here we shall discuss the mean stability of solutions of quasi-linear systems with random coefficients.

We denote by (Ω,Σ,P) a probability space with expectation operator E and a generic point $\omega\varepsilon\Omega$. If x is in R^n, then $|x|$ denotes its Euclidean norm, and if A is an n×n matrix, that is $A\varepsilon R^n \times R^n$, $|A|$ denotes its operator norm. For a random vector or a random matrix $x(\omega)$, $\|x(\cdot)\|_p$ will denote the L^p norm defined by

$$\|x(\cdot)\|_p = (E|x(\omega)|^p)^{1/p} \qquad (1 \leqq p < \infty)$$

and

$$\|x(\cdot)\|_\infty = \underset{\omega\varepsilon\Omega}{\text{ess.sup}}|x(\omega)|.$$

$L^p(\Omega)$ will denote the Banach space defined by

$$L^p(\Omega) = \{\ x(\omega) : \Omega \longrightarrow R^n\ ,\ \|x(\cdot)\|_p < \infty\ \} \qquad (1 \leqq p \leqq \infty)$$

Consider the systems

$$(1) \qquad \frac{dx}{dt} = A(t,\omega)x + f(t,x)$$

and

$$(2) \qquad \frac{dx}{dt} = A(t,\omega)x\ ,$$

where $A(t,\omega) : [0,\infty)\times\Omega \longrightarrow R^n \times R^n$ is a product measurable function such that

$$\|A(t,\cdot)\|_\infty \leqq L(t)$$

for all $t \geqq 0$ and for a locally Lebesgue integrable function L(t) and $f(t,x) : [0,\infty)\times R^n \longrightarrow R^n$ is a continuous function such that

$$|f(t,x) - f(t,x)| \leqq l(t)|x - y|$$

for $t \geqq 0$, x, $y \in R^n$ and for a locally Lebesgue integrable function $l(t)$. Then for any p, $1 \leqq p \leqq \infty$, (1) has a unique solution $x(t,\omega)$ for any initial condition $x_o(\omega) \in L^p(\Omega)$. Moreover, $\|x(t,\cdot)\|_p$ is absolutely continuous in t (see [1]).

For mean stability, we have the following lemma.

Lemma 1(Edsinger). *Let* $Y(t,\omega)$ *be a fundamental matrix solution of* (2) *which is equal to the identity matrix* I *at* $t = 0$. *Then*

(i) *the null solution of* (2) *is mean stable if and only if there exists a positive constant* K *such that*

$$\|Y(t,\cdot)\|_\infty \leqq K \qquad \textit{for} \quad t \geqq 0 \ ,$$

and

(ii) *the null solution of* (2) *is uniformly mean stable if and only if there exists a positive constant* K *such that*

$$\|Y(t,\cdot)Y^{-1}(s,\cdot)\|_\infty \leqq K \qquad \text{for} \quad 0 \leqq s \leqq t < \infty$$

Remark. Edsinger has defined $L(p,q)$–stability, which is a generalization of the mean stability, and has given necessary and sufficient conditions for the null solution of (2) to be $L(p,q)$–stable (see [2]). The above lemma corresponds to the case of $L(1,1)$–stability.

For the deterministic case, it is well known that when some stability property holds for a linear system, the same stability property holds for a quasi–linear system under suitable conditions on the perturbed term (see, for example, [3; pp. 64–67]). In the followings, we shall obtain analogous results for (1) and (2).

Theorem 1. *Let* $Y(t,\omega)$ *be a fundamental matrix solution of* (2) *which is equal to the identity matrix* I *at* $t = 0$. *Suppose that there exists a positive constant* K *such that*

$$\|Y(t,\cdot)Y^{-1}(t,\cdot)\|_\infty \leqq K \qquad \text{for} \quad 0 \leqq s \leqq t < \infty \ ,$$

and let $f(t,x)$ *in* (1) *satisfy the inequality*

$$(3) \qquad f(t,x) \leqq q(t)\ x \ ,$$

where $q(t)$ *is a nonnegative continuous function such that*

$$(4) \qquad \int_0^\infty q(s)ds < \infty ,$$

Then there exists a positive constant L such that any solution $x(t,\omega)$ of (1), which satisfies $E|x(t_1,\omega)| < \infty$ for some $t_1 \geqq 0$, is defined for $t \geqq t_1$ and satisfies

$$(5) \qquad E|x(t,\omega)| \leqq L \cdot E|x(t_1,\omega)| \qquad \text{for} \quad t \geqq t_1 .$$

Moreover, in addition, if $\|Y(t,\cdot)\|_\infty \longrightarrow 0$ as $t \longrightarrow \infty$, then $E|x(t,\omega)| \longrightarrow 0$ as $t \longrightarrow \infty$.

Let $x(t,\omega)$ be a solution of (1) which satisfies $E|x(t_1,\omega)| < \infty$ for some $t_1 \geqq 0$. Then, by the variation of constants formula,

$$(6) \qquad x(t,\omega) = Y(t,\omega)Y^{-1}(t_1,\omega)x(t_1,\omega) + \int_{t_1}^t Y(t,\omega)Y^{-1}(s,\omega)f(s,x(s,\omega))ds.$$

(5) can be shown by estimating $E|x(t,\omega)|$. Since we can verify that $\|Y^{-1}(t,\cdot)\|_\infty < \infty$ for each $t \geqq 0$, we can obtain asymptotic mean stability by estimating $E|x(t,\omega)|$.

Next we shall discuss the strong mean stability.

Definition. A solution $x_1(t,\omega)$ of (1) is strongly mean stable if given any $\varepsilon > 0$, there exists a $\delta(\varepsilon) > 0$ such that any solution $x(t,\omega)$ of (1) satisfies $E|x(t,\omega) - x_1(t,\omega)| < \varepsilon$ for $t \geqq 0$ if $E|x(t_1,\omega) - x(t_1,\omega)| \leqq \delta(\varepsilon)$ for some $t_1 \geqq 0$.

Lemma 2. Let $Y(t,\omega)$ be a fundamental matrix solution of (2) which is equal to the identity matrix I at $t = 0$. Then the null solution of (2) is strongly mean stable if and only if there exists a positive constant K such that

$$\|Y(t,\cdot)\, Y^{-1}(s,\cdot)\|_\infty \leqq K \qquad \text{for} \qquad t,s \geqq 0 .$$

The proof is done by the same method as used in [2].

Theorem 2. Let $Y(t,\omega)$ be a fundamental matrix solution of (2) which is equal to the identity matrix at $t = 0$. Suppose that there exists a positive constant K such that

$$\|Y(t,\cdot)\|_\infty \leqq K , \qquad \|Y^{-1}(t,\cdot)\|_\infty \leqq K \qquad \underline{\text{for}} \quad t \geqq 0 ,$$

and let $f(t,x)$ in (1) satisfy (3) and (4).

Then there exists a positive constant L such that any solution $x(t,\omega)$ of (1), which satisfies $E|x(t_1,\omega)| < \infty$ for some $t_1 \geqq 0$, is defined for $t \geqq 0$ and satisfies

$$E|x(t,\omega)| \leqq L \cdot E|x(t,\omega)| \qquad \underline{\text{for}} \qquad t \geqq 0 .$$

This theorem can be proved by estimating $r(t) = E|Y^{-1}(t,\omega)x(t,\omega)|$ in the relation (6).

References

[1] J.L. Strand, Random ordinary differential equations, J. Differential Eqs., 7(1970), 538-553.
[2] R.W. Edsinger, Mean stability of linear ordinary differential equations with random coefficients, J. Differential Eqs., 8(1970), 448-456.
[3] W.A. Coppel, Stability and asymptotic behavior of differential equations, Heath Mathematical Monograph, Boston, 1965.

A NONLINEAR STURM-LIOUVILLE PROBLEM

J.W. MACKI and P. WALTMAN

We consider the nonlinear problem

(1) $-y'' + q(x)y = \lambda[a(x) - f(x,y(x),y'(x))]\, y(x)$,

(2) $a_1 y(0) + a_2 y'(0) = 0$, $b_1 y(1) + b_2 y'(1) = 0$,

in relation to the associated linear problem obtained by setting $f \equiv 0$. Problems of the form (1)(2) arise in several contexts, for example the theory of buckling of plates, shells and columns ([4],[5]). Wolkowisky in [3] and Crandall and Rabinowitz in [1] used the Schauder fixed-point theorem and the theory of Leray-Schauder degree, respectively, to study the existence of eigenfunctions for problems of the form of (1)(2). In an earlier paper [2] we used a simple polar-coordinate argument to analyse certain special nonlinear eigenvalue problems. We can apply the same ideas to certain problems in bifurcation theory.

(H-1): The linear problem obtained by setting $f \equiv 0$ in (1) and keeping (2) has eigenvalues $0 < \lambda_0 < \lambda_1 < \ldots$, accumulating at $+\infty$, and the eigenfunction (unique up to a constant factor) corresponding to λ_k has exactly k simple zeros in (0,1).

(H-2): $f \in C([0,1] \times R^2)$, $f(x,0,0) \equiv 0$ and there exists a constant ρ such that $f(x,\xi,\eta) > 0$ for $x \in [0,1]$, $0 < |\xi|,|\eta| < \rho$. Furthermore, $a(x)$ and $q(x)$ are continuous on $[0,1]$ with $a(x) > 0$.

(H-2)$'$: The same as Assumption 2, except for the weaker requirement $f(x,\xi,\eta) \geq 0$ $\forall x \in [0,1]$, $0 < |\xi|,|\eta| < \rho$.

Assumptions (H-1) and (H-2) are the hypotheses $(f_1)(f_2)$ of [1]. Note that we do not assume uniqueness of solutions to initial-value problems. However, we shall need to use the fact that if P is a compact connected subset of the (y,y')-plane, and if all solutions $y(x,p)$ of (1) with initial conditions $p \in P$ exist over $[0,1]$, then $\Gamma = \bigcup_{p \in P} y(1,p)$ is connected.

Theorem 1

If (H-1) and (H-2) hold, then for each λ_j, $j = 0,1,\ldots$, *there exists an interval* $I_j = (\lambda_j, \Lambda_j)$ *such that for* $\lambda \in I_j$, *(1)(2) has at least two solutions with exactly* j *zeros in* $(0,1)$.

This is the local theorem (Theorem 3.4) of Crandall and Rabinowitz. The proof proceeds by introducing polar coordinates through the usual formulas $y(t) = r(t)\cos\theta(t)$, $y' = r(t)\sin\theta(t)$, yielding the equivalent system

$$\theta' = -\sin^2\theta - (\lambda a - q)\cos^2\theta + \lambda f\cos^2\theta$$

$$r' = r\cos\theta\sin\theta\,(1 + \lambda f - \lambda a + q)\ ,$$

$$\theta(0) = \text{Arctan}\,(-a_1/a_2) = \alpha\ ,\quad \theta(1) = \text{Arctan}\,(-b_1/b_2) - k\pi\ ,$$

where $-\pi/2 < \text{Arctan}\ \theta \leq \pi/2$ and k is any natural number. As in the standard proofs for the classical linear theory, we then show that we can hit the "target" $\theta(1,\lambda,\mu) = \beta_k \equiv \text{Arctan}\,(-b_1/b_2) - k\pi$ by varying λ and $\mu \equiv r(0,\lambda)$. The key lemmas are as follows:

I. *If (H-1) and (H-2)' hold, and if* $\lambda \in [0,\lambda^*]$, *then there exists a* $\delta(\lambda^*) > 0$ *and* $K(\lambda^*) > 0$ *such that if* $r(0,\lambda) < \delta$, *then*

$$r(x,\lambda) < Kr(0,\lambda)\ ,\quad x \in [0,1]\ .$$

In particular, solutions of (1) are continuable if $r(0,\lambda)$ *is small.*

II. *Assume (H-1) and (H-2)' hold. If* $\lambda > \lambda_j$, *then there exists* $r_1(\lambda)$ *such that*

$$0 < r(0,\lambda) < r_1 \Longrightarrow \theta(1,\lambda,\mu) < \phi(1,\lambda_j) = \beta_j\ .$$

III. *If (H-1) and (H-2) hold, then for each sufficiently small* $\mu > 0$ *there is a nondegenerate interval* $[\lambda_j, \Lambda_j(\mu)]$ *such that* $\theta(1,\lambda,\mu) > \beta_j$ *for* $\lambda \in I_j(\mu)$.

We then use Lemma I and the connectedness property mentioned above to conclude that $\theta(1,\lambda,\mu(\lambda)) = \beta_j$ for λ in some interval to the right of λ_j.

(H-3): For each λ, *there is a* $\mu(\lambda)$ *such that* $\theta(1,\lambda,\mu) \geq \beta_0$.

<u>*Theorem 2*</u> *(the global theorem)*

Assume (H-1) and (H-2) hold, and that all solutions of (1) with $a_1 u(0) + a_1 u'(0) = 0$ *are extendable over* $[0,1]$. *Then (H-3) is a necessary and sufficient condition that for each* $\lambda > \lambda_k$, $k = 0,1,2,\ldots$, *there exists a solution of (1)(2) with exactly* k *interior zeros.*

Crandall and Rabinowitz prove the existence of such solutions (without assuming extendability) if solutions of (1) (2) satisfy an a priori bound, $|y(x,\lambda)| + |y'(x,\lambda)| \leq M(\lambda)$. We can construct examples where Theorem 2 applies but there is no a priori bound.

Acknowledgment

The research for this paper was supported (in part) by the Defence Research Board of Canada, Grant Number DRB-9540-11.

Bibliography

[1] M.G. Crandall and P.H. Rabinowitz, Nonlinear Sturm-Liouville eigenvalue problems and topological degree, J. Math. Mech., 19(1970), 1083-1102.

[2] J.W. Macki and P. Waltman, A nonlinear boundary value problem of Sturm-Liouville type for a two dimensional system of ordinary differential equations, SIAM J. Appl. Math., Sept., 1971.

[3] J.H. Wolkowisky, Nonlinear Sturm-Liouville problems, Arch. Rat. Mech. and Anal., 35(1970), 299-320.

[4] ____________________, Existence of buckled states of circular plates. Comm. Pure Appl. Math. 20(1967), 549-560.

[5] ____________________, Buckling of the circular plate embedded in elastic springs. An application to geophysics, Comm. Pure Appl. Math., 22(1969), 639-667.

INITIAL VALUE PROBLEM OF AN ORDINARY DIFFERENTIAL EQUATION OF THE SECOND ORDER

Ken-iti TAKAHASI

1. Introduction. In 1949, M.J.Lighthill [1] proposed a new method for nonlinear differential equations known as PLK-method or Lighthill's method. By applying this method Y.Sibuya and K.Takahasi [2] constructed a solution of the following initial value problem of the first order:

$$(1.1)\qquad (x+\varepsilon u)\frac{du}{dx} + q(x)u = r(x), \qquad u(1) = b,$$

where ε is a small positive parameter.

On the other hand, W.Wasow [6] mentioned, for example, Lighthill's method can be justified for the following initial value problem:

$$(1.2)\qquad (x+\varepsilon u+\varepsilon^n P(x,u))\frac{du}{dx} + q(x)u = r(x), \qquad u(1) = b,$$

where the quantity n is a positive integer greater than 1, and $P(x,u)$ is a polynomial of the degree m not greater than n with respect to u. However, even if the degree m of $P(x,u)$ is greater than n, we can justify Lighthill's method under suitable assumptions [4] .

In this paper, our purpose is to show that Lighthill's method can be justified for the initial value problem of the second order:

$$(1.3)\qquad \left(x+\varepsilon\left(p(x)v+\frac{dv}{dx}\right)+\varepsilon^n P\left(x,v,\frac{dv}{dx}\right)\right)\frac{d^2v}{dx^2} + q(x)\frac{dv}{dx} + r(x)v = s(x),$$

$$v(1) = b, \qquad v'(1) = b',$$

which corresponds to Problem (1.2).

2. Theorems. First let us explain one of the results obtained

for Problem (1.1). We have the following:

THEOREM 1. *Let $u_0(x)$ be a solution of Problem* (1.1) *with* $\varepsilon = 0$. *Assume that*

(I) $q(x)$ *and* $r(x)$ *are real-valued and analytic functions of* x *for* $0 \leqq x \leqq 1$;

(II) $q_0 = q(0) > 0$;

(III) $[x^{q_0} u_0(x)]_{x=0} > 0$;

(IV) $u_0(x)q(x) - r(x) \neq 0$ for $0 < x \leqq 1$;

Then there exsist a constant ε_0 *and a function of the form*

$$x(\xi,\zeta) = \xi + \sum_{m=1}^{\infty} x_m(\xi)\zeta^m$$

such that

(a) $x(\xi,\zeta)$ *is convergent uniformly for* $0 < \xi \leqq 1, |\zeta| \leqq \delta$, *where* $\delta > 0$ *is sufficiently small*;

(b) $x(1,\zeta) = 1$, *for* $|\zeta| \leqq \delta$;

(c) *the equation* $x(\xi, \varepsilon\xi^{-s_0 q_0}) = 0$ *has a unique solution* $\hat{\xi}(\varepsilon)$ *in the interval*

$$\left[\frac{\varepsilon}{\delta}\right]^{1/q_0} \leqq \xi \leqq 1, \quad (0 < \varepsilon \leqq \varepsilon_0);$$

(d) $u = u_0(\xi)$, $x = x(\xi, \varepsilon\xi^{-s_0 q_0})$ *is a parametric representation of the solution* $u(x)$ *of Problem* (1.1);

(e) $u(0) = u_0(\hat{\xi}(\varepsilon))$.

REMARK. Let us write Equation (1.1) as follows:

$$\frac{dx}{dt} = x + \varepsilon u, \qquad \frac{du}{dt} = -q(x)u + r(x).$$

Assumption (II) shows that the critical point of this equation is a saddle point. Since Equation (1.1) with $\varepsilon = 0$ is linear, it is easily solved by the form

$$(2.1) \qquad u = x^{-q_o}(x^{q_o} w(x) + C v(x)),$$

where w(x) is a particular solution analytic for $0 \leqq x \leqq 1$, and $C x^{-q_o} v(x)$ $(v(x) > 0)$ is a general solution of the homogeneous form. If $C = (b-w(1))/v(1)$, (2.1) gives the solution $u_o(x)$ of Problem (1.1) with $\varepsilon = 0$. Therefore Assumption (III) is equivalent to $b-w(1) > 0$.

To discuss Problem (1.3), we shall consider an equivalent system:

$$(2.2) \quad \begin{cases} (x+ \varepsilon(p(x)v+u)+\varepsilon^n P(x,v,u)) \dfrac{du}{dx} + q(x)u+r(x)v = s(x), \\ \dfrac{dv}{dx} = u, \qquad u(1) = b', \quad v(1) = b. \end{cases}$$

By the same way as in [3] and [4], we have almost same results as in Theorem 1. Let us consider this problem for $\varepsilon = 0$. Putting $\varepsilon = 0$ in Problem (2.2), we obtain

$$(2.3) \quad \begin{cases} x \dfrac{du}{dx} + q(x)u+r(x)v = s(x), \\ \dfrac{dv}{dx} = u, \qquad u(1) = b', \quad v(1) = b. \end{cases}$$

Let $u_o(x)$ and $v_o(x)$ be a solution of Problem (2.3). Then we have the following:

THEOREM 2. *Suppose that the system* (2.2) *fulfils*

(I) p(x), q(x), r(x) *and* s(x) *are real-valued and analytic functions of* x *for* $0 \leqq x \leqq 1$;

(II) $q_o = q(0)$ *is a positive non-integer*;

(III) P(x,v,u) *is a polynomial of the degree m such that* $P(x,v,u) = \sum p_{ik}(x)v^i u^k$, *where* $p_{ik}(x)$ *is real-valued and analytic for* [0,1] ;

(IV) *the quantity* n *is an integer greater than* 1.

Then there exsist functions $x(\xi,\zeta)$, $u(\xi,\zeta)$ *and* $v(\xi,\zeta)$ *defined by the* power series

$$x = x(\xi,\zeta) = \xi + \sum_{m=1}^{\infty} x_m(\xi)\zeta^m,$$

$$u = u(\xi,\zeta) = u_0(\xi) + \sum_{m=1}^{\infty} u_m(\xi)\zeta^m,$$

$$v = v(\xi,\zeta) = v_0(\xi) + \sum_{m=1}^{\infty} v_m(\xi)\zeta^m$$

such that

(a) the power series $x(\xi,\zeta), u(\xi,\zeta)$ and $v(\xi,\zeta)$ are uniformly convergent for $0 < \xi \leq 1$, $|\zeta| \leq \delta$, with coefficients $x_m(\xi), u_m(\xi)$ and $v_m(\xi)$ functions real-valued and analytic for $0 < \xi \leq 1$, where δ is a sufficiently small positive constant;

(b) $x(1,\zeta)=1$, $u(1,\zeta)=b'$, $v(1,\zeta) = b$;

(c) $x = x(\xi) = x(\xi, \varepsilon\xi^{-s_0 q_0})$, $u = u(\xi) = u(\xi, \varepsilon\xi^{-s_0 q_0})$ and $v = v(\xi) = v(\xi, \varepsilon\xi^{-s_0 q_0})$ is a parametric representation of the solution $u(x), v(x)$ of Problem (2.2), where $s_0 = \max(1, m/n)$.

THEOREM 3. Assume that

(V) $m/n < 1 + 1/2q_0$;

(VI) $[x^{q_0} u_0(x)]_{x=0} > 0$,

besides the same assumptions as in Theorem 2. Then there exsists a positive constant ε_0 such that

(a) the function $x(\xi, \varepsilon\xi^{-s_0 q_0})$ is monotone increasing in

$$\left[\frac{\varepsilon}{\delta}\right]^{1/(2s_o-1)q_o} \leq \xi \leq 1, \quad (0 < \varepsilon \leq \varepsilon_o); \tag{2.4}$$

(b) the equation $x(\xi, \varepsilon\xi^{-s_0 q_0}) = 0$ has a unique solution $\hat{\xi}(\varepsilon)$ in the interval (2.4) for every ε $(0 < \varepsilon \leq \varepsilon_0)$ and the function $\hat{\xi}(\varepsilon)$ behaves asymptotically as

$$\hat{\xi}(\varepsilon) = \left[\frac{u_0(0)\,\varepsilon}{q_0+1}\right]^{1/(q_0+1)} (1+o(1)), \quad (\varepsilon \to 0);$$

(c) <u>the value of the solution</u> u(x) <u>and</u> v(x) <u>at</u> x=0 <u>is given by</u> $u(\hat{\xi}(\varepsilon))$, $v(\hat{\xi}(\varepsilon))$ <u>for</u> $0 < \varepsilon \leqq \varepsilon_0$.

Proof of Theorems 2 and 3 is based upon several lemmas stated in [5].

REMARK. It is noticed that Assumption (VI) shows that if $n/2q_0 > 1$, the degree m can be taken such that m > n.

REFERENCES

[1] M.J.Lighthill, A technique for rendering approximate solutions to physical problem uniformly valid, Philos.Mag.,40(1949),1179-1201.

[2] Y.Sibuya and K.Takahasi, On the differential equation (x+ u)du/dx +q(x)u=r(x), Funk.Ekv.,9(1966),1-17.

[3] K.Takahasi, A note on PLK-method, Proc.US-Japan Sem. Diff.Func.Eq., (1967),507-517.

[4] K.Takahasi, An application of PLK-method for second order non-linear ordinary differential equations,RIMS Rep.,67(1969),85-99.

[5] K.Takahasi, Initial value problem of an ordinary nonlinear differential equation of the second order(in Japanese),Funk.Ekv. (Internal Series)23(1970),1-10.

[6] W.Wasow, On the convergence of an approximate method of M.J.Lighthill, Jour.Rat.Mech.Anal.,4(1955),751-767, and Correction.

STOKES PHENOMENON FOR GENERAL LINEAR ORDINARY DIFFERENTIAL EQUATIONS WITH TWO SINGULAR POINTS

MITSUHIKO KOHNO

We consider the connection problem to obtain the linear relations between two fundamental sets of solutions in the neighborhood of singular points, that is the same thing, the analyses of the Stokes phenomenon occurring in the neighborhood of an irregular singular point for single n-th order linear differential equations

(1) $$t^n \frac{d^n x}{dt^n} = \sum_{\ell=1}^{n} \Big(\sum_{r=0}^{q\ell} a_{\ell,r} t^r \Big) t^{n-\ell} \frac{d^{n-\ell} x}{dt^{n-\ell}} .$$

According to the local theory of linear ordinary differential equations, there exist a fundamental set of solutions represented by the convergent power series

(2) $$x_j(t) = t^{\rho_j} \sum_{m=0}^{\infty} G_j(m) t^m \qquad (j=1,2,\ldots,n)$$

in the neighborhood of the regular singular point at the origin and also, near the irregular singular point of rank q at infinity, a fundamental set of solutions $x_S^k(t)$ defined on every sector S with vertex at the origin and central angle not exceeding $\frac{\pi}{q}$ such that

(3) $$x_S^k(t) \cong x^k(t) \quad \text{as} \quad t \to \infty \quad \text{in} \quad S \qquad (k=1,2,\ldots,n)$$

where $x^k(t)$ are the formal solutions of the differential equations (1) of the form

(4) $$x^k(t) = \exp\Big(\frac{\lambda_k}{q} t^q + \frac{\alpha_{q-1}^k}{q-1} t^{q-1} + \ldots + \alpha_1^k t\Big) t^{\mu_k} \sum_{s=0}^{\infty} h^k(s) t^{-s} .$$

In the above, the numbers $\rho_j, \alpha_1^k, \alpha_2^k, \ldots, \alpha_{q-1}^k, \lambda_k$ and μ_k are the characteristic constants determined by the constant coefficients $a_{\ell,r}$. In order to solve the global problem in detail, we would like to derive the following asymptotic relations

(5) $$x_j(t) \cong \sum_{k=1}^{n} T^{(j,k)}(S) x^k(t) \quad \text{as} \quad t \to \infty$$

in any sector S a finite number of which covers the whole neighborhood of infinity. Once we had these relations, the constant coefficients $C^{(j,k)}(\hat{S})$ of the connection problem

$$(6)\qquad x^k_{\hat{S}}(t) = \sum_{j=1}^{n} C^{(j,k)}(\hat{S})x_j(t) \quad \text{in} \quad \hat{S} \subset S$$

would be given by the formula

$$(7)\qquad \text{Matrix}\{C^{(j,k)}(\hat{S}):j,k=1,2,\ldots,n\}$$
$$= \text{Inverse matrix of matrix}\{T^{(j,k)}(S):j,k=1,2,\ldots,n\}$$

and the Stokes phenomenon could be analyzed by the following relations

$$(8)\qquad x^k_S(t) \cong \sum_{j=1}^{n} C^{(j,k)}(S) \sum_{\ell=1}^{n} T^{(j,\ell)}(S')x^\ell(t) \quad \text{as} \quad t \to \infty \quad \text{in} \quad S'.$$

For the purpose of deriving the asymptotic relations (5), we shall attempt to partition the convergent power series solutions $x_j(t)$ into functions which are somewhat easily analyzed. Such method of partition, which was first established by K. Okubo [1], are based on the detailed investigation of the coefficients $G_j(m)$ and $h^k(s)$ of power series of the convergent power series solutions (2) and of the formal solutions (4).

Here, we shall introduce the q-th order difference equations

$$(9)\qquad (m+\rho_j-\mu_k)g^{(j,k)}(m) = \alpha^k_1 g^{(j,k)}(m-1) +\ldots+ \alpha^k_{q-1}g^{(j,k)}(m-q+1)$$
$$+ \lambda_k g^{(j,k)}(m-q)$$

and, by means of a certain fundamental set of solutions of the above difference equations,

$$(10)\qquad g^{(j,k)}_1(m), g^{(j,k)}_2(m), \ldots, g^{(j,k)}_q(m),$$

we define the new functions by

$$(11)\qquad f^{(j,k)}_\ell(m) = \sum_{s=0}^{\infty} h^k(s)g^{(j,k)}_\ell(m+s)$$

for the moment formally and then, we easily see that for each j, these qn functions $f^{(j,k)}_\ell(m)$ satisfy the same qn-th order recurrence formula satisfied by the coefficient $G_j(m)$. Therefore, if we can prove the well-definedness of the functions $f^{(j,k)}_\ell(m)$ and their linear independence, the coefficients $G_j(m)$ can be represented by the linear combinations of qn functions $f^{(j,k)}_\ell(m)$,

$$(12)\qquad G_j(m) = \sum_{k=1}^{n} \sum_{\ell=1}^{n} T^{(j,k)}_\ell f^{(j,k)}_\ell(m) \qquad (j=1,2,\ldots,n)$$

where the constants $T^{(j,k)}_\ell$ are determined by the algebraic linear

equations (12), putting $m = -qn+1, -qn+2, \ldots, -1, 0$.
Thus, we have the partition of convergent power series solutions as follows:

$$(13) \qquad x_j(t) = \sum_{m=0}^{\infty} G_j(m) t^{m+\rho_j} = \sum_{\ell=1}^{q} \sum_{k=1}^{n} T_\ell^{(j,k)} h^k(s) X_\ell^{(j,k)}(t,s)$$

where we put

$$(14) \qquad X_\ell^{(j,k)}(t,s) = \sum_{m=0}^{\infty} g_\ell^{(j,k)}(m+s) t^{m+\rho_j}.$$

Here we remark that the proof of the well-definedness and the linear independence of the functions $f_\ell^{(j,k)}(m)$ will be done by the method of ε-parameter [1] or by the estimation method [2] or so, although we have not yet showed them in this stage.
Now, returning to the newly defined functions (14), we find out that the functions $X_\ell(t,s)$, dropping the indices (j,k), satisfy the first order nonhomogeneous differential equations and then, by the guadrature, we have the following integral representations

$$\begin{aligned}(15)\ X_\ell(t,s) = {} & [\lambda g_\ell(s-1)] t^{\rho+q-1} \hat{X}(\lambda^{\frac{1}{q}} t : s-\mu+\rho+q-1) \\ & + [\lambda g_\ell(s-2) + \alpha_{q-1} g_\ell(s-1)] t^{\rho+q-2} \hat{X}(\lambda^{\frac{1}{q}} t : s-\mu+\rho+q-2) \\ & + \\ & \vdots \\ & + [\lambda g_\ell(s-q+1) + \alpha_{q-1} g_\ell(s-q+2) + \ldots + \alpha_2 g_\ell(s-1)] t^{\rho+1} \hat{X}(\lambda^{\frac{1}{q}} t : s-\mu+\rho+1) \\ & + [\lambda g_\ell(s-q) + \alpha_{q-1} g_\ell(s-q+1) + \ldots + \alpha_2 g_\ell(s-2) + \alpha_1 g_\ell(s-1)] t^{\rho} \\ & \qquad \times \hat{X}(\lambda^{\frac{1}{q}} t : s-\mu+\rho)\end{aligned}$$

where

$$\begin{aligned}(16) \qquad \hat{X}(t:\nu) = \int_0^1 \exp[& \tfrac{1}{q} t^q (1-\tau^q) + \frac{\alpha_{q-1}}{q-1} \lambda^{-\frac{q-1}{q}} t^{q-1} (1-\tau^{q-1}) \\ & + \ldots + \alpha_1 \lambda^{-\frac{1}{q}} t (1-\tau)] \tau^{\nu-1} d\tau.\end{aligned}$$

Hence, the analyses of $x_j(t)$ are converted into that of $\hat{X}(t:\nu)$.
At first, if we investigate the behaviours of the integral $\hat{X}(t:\nu)$ for sufficiently large values of t, we have the following theorem.

<u>Theorem 1.</u> <u>If we choose an appropriate fundamental set of solutions (10) of the difference equations (9), then we have</u>

(17) $\hat{X}(\lambda^{\frac{1}{q}}t:s-\mu+\rho) \cong E_{j+1}(s)\exp(\frac{\lambda}{q}t^q+\frac{\alpha_{q-1}}{q-1}t^{q-1}+\ldots+\alpha_1 t)t^{-s+\mu-\rho}+O(1)$

for sufficiently large values of t in the sector

(18) $S_j(\lambda):\ -\frac{\pi}{q}+\frac{2\pi}{q}j \leqq \arg\lambda^{\frac{1}{q}}t \leqq \frac{\pi}{q}+\frac{2\pi}{q}j \qquad (j=0,1,\ldots,q-1).$

Here the coefficients $E_j(s)$ are of the form

(19) $$E_j(s) = \frac{D_j(s-1)}{\lambda C(s-1)} \qquad (j=1,2,\ldots,q)$$

where $C(s-1)$ is the Casorati determinant constructed by the fundamental set of solutions $g_1(s-1), g_2(s-1), \ldots, g_q(s-1)$ and $D_j(s-1)$ are the cofactors of elements $g_j(s-q)$ of the Casorati determinant $C(s-1)$. The important part of Theorem 1 is the fact that the coefficients $E_j(s)$, appearing in the asymptotic behaviours of $\hat{X}(t:\nu)$ and varying with the sector $S_j(\lambda)$, have the beautiful form (19).

We have the relations with respect to the coefficients $E_j(s)$.

Theorem 2. It holds that

(20) $$\lambda E_j(s+q-k)+\alpha_{q-1}E_j(s+q-1-k)+\ldots+\alpha_{k+1}E_j(s+1)+\alpha_k E_j(s) = \frac{\Delta_j^k(s-1)}{C(s-1)}$$

$(j,k=1,2,\ldots,q)$

where $\Delta_j^k(s-1)$ are the cofactors of the elements $g_j(s-k)$ of the Casorati determinant $C(s-1)$.

Now, using Theorem 1, Theorem 2, the relations (15) and (13), we have the final results.

Theorem 3. Suppose that t is sufficiently large. Then t necessarily lies in some sector

(21) $$S(\ell_1,\ell_2,\ldots,\ell_n) = S_{\ell_1-1}(\lambda_1) \cap S_{\ell_2-1}(\lambda_2) \cap\ldots\cap S_{\ell_n-1}(\lambda_n)$$

where $1 \leqq \ell_1,\ell_2,\ldots,\ell_n \leqq q$, and hence, it holds that

(22) $$x_j(t) \cong \sum_{k=1}^{n} T_{\ell_k}^{(j,k)}x^k(t)+O(t^{\rho_j+q-1}) \text{ as } t \to \infty \text{ in } S(\ell_1,\ell_2,\ldots,\ell_n).$$

REFERENCES

1. Okubo, K., A global representation of a fundamental set of solutions and a Stokes phenomenon for a system of linear ordinary differential equation, J. Math. Soc. Japan, 15(1963).

2. Kohno, M., A two point connection problem for n-th order single linear ordinary differential equations with an irregular singular point of rank two, Japanese Journal of Math., 40(1970), 11-62.

ON AN ASYMPTOTIC EXPANSION OF SOLUTIONS OF ORR-SOMMERFELD TYPE EQUATION

Minoru NAKANO and Toshihiko NISHIMOTO

1. We consider a 4-th order linear ordinary differential equation

(1) $$\varepsilon^2 y^{(4)} - p_3(x,\varepsilon)y'' - p_2(x,\varepsilon)y' - p_1(x,\varepsilon)y = 0 ,$$

where $p_i(x,\varepsilon)$'s are holomorphic in $\varepsilon : 0 < \varepsilon \leqq \varepsilon_0 < 1$ and $x: |x| < \infty$, and supposed to be expanded asymptotically in $\sum_{r=0}^{\infty} p_{ir}(x)\varepsilon^r$ as ε tends to 0 for $x: |x| < M$.

If we write in vectorial form we have a 1st order linear system

(2) $$\varepsilon Y' = \begin{bmatrix} 0 & \varepsilon & 0 & 0 \\ 0 & 0 & \varepsilon & 0 \\ 0 & 0 & 0 & 1 \\ p_1(x,\varepsilon) & p_2(x,\varepsilon) & p_3(x,\varepsilon) & 0 \end{bmatrix} Y ,$$

where $$Y = \begin{bmatrix} y \\ y' \\ y'' \\ \varepsilon y''' \end{bmatrix} .$$

We call zeros of the function $p_3(x,0)=p_{30}(x)$ turning points of the differential equation (1). A reduced equation becomes 2nd order, and it has singularities at turning points. We suppose that all the singular points of the reduced equation are regular singular points (cf.[4],[9]).

A special case of a 4-th order equation

(3) $$\varepsilon^2 y^{(4)} - [i(w(x)-c)+2\alpha^2\varepsilon^2]y'' - [i\alpha^2(w(x)-c)-w''(x)-\alpha^4\varepsilon^2]y = 0$$

is the so-called Orr-Sommerfeld equation, which appears in the theory of hydrodynamic linear stability. Here α is real and c is complex. If we put

$$w(x) = 1-x^2,$$

$x = \pm(1-c)^{1/2}$ are simple turning points and the reduced equation has regular singular points at these two points.

2. About the equation (3), there are many papers by many authors ([2],[3],[5]-[8],[10],[11]). Among them, the papers of Lin and Wasow are very important. Wasow's works are mostly concerned with the local theory around one simple turning point using the related equation method.

The linear stability theory has been almost completely solved, but there are many difficult problems still unsolved mathematically in this type of equations. Recently much attention has been paid to this equation relating to the non-linear stability theory. On the other hand, the papers treated global theory are very few. We know the papers of Foote and Lin [2] and of Wasow [11]. They treat the plane Poiseuille flow case and the plane Couette flow case respectively.

For the nearly parallel flow, we need not restrict the basic function $w(x)$ to be at most quadratic function. Thus, our purpose of this paper is to make a systematical treatment about asymptotic expansions of the differential equations (1) and (3).

3. We calculate formal solutions and we can prove that these formal solutions approximate asymptotically true solutions in some x-region.

After heavy computation, we get formal solutions. We write the fundamental solution

$$Y=\begin{bmatrix} Y_{11} & Y_{12} \\ Y_{21} & Y22 \end{bmatrix},$$

where Y_{ij}'s are 2-by-2 matrices, and we expand it in power series of ε, and we write only upper two rows of first three terms for the special case (3), because y and y' are very necessary for some boundary value problem.

$$Y_{11}=[\, U_o(x)+\varepsilon U_1(x) + \varepsilon^2 U_2(x)] \cdot \begin{bmatrix} (x-x_i)^{\lambda} & C(x-x_i)^{\lambda}\log(x-x_i) \\ 0 & (x-x_i)^{\mu} \end{bmatrix} + \varepsilon^3(\cdots),$$

$$Y_{12}=\left\{\varepsilon\begin{bmatrix} 0 & 0 \\ 1/(p_{30})^{1/2} & 1/(p_{30})^{1/2} \end{bmatrix} + \varepsilon^2\begin{bmatrix} 1/p_{30} & -1/p_{30} \\ m & n \end{bmatrix}\right\} \times$$

$$\times \left\{ (p_{30})^{-1/4} \exp\left[\frac{1}{\varepsilon}\int^x (p_{30})^{1/2}dx \begin{bmatrix} 1 & 0 \\ 0 & -1 \end{bmatrix}\right]\right\} + \varepsilon^3(\cdots),$$

where U_0 is a solution of the reduced equation and λ and μ are roots of the characteristic equation associated to the reduced equation. The constant C vanishes if $\lambda - \mu$ is equal to no integers. Two functions m and n are given by

$$m= 3p_{30}'/(4p_{30}^2)+1/(p_{30})^{1/2}\int^x [p_{32}/2(p_{30})^{1/2}+1/2(p_{30})^{3/2}(p_{10}+p''_{30}/4)+$$

$$+ 3(p'_{30})^2/32(p_{30})^{5/2}]dx,$$

$$n= -p'_{30}/(p_{30})^2+1/(p_{30})^{1/2}\int^x [-p_{32/2}(p_{30})^{1/2}+1/2(p_{30})^{3/2}(-p_{10}+p''_{30/4})+$$

$$+5\,(p'_{30})^2/32(p_{30})^{5/2}]dx.$$

General case has the same forms, but much more complicated.

4. We consider some x-region in order to show asymptoticity of solutions. This region is unbounded and surrounded by Stokes curves. It is called a canonical region by a Russian mathematician Fedoryuk [1].

Stokes curves are defined by

$$\mathrm{Re}\int_{x_i}^{x} (p_{30}(x))^{1/2}dx = 0,$$

where x_i is a turning point or a zero of the function $p_{30}(x)$.

For one differential equation there are several canonical regions. It is not easy to draw canonical regions for general equations, but for a special case we can draw all Stokes curves and get canonical regions.

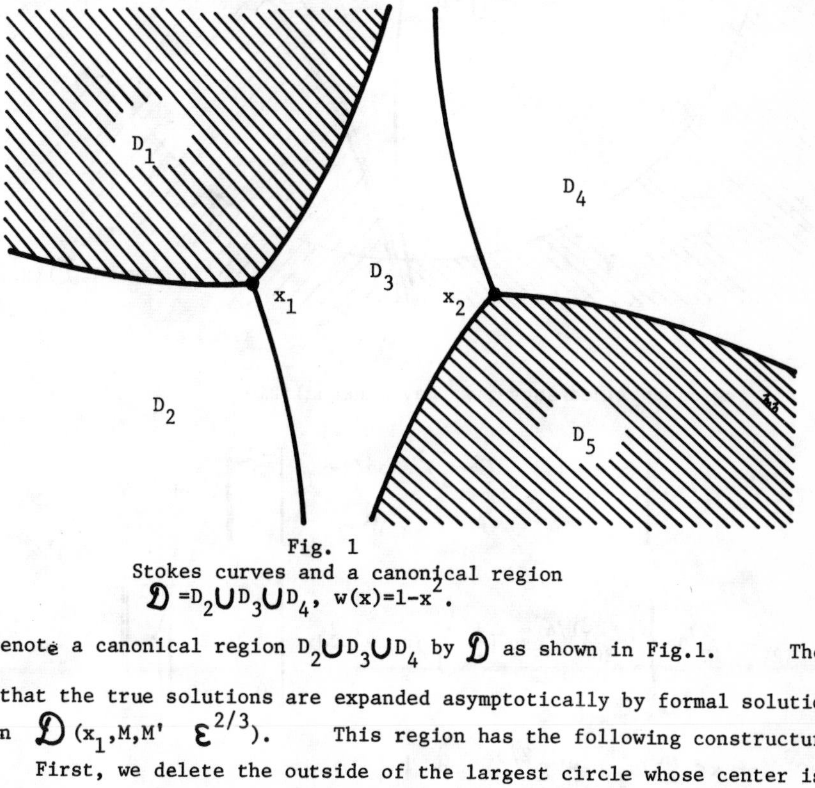

Fig. 1
Stokes curves and a canonical region
$\mathcal{D} = D_2 \cup D_3 \cup D_4$, $w(x)=1-x^2$.

We denote a canonical region $D_2 \cup D_3 \cup D_4$ by $\mathcal{D}$ as shown in Fig.1. Then we can prove that the true solutions are expanded asymptotically by formal solutions in the region $\mathcal{D}(x_1, M, M'\varepsilon^{2/3})$. This region has the following constructure (see Fig.2): First, we delete the outside of the largest circle whose center is situated at the turning point x_1 and a radius is M an arbitrarily large constant. Then we delete the inside of the smallest circle whose center is situated at x_1 and a radius is $M'\varepsilon^{2/3}$, where M' is an arbitrarily large constant, and delete the inside of the circle whose center is situated at x_2 and a radius is δ an arbitrarily small constant. And then, we remove neighbourhoods of Stokes curves as shown in Fig.2.

Thus, we get the region $\mathfrak{D}(x_1,M,M'\varepsilon^{2/3})$ which is a white part in the following figure.

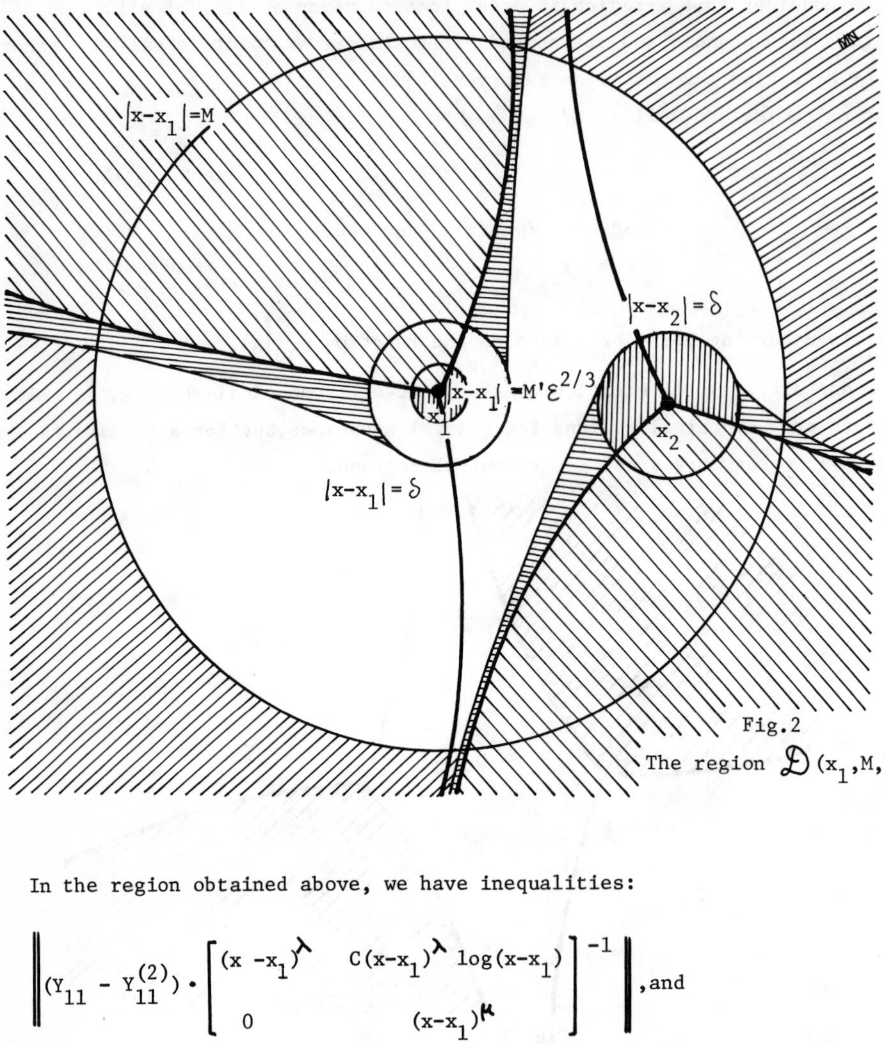

Fig.2
The region $\mathfrak{D}(x_1,M,M'\varepsilon^{2/3})$

5. In the region obtained above, we have inequalities:

$$\left\| (Y_{11} - Y_{11}^{(2)}) \cdot \begin{bmatrix} (x-x_1)^{\lambda} & C(x-x_1)^{\lambda}\log(x-x_1) \\ 0 & (x-x_1)^{\mu} \end{bmatrix}^{-1} \right\|, \text{and}$$

$$\left\| (Y_{12} - Y_{12}^{(2)}) \cdot \left\{ (p_{30})^{-1/4} \exp\left[\frac{1}{\varepsilon}\int^{x} (p_{30})^{1/2} dx \begin{bmatrix} 1 & 0 \\ 0 & -1 \end{bmatrix}\right]\right\}^{-1} \right\|$$

$$\leqq \begin{cases} K\varepsilon^{3} \text{ for } x \in \mathfrak{D}(x_1,M,M'\varepsilon^{2/3}) \text{ and } |x| \geqq \delta, \\ K(|x-x_1|\ \varepsilon^{-3/2})^{3} \text{ for } x \in \mathfrak{D}(x_1,M,M'\varepsilon^{2/3}) \text{ and } \delta \geqq |x| \geqq M'\varepsilon^{2/3}, \end{cases}$$

where $Y_{1i}^{(2)}$'s are sums of terms up to degree 2 with respect to ε of formal solutions.

In order to prove the above inequalities, we must precisely study integral equations whose integral paths have to be taken so that the integrals converge. We can take easily integral paths between the largest circle and the intermediate one by considering canonical regions. However integral paths between the intermediate and the smallest circles are very complicated because the integrals are influenced by the turning point.

Foote and Lin have gained same asymptoticity but in the region $\mathfrak{D}(x_1, M, \delta)$. We should notice δ is a constant.

As the boundary near the turning point x_1 depends on $\varepsilon^{2/3}$ it shrinks to the turning point x_1 as ε tends to zero. But the circle around the other turning point x_2 does not depend on ε. Our region is therefore larger than one of them.

For the general case (1), asymptoticity can be proved in similar way.

6. Other canonical regions are, for instance, $D_1 \cup D_2$, $D_4 \cup D_5$ and $D_1 \cup D_3 \cup D_5$. As we are considering a linear differential equation, the solutions in $\mathfrak{D}$ and other canonical regions are connected linearly. The connection relation may be represented by constant matrices.

REFERENCES

[1] Evgrafov, E.M., and M.V. Fedoryuk, Russian Math. Surveys. 21 (1966), 1-48.

[2] Foote, J.R., and C.C. Lin, Quart. Appl. Math. 8 (1950), 265-280.

[3] Graebel, W.P., J. Fluid Mech. 24 (1966), 497-508.

[4] Iwano, M., and Y. Shibuya, Kodai Math. Sem. Rep. 15 (1963), 1-28.

[5] Kuentam, K., J. Fluid Mech. 34 (1968), 145-158.

[6] Langer, R.E., Trans. Amer. Math. Soc. 84 (1957), 144-158.

[7] Lin, C.C., Theory of hydrodynamic stability. Cambridge Univ. Press (1955).

[8] Lin, C.C., and A.L. Rabenstein, Trans. Amer. Math. Soc. 94 (1960), 24-57.

[9] Nishimoto, T., Kodai Math. Sem. Rep. 20 (1968), 218-256.

[10] Wasow, W., Ann. of Math. 58 (1953), 222-252.

[11] Wasow, W., J. Res. Nat. Bur. Standards B51 (1953), 195-202.

ON PERIODS OF PERIODIC SOLUTIONS OF A CERTAIN NONLINEAR DIFFERENTIAL EQUATION

Shigeru FURUYA

Prof. T. Ōtsuki has proposed a question ([1]) concerning periods of periodic solutions of the differential equation

$$(1) \qquad nx(1-x^2)\frac{d^2x}{dt^2} + \left(\frac{dx}{dt}\right)^2 + (1-x^2)(nx^2-1) = 0,$$

where n is a positive integer not equal to one.

Since the equation does not involve t explicitly, by taking $y = \frac{dx}{dt}$ (1) is transformed into the first order differential equation:

$$(2) \qquad nx(1-x^2)y\frac{dy}{dx} + y^2 + (1-x^2)(nx^2-1) = 0$$

The general solution of (2) is given by

$$(3) \qquad y^2 = 1-x^2-c\left(\frac{|1-x^2|}{x^2}\right)^{1/n}$$

where c is an intergral constant.

It is easily seen from behaviors of curves (3) that the solution $x(t)$ of (1) with the initial condition

$$0 < x(0)\ , \qquad x^2(0) + \left(\frac{dx}{dt}(0)\right)^2 < 1$$

is periodic and it holds

$$x^2(t) + \left(\frac{dx}{dt}(t)\right)^2 < 1\ , \qquad t \geqq 0$$

Let Λ be the supremum of periods of these periodic solutions. Then Ōtsuki's problem is to decide whether $\Lambda < 2\pi$ or $\Lambda \geqq 2\pi$. Here we shall show that $\Lambda < 2\pi$.

The period T of a periodic solution of our problem is given by

$$T = \int_{x_1}^{x^2} \frac{dx}{\sqrt{1-x^2-c(\frac{1-x^2}{x^2})^{1/n}}}$$

where $0<x_1<\frac{1}{\sqrt{n}}< x_2<1$, $x_i^2(1-x_i^2)^{n-1} = c^n \quad (i=1,2)$

Our problem is reduced to the problem of estimating the value of the above integral.

We set

$$u=u(x)=x(1-x^2)^{\frac{n-1}{2}}, \qquad \alpha=u(\frac{1}{\sqrt{n}})=\frac{1}{\sqrt{n}}(1-\frac{1}{n})^{\frac{n-1}{2}} .$$

By changing the variable from x to u, we get

$$T = 2(I_1+I_2),$$

$$I_1 = \int_{c^{n/2}}^{\alpha} \frac{u^{1/n}}{\sqrt{u^{2/n}-c}} \quad \frac{(1-x_1^2(u))^{1-\frac{n}{2}}}{1-nx_1^2(u)} du,$$

$$I_2 = \int_{c^{n/2}}^{\alpha} \frac{u^{1/n}}{\sqrt{u^{2/n}-c}} \quad \frac{(1-x_2^2(u))^{1-\frac{n}{2}}}{nx_2^2(u)-1} du,$$

where $x_1(u)$, $x_2(u)$ are inverse functions of $u=u(x)$ on

$[x_1, \frac{1}{\sqrt{n}}]$, $[\frac{1}{\sqrt{n}}, x_2]$ respectively.

First, we consider the integral I_1. For simplicity we write x in stead of $x_1(u)$ and let

$$y = x^{\frac{2}{n-1}}, \qquad p = (\frac{1}{\sqrt{n}})^{\frac{2}{n-1}} .$$

Then we have

$$(4) \quad 1-nx^2=n(p^{n-1} - y^{n-1})=n(p-y)A, \quad A=p^{n-2}+ yp^{n-3}+\ldots+y^{n-2}$$

and

$$(5) \quad \alpha^{\frac{2}{n-1}} - u^{\frac{2}{n-1}} =(p-y)-(p^n-y^n)=(p-y)^2B,$$

$$B = (n-1)p^{n-2}+(n-2)p^{n-3}y+\ldots+y^{n-2}.$$

Let $v = u^{\frac{2}{n(n-1)}}$, $\beta = \alpha^{\frac{2}{n(n-1)}}$.

Then

$$(6) \quad \alpha^{\frac{2}{n(n-1)}} - u^{\frac{2}{n(n-1)}} = \beta^n - v^n = C(\beta^{n-1} - v^{n-1}),$$

$$C = (\beta^n - v^n)(\beta^{n-1} - v^{n-1})^{-1}.$$

From (4), (5) and (6) we have

$$(7) \quad p-y = \sqrt{(C/B)(\beta^{n-1} - v^{n-1})} = \sqrt{C/B}\ \sqrt{\alpha^{2/n} - u^{2/n}} .$$

Since $(1-x^2)^{1-\frac{n}{2}} = y^{n-2}\, u^{-\frac{n-2}{n-1}}$, after noticing (4) and (7), we obtain

$$(8) \quad I_1 = \frac{1}{2}\int_{c^{n/2}}^{\alpha} \frac{(2/n)u^{\frac{2}{n}-1}}{\sqrt{(u^{2/n}-c)(\alpha^{2/n}-u^{2/n})}}\ \frac{\sqrt{vy^{n-2}B}}{A\sqrt{C}}\, du.$$

Next, we consider the integral I_2. Also in this case we write x instead of $x_2(u)$. Similarly as above we get

$$nx^2-1=n(y-p)A , \qquad \alpha^{\frac{2}{n-1}} - u^{\frac{2}{n-1}} = (y-p)^2 B ,$$

$$y-p = \sqrt{(C/B)(\alpha^{2/n} - u^{2/n})}$$

so that

$$I_2 = \frac{1}{2}\int_{c^{n/2}}^{\alpha} \frac{(2/n)u^{\frac{2}{n}-1}}{\sqrt{(u^{2/n}-c)(\alpha^{2/n}-u^{2/n})}}\ \frac{\sqrt{vy^{n-2}B}}{A\sqrt{C}}\, du .$$

In both cases we can show that

$$(9) \qquad vy^{n-2}B \leqq A^2C\,\frac{n-1}{n} .$$

It is easy to see

$$A^2 = \Big(\sum_{i=0}^{n-2} p^i y^{n-2-i}\Big)\Big(\sum_{j=0}^{n-2} p^j y^{n-2-j}\Big) = \sum_{i+j\leqq n-2} + \sum_{i+j>n-2} =$$

$y^{n-2}B + \sum_{i+j>n-2} > y^{n-2}B$.

Let $P=\beta^{n-2}v + \beta^{n-3}v^2 + \ldots + v^{n-1}$.

Then, from $v \leqq \beta$ it follows $P \leqq (n-1)\beta^{n-1}$, so that

$$\frac{C}{v} = \frac{P+\beta^{n-1}}{P} = 1 + \frac{\beta^{n-1}}{P} \geqq 1 + \frac{1}{n-1} = \frac{n}{n-1} .$$

Hence the inequality (9) is proved.
Thus we obtain

$$I_i \leqq \frac{1}{2} \int_{c^{n/2}}^{\alpha} \frac{(2/n)u^{\frac{1}{n}-1}}{\sqrt{(u^{2/n} - c)(\alpha^{2/n} - u^{2/n})}} \sqrt{\frac{n-1}{n}}\, du \qquad (i=1,2)$$

$$= \frac{1}{2}\sqrt{\frac{n-1}{n}} \int_{c}^{\alpha^{2/n}} \frac{dw}{\sqrt{(w-c)(\alpha^{2/n} - w)}} = \frac{\pi}{2}\sqrt{\frac{n-1}{n}} .$$

Therefore we conclude that

$$T = 2(I_1 + I_2) \leqq 2\sqrt{\frac{n-1}{n}}\,\pi .$$

This completes the proof of $\Lambda < 2\pi$.

Professor Urabe conjectured after numerical calculations that $\Lambda = \sqrt{2}\,\pi$. But the author cannot prove it as yet.

BIBLIOGRAPHY

[1] T. Ōtsuki, Minimal hypersurfaces in a Riemannian manifold of constant curvature, Amer. Jour. Math., 92(1970).

A BIBASIC FUNCTIONAL EQUATION

Wazir Hasan Abdi (Adelaide)

1. Basic difference equations are defined by the operator

$$\theta f(z) = \frac{f(qz)-f(z)}{z(q-1)}, \ldots, \theta^{(n+1)} f(z) = \theta(\theta^{(n)} f(z)),$$

where the base q is fixed and f may be defined on any domain D of the z-plane or on the set $\{z | z=z_0 q^n;\ z_0 \text{ fixed},\ n \in \underline{N}\}$. The general form of the linear basic equation is

$$\sum_{m=0}^{n} a_m(z)\theta^{(m)} f(z) = g(z) \tag{1}$$

or

$$\sum_{m=0}^{n} b_m(z) f(q^m z) = h(z). \tag{2}$$

The theory of such equations has been extensively studied (amongst others) by Adams, Jackson, Birkhoff, Tritjzinsky, Hahn* and the author [1,2]. As with $q \to 1$ $\theta f \to \frac{df}{dz}$ and $\theta^{(m)} f \to \frac{d^m f}{dz^m}$, (1) reduces to a linear O.D.E. of order n. Also (2) is especially useful in handling ordinary difference equations of the type

$$\sum_{m=0}^{n} a_m(z) F(z+m\omega) = g(z) \tag{3}$$

because with $F(z) \to f(e^z)$, $e^z \to z$, (3) reduces to (2) which has the advantage of using power series. But the unibasic theory is not suitable for difference-differential and similar equations. So we prescribe two bases p,q and without loss of generality assume $0<p<q$, $p \neq 1$, $q \neq 1$.

Consider the homogeneous equation

$$a(z)f(pz) + b(z)f(qz) + c(z)f(z) = 0 \tag{4}$$

where a,b,c are given functions regular in some region of the z-plane enclosing the origin. In another form it becomes

$$\frac{f(pz)-f(z)}{z(p-1)} + A(z)f(qz) + B(z)f(z) = 0 \tag{5}$$

which in the limit $p \to 1$ yields a difference-differential equation

$$\frac{dF}{d\xi} + \alpha(\xi)F(\xi+\omega) + \beta(\xi)F(\xi) = 0 \tag{6}$$

by the transformation $e^z = \xi$, $f(z) = F(\xi)$, $q = e^\omega$.

2. Let the coefficients of z^n in the Taylor Series representation of a,b,c at the origin be a_n, b_n, c_n respectively. Suppose the desired solution of (4) is $\sum_{n=0}^{\infty} A_n z^{n+\sigma}$. Then by direct substitution and equating the coefficients to zero

$$a_0 p^\sigma + b_0 q^\sigma + c_0 = 0 \tag{7}$$

*For bibliography see Kuczma [3].

$$\sum_{m=0}^{n}(a_m p^{\sigma+n-m}+b_m q^{\sigma+n-m}+c_m)A_{n-m} = 0. \qquad (8)$$

As usual (7) is called the characteristic equation, and (8) the recurrence relation which determines the coefficients A_n uniquely upto a multiplicative constant A_0. In general, the characteristic has an infinity of complex zeros and hence the bibasic functional equation itself will have an infinite number of solutions. However, we are concerned here with the real indices, so we stipulate further that ka_0, kb_0, kc_0 must be real for some non-zero complex (or real) constant k.

3. If p and q are connected, i.e. for some integers, $m,n,p^m=q^n$, then the characteristic equation becomes

$$a_0 r^{\sigma n} + b_0 r^{\sigma m} + c_0 = 0 \qquad (9)$$

which is a polynomial equation in r^{σ} of degree $\max(|m|,|n|,|m-n|)$. The solution leads to the unibasic theory with base

$$r = p^{\frac{1}{n}} = q^{\frac{1}{m}} .$$

But when the bases are unconnected, the transcendental equation (7) is written as

$$a_0 e^{(\ln p)\sigma} + b_0 e^{(\ln q)\sigma} + c_0 = 0,$$

or re-arranging

$$C(1+Ae^{\alpha\sigma} + Be^{\beta\sigma}) = 0, \qquad 0 < \alpha < \beta.$$

Let $K(\sigma) = 1 + Ae^{\alpha\sigma} + Be^{\beta\sigma}$. Then the equation

$$K(\sigma) = 0 \qquad (10)$$

is called the <u>canonical form of the characteristic equation</u>.

Now

$$\left.\begin{aligned} &\lim_{\sigma\to-\infty} K(\sigma) = 1 \\ &\lim_{\sigma\to+\infty} K(\sigma) = (\operatorname{sgn} B).\infty . \end{aligned}\right\} \qquad (11)$$

Also, the turning point will be t where $t = \frac{1}{\beta-\alpha}\ln\left(-\frac{A\alpha}{B\beta}\right)$. As $\alpha,\beta > 0$, t will not exist if sgn A = sgn B. If t exists, sgn K(t) will determine the number of real finite zeros of $K(\sigma)$ according to the following table:

TABLE

sgn A	sgn B	sgn K(t)	Number
-	-	non-existent	1
-	+	+	Nil
-	+	0	1,repeated
-	+	-	2,distinct
+	+	non-existent	Nil
+	-	immaterial	1.

4. Now, suppose, without loss of generality, that a,b,c are analytic on the domain $|z| < \rho \leqslant 1$. Then

$$|a_{n-m}p^{\sigma+m}+b_{n-m}q^{\sigma+m}+c_{n-m}| \leqslant M\,\frac{\mu^m}{\rho^n} \tag{12}$$

and

$$|a_0 p^{\sigma+n} + b_0 q^{\sigma+n} + c_0| \geqslant |c_0(1-kq^n)|, \tag{13}$$

where M,k are suitable positive constants and $\mu = \rho \max(q,1)$. Hence for the recurrence relation (8),

$$|A_n| \leqslant \frac{M\sum_{m=0}^{n-1}\mu^m|A_m|}{\rho^n|c_0(1-kq^n)|}, \quad c_0 \neq 0. \tag{14}$$

Writing A_n^* for the right hand side of (14), we have

$$\lim_{n\to\infty}\frac{A_n^*}{A_{n+1}^*} = \begin{cases} \rho|c_0|/(M+|c_0|) & q<1 \\ \rho|c_0|qk/(M+k|c_0|) & q>1. \end{cases} \tag{15}$$

It follows therefore that

$$\sum_{n=0}^{\infty} A_n z^n$$

has a non-zero radius of convergence. In case c_0=o, the inequality (13) reduces to $|a_0 p^{\sigma+n} + b_0 q^{\sigma+n}| \geqslant kp^n$ and still the limit (15) is non-zero.

Summing up, we obtain the following

Theorem

If the coefficients A,B in the canonical form of the characteristic are both positive no real-index solution of the homogeneous bibasic function equation (4) with unconnected bases exists. Also when A<0, B>0 and K(t) > 0 there is no solution of this type. In all other cases one and at most two solutions $z^{\sigma}F(z;p,q)$ exist where F is analytic in some non-zero disc around the origin.

When the two indices differ by an integer or zero, one of the solutions may involve logarithms.

5. To illustrate, consider a difference-differential equation with exponential coefficients:

$$\frac{df}{dz} + (a+be^z)f(z+\omega) - (a-ce^z)f(z) = 0 \tag{16}$$

where a,b,c,ω are real constants. With $f(x) \to F(t)$, $e^z \to t$, $\omega \to \ln q$, the bibasic equation is

$$F(pt)+(p-1)(a+bt)F(qt)+\{(p-1)(-a+ct)-1\}F(t) = 0.$$

The characteristic equation is

$$p^{\sigma}+a(p-1)q^{\sigma}-(1+ap-a) = 0$$

so the exponent σ=0 yields the solution

$$F(t) = A_0 + A_0 \sum_{n=1}^{\infty} (-)^n \frac{(c+b)(c+qb)\ldots(c+q^{n-1}b)}{\prod_{m=0}^{n-1} \left\{ a(q^{m+1}-1) + \frac{1-p^{m+1}}{1-p} \right\}} t^n \tag{17}$$

or in the limit $p \to 1$ a solution of (16) is

$$f(z) = A_0 + A_0 \sum_{n=0}^{\infty} (-)^n \prod_{m=0}^{n-1} \left\{ \frac{c + be^{m\omega}}{a(e^{(m+1)\omega}-1)+m+1} \right\} e^{zn}$$

which is entire when $\mathcal{R}\ell\omega<0$ and convergent for $\mathcal{R}\ell z < \mathcal{R}\ell\omega + \ln\left|\frac{a}{b}\right|$ when $\mathcal{R}\ell\omega > 0$.

REFERENCES

[1] Abdi, W.H., Proc.Nat.Inst.Sc. (India) 28A (1962)1-15

[2] Abdi, W.H., Rend.Cir.Mat.Palermo.11 (1962)1-13

[3] Kuczma, M., Functional Equations in a Single Variable (1968)

LIST OF PARTICIPANTS

Egawa, Jiro

Faculty of Engineering, Kobe University, Kobe Japan

Erbe, Lynn H.

Department of Mathematics, The University of Alberta, Edmonton 7, Alberta, Canada

Frederickson, Paul O.

Department of Mathematics, Lakehead University, Postal Station P., Thunder Bay, Ontario, Canada

Furumochi, Tetsuo

Mathematical Institute, Faculty of Science, Tohoku University, Sendai, Japan

Furuya, Shigeru

Department of Mathematics, Faculty of Science, The University of Tokyo, Tokyo, Japan

Harris, William A. Jr.

Department of Mathematics, The University of Southern California, Los Angeles, California 90007, USA

Hartman, Philip

Department of Mathematics, Johns Hopkins University, Baltimore, Maryland 21218, USA

Hino, Yoshiyuki

Faculty of Education, Iwate University, Morioka, Japan

Hukuhara, Masuo

Tsuda College, Kodaira-shi, Tokyo-to, Japan

Infante, E. F.

Center for Dynamical Systems, Division of Applied Mathematics, Brown University, Providence, Rhode Island 02912, USA

Iwano, Masahiro
Department of Mathematics, Faculty of Science, Tokyo Metropolitan University, Tokyo, Japan

Kato, Junji
Mathematical Institute, Faculty of Science, Tohoku University, Sendai, Japan

Kikuchi, Norio
Department of Mathematics, Faculty of Science, Kobe University, Kobe, Japan

Kimura, Tosihusa
Department of Mathematics, Faculty of Science, The University of Tokyo, Tokyo, Japan

Kohno, Mitsuhiko
Department of Mathematics, Faculty of Science, Hiroshima University, Hiroshima, Japan

Kurihara, Mitsunobu
Faculty of Engineering, Yamanashi University, Kofu, Japan

Lee, Richard W. M.
Department of Mathematics, The University of New Brunswick, Fredericton, N 13, Canada

Levin, Jacob J.
Department of Mathematics, The University of Wisconsin, Madison, Wisconsin 53706, USA

Macki, Jack W.
Department of Mathematics, The University of Alberta, Edmonton 7, Alberta, Canada

Muldowney, James S.
Department of Mathematics, The University of Alberta, Edmonton 7, Alberta, Canada

Naito, Toshiki
Mathematical Institute, Faculty of Science, Tohoku University, Sendai, Japan

Nakajima, Fumio
Mathematical Institute, Faculty of Science, Tohoku University, Sendai, Japan

Nakano, Minoru
Department of Mathematics, Faculty of Science, Tokyo Institute of Technology, Tokyo, Japan

Nishimoto, Toshihiko
Department of Mathematics, Faculty of Science, Chiba University, Chiba, Japan

Nohel, John. A.
Department of Mathematics, The University of Wisconsin, Madison, Wisconsin 53706, USA

Okamoto, Kazuo
Department of Mathematics, Faculty of Science, The University of Tokyo, Tokyo, Japan

Okubo, Kenjiro
Department of Mathematics, Faculty of Science, Tokyo Metropolitan University, Tokyo, Japan

Rabinowitz, Paul
Department of Mathematics, The University of Wisconsin, Madison, Wisconsin 53706, USA

Saito, Kaoru
Department of Mathematics, Kyushu Institute of Technology, Kitakyushu-shi, Japan

Saito, Tosiya
Department of Mathematics, College of General Education, The University of Tokyo, Tokyo, Japan

Sakawa, Yoshiyuki
Faculty of Engineering Science, Osaka University, Toyonaka, Japan

Sato, Yukichi
Faculty of Science and Engineering, Saitama University, Urawa, Japan

Seifert, George
Department of Mathematics, Iowa State University, Ames, Iowa 50010, USA

Sibuya, Yasutaka
School of Mathematics, The University of Minnesota, Minneapolis, Minnesota 55455, USA

Sugiyama, Shohei
Department of Mathematics, Faculty of Science and Engineering, Waseda University, Tokyo, Japan

Takahashi, Ken-iti
Tokyo Institute of Agriculture and Technology, Tokyo, Japan

Takano, Kyoichi
Department of Mathematics, Faculty of Science, The University of Tokyo, Tokyo, Japan

Ura, Taro
Department of Mathematics, Faculty of Science, Kobe University, Kobe, Japan

Urabe, Minoru
Research Institute for Mathematical Sciences, Kyoto University, Kyoto, Japan

Yorke, James A.
Institute for Fluid Dynamics, The University of Maryland, College Park, Maryland 20740, USA

Yoshizawa, Taro
Mathematical Institute, Faculty of Science, Tohoku University, Sendai, Japan

Lecture Notes in Mathematics

Comprehensive leaflet on request

Vol. 38: R. Berger, R. Kiehl, E. Kunz und H.-J. Nastold, Differentialrechnung in der analytischen Geometrie IV, 134 Seiten. 1967 DM 12,-

Vol. 39: Séminaire de Probabilités I. II, 189 pages. 1967. DM 14,-

Vol. 40: J. Tits, Tabellen zu den einfachen Lie Gruppen und ihren Darstellungen. VI, 53 Seiten. 1967. DM 6.80

Vol. 41: A. Grothendieck, Local Cohomology. VI, 106 pages. 1967. DM 10,-

Vol. 42: J. F. Berglund and K. H. Hofmann, Compact Semitopological Semigroups and Weakly Almost Periodic Functions. VI, 160 pages. 1967. DM 12,-

Vol. 43: D. G. Quillen, Homotopical Algebra. VI, 157 pages. 1967. DM 14,-

Vol. 44: K. Urbanik, Lectures on Prediction Theory. IV, 50 pages. 1967. DM 5,80

Vol. 45: A. Wilansky, Topics in Functional Analysis. VI, 102 pages. 1967. DM 9,60

Vol. 46: P. E. Conner, Seminar on Periodic Maps. IV, 116 pages. 1967. DM 10,60

Vol. 47: Reports of the Midwest Category Seminar I. IV, 181 pages. 1967. DM 14,80

Vol. 48: G. de Rham, S. Maumary et M. A. Kervaire, Torsion et Type Simple d'Homotopie. IV, 101 pages. 1967. DM 9,60

Vol. 49: C. Faith, Lectures on Injective Modules and Quotient Rings. XVI, 140 pages. 1967. DM 12,80

Vol. 50: L. Zalcman, Analytic Capacity and Rational Approximation. VI, 155 pages. 1968. DM 13.20

Vol. 51: Séminaire de Probabilités II. IV, 199 pages. 1968. DM 14,-

Vol. 52: D. J. Simms, Lie Groups and Quantum Mechanics. IV, 90 pages. 1968. DM 8,-

Vol. 53: J. Cerf, Sur les difféomorphismes de la sphère de dimension trois (Γ_4= O). XII, 133 pages. 1968. DM 12,-

Vol. 54: G. Shimura, Automorphic Functions and Number Theory. VI, 69 pages. 1968. DM 8,-

Vol. 55: D. Gromoll, W. Klingenberg und W. Meyer, Riemannsche Geometrie im Großen. VI, 287 Seiten. 1968. DM 20,-

Vol. 56: K. Floret und J. Wloka, Einführung in die Theorie der lokalkonvexen Räume. VIII, 194 Seiten. 1968. DM 16,-

Vol. 57: F. Hirzebruch und K. H. Mayer, O (n)-Mannigfaltigkeiten, exotische Sphären und Singularitäten. IV, 132 Seiten. 1968. DM 10,80

Vol. 58: Kuramochi Boundaries of Riemann Surfaces. IV, 102 pages. 1968. DM 9,60

Vol. 59: K. Jänich, Differenzierbare G-Mannigfaltigkeiten. VI, 89 Seiten. 1968. DM 8,-

Vol. 60: Seminar on Differential Equations and Dynamical Systems. Edited by G. S. Jones. VI, 106 pages. 1968. DM 9,60

Vol. 61: Reports of the Midwest Category Seminar II. IV, 91 pages. 1968. DM 9,60

Vol. 62: Harish-Chandra, Automorphic Forms on Semisimple Lie Groups X, 138 pages. 1968. DM 14,-

Vol. 63: F. Albrecht, Topics in Control Theory. IV, 65 pages. 1968. DM 6,80

Vol. 64: H. Berens, Interpolationsmethoden zur Behandlung von Approximationsprozessen auf Banachräumen. VI, 90 Seiten. 1968. DM 8,-

Vol. 65: D. Kölzow, Differentiation von Maßen. XII, 102 Seiten. 1968. DM 8,-

Vol. 66: D. Ferus, Totale Absolutkrümmung in Differentialgeometrie und -topologie. VI, 85 Seiten. 1968. DM 8,-

Vol. 67: F. Kamber and P. Tondeur, Flat Manifolds. IV, 53 pages. 1968. DM 5,80

Vol. 68: N. Boboc et P. Mustatǎ, Espaces harmoniques associés aux opérateurs différentiels linéaires du second ordre de type elliptique. VI, 95 pages. 1968. DM 8,60

Vol. 69: Seminar über Potentialtheorie. Herausgegeben von H. Bauer. VI, 180 Seiten. 1968. DM 14,80

Vol. 70: Proceedings of the Summer School in Logic. Edited by M. H. Löb. IV, 331 pages. 1968. DM 20,-

Vol. 71: Séminaire Pierre Lelong (Analyse), Année 1967 – 1968. VI, 190 pages. 1968. DM 14,

Vol. 72: The Syntax and Semantics of Infinitary Languages. Edited by J. Barwise. IV, 268 pages. 1968. DM 18,-

Vol. 73: P. E. Conner, Lectures on the Action of a Finite Group. IV, 123 pages. 1968. DM 10,-

Vol. 74: A. Fröhlich, Formal Groups. IV, 140 pages. 1968. DM 12,-

Vol. 75: G. Lumer, Algèbres de fonctions et espaces de Hardy. VI, 80 pages. 1968. DM 8,-

Vol. 76: R. G. Swan, Algebraic K-Theory. IV, 262 pages. 1968. DM 18,-

Vol. 77: P.-A. Meyer, Processus de Markov: la frontière de Martin. IV, 123 pages. 1968. DM 10,-

Vol. 78: H. Herrlich, Topologische Reflexionen und Coreflexionen. XVI, 166 Seiten. 1968. DM 12,-

Vol. 79: A. Grothendieck, Catégories Cofibrées Additives et Complexe Cotangent Relatif. IV, 167 pages. 1968. DM 12,-

Vol. 80: Seminar on Triples and Categorical Homology Theory. Edited by B. Eckmann. IV, 398 pages. 1969. DM 20,-

Vol. 81: J.-P. Eckmann et M. Guenin, Méthodes Algébriques en Mécanique Statistique. VI, 131 pages. 1969. DM 12,-

Vol. 82: J. Wloka, Grundräume und verallgemeinerte Funktionen. VIII, 131 Seiten. 1969. DM 12,-

Vol. 83: O. Zariski, An Introduction to the Theory of Algebraic Surfaces. IV, 100 pages. 1969. DM 8,-

Vol. 84: H. Lüneburg, Transitive Erweiterungen endlicher Permutationsgruppen. IV, 119 Seiten. 1969. DM 10.-

Vol. 85: P. Cartier et D. Foata, Problèmes combinatoires de commutation et réarrangements. IV, 88 pages. 1969. DM 8,-

Vol. 86: Category Theory, Homology Theory and their Applications I. Edited by P. Hilton. VI, 216 pages. 1969. DM 16,-

Vol. 87: M. Tierney, Categorical Constructions in Stable Homotopy Theory. IV, 65 pages. 1969. DM 6,-

Vol. 88: Séminaire de Probabilités III. IV, 229 pages. 1969. DM 18,-

Vol. 89: Probability and Information Theory. Edited by M. Behara, K. Krickeberg and J. Wolfowitz. IV, 256 pages. 1969. DM 18,-

Vol. 90: N. P. Bhatia and O. Hajek, Local Semi-Dynamical Systems. II, 157 pages. 1969. DM 14,-

Vol. 91: N. N. Janenko, Die Zwischenschrittmethode zur Lösung mehrdimensionaler Probleme der mathematischen Physik. VIII, 194 Seiten. 1969. DM 16,80

Vol. 92: Category Theory, Homology Theory and their Applications II. Edited by P. Hilton. V, 308 pages. 1969. DM 20,-

Vol. 93: K. R. Parthasarathy, Multipliers on Locally Compact Groups. III, 54 pages. 1969. DM 5,60

Vol. 94: M. Machover and J. Hirschfeld, Lectures on Non-Standard Analysis. VI, 79 pages. 1969. DM 6,-

Vol. 95: A. S. Troelstra, Principles of Intuitionism. II, 111 pages. 1969. DM 10,-

Vol. 96: H.-B. Brinkmann und D. Puppe, Abelsche und exakte Kategorien, Korrespondenzen. V, 141 Seiten. 1969. DM 10,-

Vol. 97: S. O. Chase and M. E. Sweedler, Hopf Algebras and Galois theory. II, 133 pages. 1969. DM 10,-

Vol. 98: M. Heins, Hardy Classes on Riemann Surfaces. III, 106 pages. 1969. DM 10,-

Vol. 99: Category Theory, Homology Theory and their Applications III. Edited by P. Hilton. IV, 489 pages. 1969. DM 24,-

Vol. 100: M. Artin and B. Mazur, Etale Homotopy. II, 196 Seiten. 1969. DM 12,-

Vol. 101: G. P. Szegö et G. Treccani, Semigruppi di Trasformazioni Multivoche. VI, 177 pages. 1969. DM 14,-

Vol. 102: F. Stummel, Rand- und Eigenwertaufgaben in Sobolewschen Räumen. VIII, 386 Seiten. 1969. DM 20,-

Vol. 103: Lectures in Modern Analysis and Applications I. Edited by C. T. Taam. VII, 162 pages. 1969. DM 12,-

Vol. 104: G. H. Pimbley, Jr., Eigenfunction Branches of Nonlinear Operators and their Bifurcations. II, 128 pages. 1969. DM 10,-

Vol. 105: R. Larsen, The Multiplier Problem. VII, 284 pages. 1969. DM 18,-

Vol. 106: Reports of the Midwest Category Seminar III. Edited by S. Mac Lane. III, 247 pages. 1969. DM 16,-

Vol. 107: A. Peyerimhoff, Lectures on Summability. III, 111 pages. 1969. DM 8,-

Vol. 108: Algebraic K-Theory and its Geometric Applications. Edited by R. M. F. Moss and C. B. Thomas. IV, 86 pages. 1969. DM 6,-

Vol. 109: Conference on the Numerical Solution of Differential Equations. Edited by J. Ll. Morris. VI, 275 pages. 1969. DM 18,-

Vol. 110: The Many Facets of Graph Theory. Edited by G. Chartrand and S. F. Kapoor. VIII, 290 pages. 1969. DM 18,-

Vol. 111: K. H. Mayer, Relationen zwischen charakteristischen Zahlen. III, 99 Seiten. 1969. DM 8,–

Vol. 112: Colloquium on Methods of Optimization. Edited by N. N. Moiseev. IV, 293 pages. 1970. DM 18,–

Vol. 113: R. Wille, Kongruenzklassengeometrien. III, 99 Seiten. 1970. DM 8,–

Vol. 114: H. Jacquet and R. P. Langlands, Automorphic Forms on GL (2). VII, 548 pages. 1970. DM 24,–

Vol. 115: K. H. Roggenkamp and V. Huber-Dyson, Lattices over Orders I. XIX, 290 pages. 1970. DM 18,–

Vol. 116: Sèminaire Pierre Lelong (Analyse) Année 1969. IV, 195 pages. 1970. DM 14,–

Vol. 117: Y. Meyer, Nombres de Pisot, Nombres de Salem et Analyse Harmonique. 63 pages. 1970. DM 6.–

Vol. 118: Proceedings of the 15th Scandinavian Congress, Oslo 1968. Edited by K. E. Aubert and W. Ljunggren. IV, 162 pages. 1970. DM 12,–

Vol. 119: M. Raynaud, Faisceaux amples sur les schémas en groupes et les espaces homogènes. III, 219 pages. 1970. DM 14,–

Vol. 120: D. Siefkes, Büchi's Monadic Second Order Successor Arithmetic. XII, 130 Seiten. 1970. DM 12,–

Vol. 121: H. S. Bear, Lectures on Gleason Parts. III, 47 pages. 1970. DM 6,–

Vol. 122: H. Zieschang, E. Vogt und H.-D. Coldewey, Flächen und ebene diskontinuierliche Gruppen. VIII, 203 Seiten. 1970. DM 16,–

Vol. 123: A. V. Jategaonkar, Left Principal Ideal Rings. VI, 145 pages. 1970. DM 12,–

Vol. 124: Séminare de Probabilités IV. Edited by P. A. Meyer. IV, 282 pages. 1970. DM 20,–

Vol. 125: Symposium on Automatic Demonstration. V, 310 pages. 1970. DM 20,–

Vol. 126: P. Schapira, Théorie des Hyperfonctions. XI, 157 pages. 1970. DM 14,–

Vol. 127: I. Stewart, Lie Algebras. IV, 97 pages. 1970. DM 10,–

Vol. 128: M. Takesaki, Tomita's Theory of Modular Hilbert Algebras and its Applications. II, 123 pages. 1970. DM 10,–

Vol. 129: K. H. Hofmann, The Duality of Compact Semigroups and C^*-Bigebras. XII, 142 pages. 1970. DM 14,–

Vol. 130: F. Lorenz, Quadratische Formen über Körpern. II, 77 Seiten. 1970. DM 8,–

Vol. 131: A Borel et al., Seminar on Algebraic Groups and Related Finite Groups. VII, 321 pages. 1970. DM 22,–

Vol. 132: Symposium on Optimization. III, 348 pages. 1970. DM 22,–

Vol. 133: F. Topsøe, Topology and Measure. XIV, 79 pages. 1970. DM 8,–

Vol. 134: L. Smith, Lectures on the Eilenberg-Moore Spectral Sequence. VII, 142 pages. 1970. DM 14,–

Vol. 135: W. Stoll, Value Distribution of Holomorphic Maps into Compact Complex Manifolds. II, 267 pages. 1970. DM 18,–

Vol. 136: M. Karoubi et al., Séminaire Heidelberg-Saarbrücken-Strasbuorg sur la K-Théorie. IV, 264 pages. 1970. DM 18,–

Vol. 137: Reports of the Midwest Category Seminar IV. Edited by S. MacLane. III, 139 pages. 1970. DM 12,–

Vol. 138: D. Foata et M. Schützenberger, Théorie Géométrique des Polynômes Eulériens. V, 94 pages. 1970. DM 10,–

Vol. 139: A. Badrikian, Séminaire sur les Fonctions Aléatoires Linéaires et les Mesures Cylindriques. VII, 221 pages. 1970. DM 18,–

Vol. 140: Lectures in Modern Analysis and Applications II. Edited by C. T. Taam. VI, 119 pages. 1970. DM 10,–

Vol. 141: G. Jameson, Ordered Linear Spaces. XV, 194 pages. 1970. DM 16,–

Vol. 142: K. W. Roggenkamp, Lattices over Orders II. V, 388 pages. 1970. DM 22,–

Vol. 143: K. W. Gruenberg, Cohomological Topics in Group Theory. XIV, 275 pages. 1970. DM 20,–

Vol. 144: Seminar on Differential Equations and Dynamical Systems, II. Edited by J. A. Yorke. VIII, 268 pages. 1970. DM 20,–

Vol. 145: E. J. Dubuc, Kan Extensions in Enriched Category Theory. XVI, 173 pages. 1970. DM 16,–

Vol. 146: A. B. Altman and S. Kleiman, Introduction to Grothendieck Duality Theory. II, 192 pages. 1970. DM 18,–

Vol. 147: D. E. Dobbs, Čech Cohomological Dimensions for Commutative Rings. VI, 176 pages. 1970. DM 16,–

Vol. 148: R. Azencott, Espaces de Poisson des Groupes Localement Compacts. IX, 141 pages. 1970. DM 14,–

Vol. 149: R. G. Swan and E. G. Evans, K-Theory of Finite Groups and Orders. IV, 237 pages. 1970. DM 20,–

Vol. 150: Heyer, Dualität lokalkompakter Gruppen. XIII, 372 Seiten. 1970. DM 20,–

Vol. 151: M. Demazure et A. Grothendieck, Schémas en Groupes I. (SGA 3). XV, 562 pages. 1970. DM 24,–

Vol. 152: M. Demazure et A. Grothendieck, Schémas en Groupes II. (SGA 3). IX, 654 pages. 1970. DM 24,–

Vol. 153: M. Demazure et A. Grothendieck, Schémas en Groupes III. (SGA 3). VIII, 529 pages. 1970. DM 24,–

Vol. 154: A. Lascoux et M. Berger, Variétés Kähleriennes Compactes. VII, 83 pages. 1970. DM 8,–

Vol. 155: Several Complex Variables I, Maryland 1970. Edited by J. Horváth. IV, 214 pages. 1970. DM 18,–

Vol. 156: R. Hartshorne, Ample Subvarieties of Algebraic Varieties. XIV, 256 pages. 1970. DM 20,–

Vol. 157: T. tom Dieck, K. H. Kamps und D. Puppe, Homotopietheorie. VI, 265 Seiten. 1970. DM 20,–

Vol. 158: T. G. Ostrom, Finite Translation Planes. IV. 112 pages. 1970. DM 10,–

Vol. 159: R. Ansorge und R. Hass. Konvergenz von Differenzenverfahren für lineare und nichtlineare Anfangswertaufgaben. VIII, 145 Seiten. 1970. DM 14,–

Vol. 160: L. Sucheston, Constributions to Ergodic Theory and Probability. VII, 277 pages. 1970. DM 20,–

Vol. 161: J. Stasheff, H-Spaces from a Homotopy Point of View. VI, 95 pages. 1970. DM 10,–

Vol. 162: Harish-Chandra and van Dijk, Harmonic Analysis on Reductive p-adic Groups. IV, 125 pages. 1970. DM 12,–

Vol. 163: P. Deligne, Equations Différentielles à Points Singuliers Reguliers. III, 133 pages. 1970. DM 12,–

Vol. 164: J. P. Ferrier, Seminaire sur les Algebres Complètes. II, 69 pages. 1970. DM 8,–

Vol. 165: J. M. Cohen, Stable Homotopy. V, 194 pages. 1970. DM 16,–

Vol. 166: A. J. Silberger, PGL_2 over the p-adics: its Representations, Spherical Functions, and Fourier Analysis. VII, 202 pages. 1970. DM 18,–

Vol. 167: Lavrentiev, Romanov and Vasiliev, Multidimensional Inverse Problems for Differential Equations. V, 59 pages. 1970. DM 10,–

Vol. 168: F. P. Peterson, The Steenrod Algebra and its Applications: A conference to Celebrate N. E. Steenrod's Sixtieth Birthday. VII, 317 pages. 1970. DM 22,–

Vol. 169: M. Raynaud, Anneaux Locaux Henséliens. V, 129 pages. 1970. DM 12,–

Vol. 170: Lectures in Modern Analysis and Applications III. Edited by C. T. Taam. VI, 213 pages. 1970. DM 18,–

Vol. 171: Set-Valued Mappings, Selections and Topological Properties of 2^X. Edited by W. M. Fleischman. X, 110 pages. 1970. DM 12,–

Vol. 172: Y.-T. Siu and G. Trautmann, Gap-Sheaves and Extension of Coherent Analytic Subsheaves. V, 172 pages. 1971. DM 16,–

Vol. 173: J. N. Mordeson and B. Vinograde, Structure of Arbitrary Purely Inseparable Extension Fields. IV, 138 pages. 1970. DM 14,–

Vol. 174: B. Iversen, Linear Determinants with Applications to the Picard Scheme of a Family of Algebraic Curves. VI, 69 pages. 1970. DM 8,–

Vol. 175: M. Brelot, On Topologies and Boundaries in Potential Theory. VI, 176 pages. 1971. DM 18,–

Vol. 176: H. Popp, Fundamentalgruppen algebraischer Mannigfaltigkeiten. IV, 154 Seiten. 1970. DM 16,–

Vol. 177: J. Lambek, Torsion Theories, Additive Semantics and Rings of Quotients. VI, 94 pages. 1971. DM 12,–

Vol. 178: Th. Bröcker und T. tom Dieck, Kobordismentheorie. XVI, 191 Seiten. 1970. DM 18,–

Vol. 179: Seminaire Bourbaki – vol. 1968/69. Exposés 347-363. IV. 295 pages. 1971. DM 22,–

Vol. 180: Séminaire Bourbaki – vol. 1969/70. Exposés 364-381. IV, 310 pages. 1971. DM 22,–

Vol. 181: F. DeMeyer and E. Ingraham, Separable Algebras over Commutative Rings. V, 157 pages. 1971. DM 16.–